SHAPED BY THE WEST WIND

INUKSHUKS

Gitchi Manitou
spirit of nature

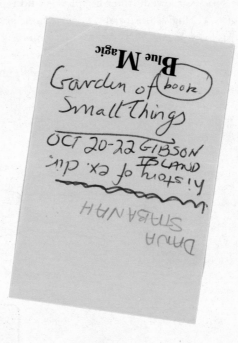

Shaped by the West Wind is the first volume in the Nature/History/Society series. The series is devoted to the publication of high-quality scholarship in environmental history and allied fields. Its broad compass is signalled by its title: *nature* because it takes the natural world seriously; *history* because it aims to foster work that has temporal depth; and *society* because its essential concern is with the interface between nature and society, broadly conceived. The series is avowedly interdisciplinary and is open to the work of anthropologists, ecologists, historians, geographers, literary scholars, political scientists, sociologists, and others whose interests resonate with its mandate. It offers a timely outlet for lively, innovative, and well-written work on the interactions of people and nature through time in North America.

General Editor: Graeme Wynn, University of British Columbia

NATURE | HISTORY | SOCIETY

SHAPED BY THE WEST WIND:

Nature and History in Georgian Bay

CLAIRE ELIZABETH CAMPBELL

UBCPress · Vancouver · Toronto

15 14 13 12 11 10 09 08   5 4 3 2

Printed in Canada on acid-free paper that is 100% post-consumer recycled, processed chlorine-free, and printed with vegetable-based, low-VOC inks.

Library and Archives Canada Cataloguing in Publication

Campbell, Claire Elizabeth, 1974-
   Shaped by the west wind: nature and history in Georgian Bay /
Claire Elizabeth Campbell.

(Nature, history, society)
Includes bibliographical references and index.
ISBN 0-7748-1098-X (bound); ISBN 0-7748-1099-8

   1. Georgian Bay Region (Ont.) – History. 2. Landscape – Symbolic aspects – Ontario – Georgian Bay Region. 3. Georgian Bay Region (Ont.) – Environmental conditions. 4. Georgian Bay (Ont.: Bay) I. Title. II. Series.

FC3095.G34C33 2005   971.3'15   C2004-906323-5

Canadä

UBC Press gratefully acknowledges the financial support for our publishing program of the Government of Canada through the Book Publishing Industry Development Program (BPIDP), and of the Canada Council for the Arts, and the British Columbia Arts Council.

This book has been published with the help of a grant from the Canadian Federation for the Humanities and Social Sciences, through the Aid to Scholarly Publications Programme, using funds provided by the Social Sciences and Humanities Research Council of Canada.

Printed and bound in Canada by Friesens
Set in New Baskerville by Artegraphica Design Co. Ltd
Text designer: Irma Rodriguez
Copy editor: Judy Phillips
Proofreader: Lenore Heitkamp
Indexer: Noeline Bridge
Cartographer: Eric Leinberger

UBC Press
The University of British Columbia
2029 West Mall
Vancouver, BC V6T 1Z2
604-822-5959 / Fax: 604-822-6083
www.ubcpress.ca

For my family

and "Bay people,"

especially my father

Contents

Illustrations

MAPS

Of Canoes and Pines and Rock-Bound Gardens

Graeme Wynn

Spending a quiet winter evening at home several years ago, browsing through some of the beautifully illustrated books on gardens and gardening that have come to fill an ample shelf in an increasingly book-filled house, I came across a set of striking images that imprinted themselves on my memory. Tucked away in the middle of what can only be described as a coffee-table book with magnificent pictures displaying the glories of private gardens across Canada, I found a handful of photographs that seemed – to me – to offer a wonderful metaphor for the Canadian settlement experience.

Set in a book intended to show that "a Canadian garden is not a contradiction in terms," these illustrations appear to do almost exactly the opposite. They depict Nicole Eaton's garden on Little Jane Island in Ontario.[1] One of them, "an aerial view of the vegetable and herb gardens," is particularly striking. The carefully composed frame is bordered on one side by water and on another by a ribbon of small trees. At least half, perhaps two-thirds, of its area is occupied by rock – glaciated, striated, and essentially bare pre-Cambrian rock, variegated by patches of lichen and bits and pieces of grass and bush growing where thin soil accumulated in the cracks and hollows of the granite and gneiss. On this windswept outcrop, the kitchen gardens stand out, not only for the type, density, and lushness of the growth within them, but also by virtue of the boulders stacked around their perimeter. Other photographs show how clearly these boulders mark the barren from the sown, the bleak, inhospitable, unproductive rock of the country from the fertile creation of human endeavour.

Imposed upon the bedrock, these gardens seem remarkable, unlikely, preposterous, fanciful ... and impermanent. They speak of labour, and dreams and conviction, of expectations as well as seedlings transplanted to this remote and forbidding spot. They signal the human desire to tame the wild, the human capacity to alter environments, and the human longing to shape places to cultural norms. "Soil," says the caption to one of these photographs, "had to be brought in by barge to create this raised ... garden." Roses and lavender bloom among the parsley, chives, basil, lettuces, and tomatoes. Elsewhere, lilies and delphiniums, zinnias and cosmos, and "masses of phlox for August" burst into bright seasonal flower.

All of this is against great odds. Each spring, the maker of this place returns in trepidation, dreading "to see what the bears and beavers have nibbled, trampled or simply done away with altogether." The wind and the winter are also hard on gardening dreams. But the survival of roses among the "boulders of the shoreline ... encourages [the gardener's] foolhardiness to plant more." Little Jane Island is not ideal gardening territory, reflects Ms. Eaton, who professes a liking for symmetry and classical statuary, clipped hedges, and ordered vistas, but the fragrance of the flowers borne into the house by breezes off the water is "wonderful" and the "first ripe tomato of the season" is a reminder that these patches of good earth are "a triumph of greed over nature."

Contemplating these words and images, I thought of innumerable others, less fortunate than the owners of Little Jane Island, who had struggled similarly to establish themselves – and to create their own particular "gardens" – across the vast and so often niggardly expanse of Canadian space. The parade of characters and circumstances was long and varied: fisherfolk clinging to the shores of Newfoundland and Nova Scotia, creating rough and tiny patches of cultivation between the grey and tempestuous sea and the barrens at their back; *Canadien* habitants, whose seigneurial rotures (the "long narrow farms" of countless textbooks) joined the eel fisheries and marsh-hay of the St. Lawrence with the forests of the pre-Cambrian shield; middle-class English woman Susanna Moodie and her husband who, in poet Margaret Atwood's re-telling at least, clung to the illusion that the future lay in the stumpy patch of beans and potatoes that they tended in the forest; the shipyard worker in Saint John, New Brunswick, whose calloused hands found tender joy in the flowers his careful ministrations coaxed from the rocky, infertile earth; the mining families, whose houses perched, in Yvonne McKague Housser's emblematic paintings, upon the hard bare rock that yielded the ore that dictated the creation of settlements in such unpropitious surroundings; the Canadian Pacific Railroad employees who seeded and watered and weeded flowerbeds at stations along the line to bring colour to the prairie and comfort to new immigrants far from home; Daisy Phillips, whose memories of English landscapes helped sustain her efforts to establish a home in the Windermere valley of British Columbia before the First World War.[2] No barges of new soil every few years for them. No clipped hedges and statuary where they struggled to find a foothold in the new world. No luxury of retreat during the winter, either, no matter what havoc beasts and ice might wreak in one's absence. But at least most of those of whom I thought shared with Ms. Eaton the possibly foolhardy conviction that it was important to persist, and to plant more, despite setbacks and disappointments, frustration, and despair. Thus were islands of settlement made. Thus were imported hopes, perceptions, and aspirations, reconciled with local circumstances. Thus did necessity (more often, I think, than greed) triumph over nature. Thus were the foundations of a country laid. In this light, and constrained, hard-won and limited though they may be, Canadian gardens are not oxymorons.

I am prompted to these reflections by the fact that Little Jane is one among the "thirty thousand islands" scattered along the rugged north shore of the Georgian Bay to which Claire Campbell draws attention in this book. Revealingly, this handsome volume – as fresh and invigorating as the west wind that provides its title – has relatively little to say about Nicole Eaton's garden. Its author, a lively and perceptive student whose affection for the flat pink granite, cold turquoise water, pale straw-coloured grasses, and stunted white pines of her chosen territory shines through every page of her story, judges that foolhardy creation "painfully *out* of place." Her interests in the Bay and its islands – which we come to know, collectively and in dozens of surprising ways, as a result of her labours – carry her inquiries into other channels; she has limited sympathy for this recent example of imperious colonization. Far from epitomizing facets of a broader Canadian experience with the land, each of Ms. Eaton's boulder-edged plots of soil on gneiss remind Claire Campbell of "an ill-fitting toupee" that threatens to slide from its appointed place with the first winter rains. Her focus is local and regional. She draws attention to a part of North America that, she argues, has been too-long overlooked. She writes of a specific place. She seeks to contextualize landscapes immortalized by members of Canada's most famous artistic movement, the Group of Seven, and thus "recognized" far beyond the Bay by people who have never been there. She aspires to understand what the poet Douglas LePan called a "Rough Sweet Land," the space he knew as the "Islands of Summer." Her Georgian Bay is surely a place of innocent pleasures – "lots of leaky canoes, and the smell of pine needles" as John Irving had it in *A Prayer for Owen Meany* – but it is much more besides. She wants to expose the irony in the words that novelist Katherine Govier has a Canadian in England speak of this place: "If you went, you'd say there was nothing there. No history, not like here." She aims to give this familiar but neglected spot a usable past, to rescue it from its common characterization as a place of escape, *outside* history, to forge a "community of interest" between academic scholarship and local concern, to engage us with the past and present of this extraordinary arc of rock and water, where (to paraphrase Govier) a surfeit of energy beats against the shore.[3]

But do not be misled by this. Claire Campbell is no parish-pump historian devoted to chronicling the arcane and obscure. For all its fine-grained local detail, *Shaped by the West Wind* has much to say to readers everywhere. Canadians, especially, should heed Dr. Campbell's well-told story, both because Georgian Bay has occupied a pivotal place in the long history of the country, and because this fine account of its evolution offers new ways of seeing, and thinking about, Canada. The chronological sweep of the book is remarkable. It ranges over several hundred years, carrying readers into the Bay, metaphorically, in the canoes of Samuel de Champlain early in the seventeenth century, and giving them insight into the conflicts, some three hundred and fifty years later, between cottagers (who want to "save" the Bay) and locals (who live and need to work there). Cartographers, native peoples, painters, poets, novelists, loggers, fishers, engineers, environmentalists, cottagers, and bureaucrats enter the account in various ways in the intervening years – but they never entirely dominate the story, because the land, the environment, nature are a constant foil to their cultural and economic ambitions. In Claire Campbell's telling, the history of Georgian Bay is the tale of a middle ground. She sees her chosen territory as a place of encounter between people and nature. It appears, in her skillfully crafted prose, as a locale shaped and transformed by the incursion and retreat of successive waves of human actors following somewhat different scripts. It is, to borrow and extend one of her many telling metaphors, "akin to an island shoreline" enveloped, exposed, and almost imperceptibly sculpted over time by the storm surges that wash against the rock.

Writing this sort of account is no easy task. The effort to hold space and time, geography and history, in hand together has vexed generations of historical geographers and environmentally minded historians. It has also produced a variety of more-or-less unsatisfactory solutions. These range from geographers' attempts to combine (spatial) cross-sections with (chronological) vertical themes, to Fernand Braudel's grand division of time into three, with the slow play of geographical forces making up history in the *longue durée*, the history of social forms (*conjonctures*) constituting the *moyenne durée*, and the evanescent acts of politics forming the stuff of *histoire evenementielle*. Claire Campbell develops her own way around this conundrum. It gives her study an unusual quality when judged by the standards of much earlier Canadian historical writing. Recognizing that human encounters with, and experiences of, the Bay have differed over time and been refracted through cultural, economic, social, and other lenses that differ one from another, she presents a series of somewhat discrete chapters, each of which focuses on a different type of encounter with the Bay. Indeed, it seems fair to say that this work is organized much as people relate to the natural world – by mapping and measuring, exploiting and managing, mythologizing and representing it. This does some violence to time's arrow, which appears to circle back on itself in successive chapters, but there is order in the kaleidoscope, and the different perspectives are tied loosely to a broad chronological arc that places Champlain comfortably at the beginning and recent debates – over recreational use and ecological integrity in planning a park for the area – appropriately near the end of the account.

More than this, however, Dr. Campbell's efforts to deal with the tensions between change and continuity, and with those between nature and culture, are inflected by her training as a public historian, and by her desire to make her work relevant and accessible to a wide audience. These influences – and the intricate and varied histories of the Bay itself – have yielded a remarkable book. Coherent, lucid, and compelling, it not only ranges across several seemingly discrete fields, it integrates the best of them into a work that is at once a contribution to environmental, cultural, conservation, administrative, and public history, as well as historical geography. Yet even to note this is to risk denting Claire Campbell's achievement, for her book is more than the sum of its parts. By making the landscape the touchstone of her concerns, by refusing to embrace the analysis of representations and imagery to the point that the physical, material actuality of places seems irrelevant, by recognizing that nature and culture are inextricably bound together in the landscape, and by insisting that a focus on landscape invites a broad view of history, she has provided us with a powerful and satisfying account of Georgian Bay that should intrigue and inspire others to explore the fascinating (environmental) history of Canada. By writing her story from a regional perspective and "from the ground up" – and I borrow her phrases to make these points – she shows how ideas shaped, and were shaped by, encounters with specific places, and demonstrates that theoretical concepts ultimately derive their relevance from the local experiences and local sources that anchor them to a specific place and time. Here, in this history, nature and society are treated as inextricably interrelated and dialectical.

This is no mean achievement. For the better part of a century, Georgian Bay has held a special place in the Canadian imagination. It has been portrayed as a site of escape, treated as a generic wilderness symbolic of the country, and taken to embody the essence of Canada. In the process, the region has been flattened into an archetype. Elided with neighbouring Muskoka, and characterized as part of an ill-defined "north country," it has been the locus of a great deal of a-historical and in some sense dis-located sentiment. Examples abound. So the poet, Douglas LePan, as fond of this place as anyone, describes it as existing in a "time without tense." Recall the iconic works of the Group of Seven, almost all of which erase every sign of native peoples and industrial activity (in other words history) from the landscape. Remember Hana, the Canadian nurse in the war-torn Italy of Michael Ondaatje's *The English Patient*, who yearns for home and a "small cabin and pink rock in Georgian Bay." Think of expatriate Australian academic Jill Ker Conway, who memorialized her time in Toronto under the title *True North* and thought that Canadians derived their sense of place from lives lived against such backdrops as "the sparkling water and black rocks of Georgian Bay." Or consider – as Claire Campbell has done elsewhere – homesick Canadian artist Doris McCarthy sitting in London's Hyde Park, feeling the sharp sting of exile and crying "with longing for the feel of a paddle in my hand, the warmth of flat Georgian Bay rocks, [and] the cool ripple of blue water."[4] Inattentive to time and place, all of this places Georgian Bay beyond history and geography and makes it a synecdoche for the nation.

By contrast, *Shaped by the West Wind* insists on the complex regional variety of the Bay and on the differences among those who lived there over time. In doing so, it helps pull back the curtain on the comforting deception that rock-and-pine define the "natural soul" of the country.[5] Between the particular contours of the Canadian wilderness myth that have made it easy for Canadians – or at least those who accept the cultural influence of Ontario – to embrace this view, Claire Campbell maps a new and important topography given form and substance by the long and local, intricate, and contested *environmental* history of Georgian Bay itself. By insisting that the lives of those who moved through and lived in this area (even if only in the summer time – and even if to indulge their own fanciful and foolhardy, but ultimately widely shared ambitions to create a garden in the Canadian "wilderness") are worth the telling, she adds depth and texture to the nationalist façade. By attending to the ways in which generations of people struggled, variously, to make sense of this place, to tame it, to exploit it, to enjoy it, and to live with its limitations, she animates the still life images created by artists' brushes and nostalgic memories. By skillfully weaving land and life together, by paying focused attention to people and place, this book makes it impossible to think of those familiar nationalist images of blue lake and rocky shore in quite the same way again.

In the end, Claire Campbell's book carries my thoughts back to a long-ago essay by anthropologist Clifford Geertz in which he argues that small things raise large questions, and that "winks speak to epistemology and sheep marks to revolution because they are made to."[6] Here Geertz is concerned to establish social anthropology (loosely conceived as ethnography) as an interpretive pursuit in search of meaning rather than an experimental science in search of laws. He insists that ethnographers (and we might add historians here) characteristically construct their understanding on the basis of "exceedingly extended acquaintances with extremely small matters." This habit of mind poses the problem of moving from local truths to grander visions. He finds the two most common ways of addressing this dilemma – finding heaven in a grain of sand or insisting that one's micro-study is a test case that defines "the farther shores of possibility" – to be false and pernicious. To avoid them, he draws a useful distinction between inscription and specification to identify a twofold challenge for interpretive scholarship. The first, inscription or thick description, reveals the ways in which the subjects of investigation conceptualize their lives. The second, diagnosis or specification, deciphers what this demonstrates about their societies. By offering new ways of seeing Georgian Bay and new ways of thinking about Canada, *Shaped by the West Wind* moves Canadian historical scholarship forward on both fronts and fulfils one of the important purposes of Geertz's mandate for interpretive, humanistic inquiry: that of enlarging the discourse of human experience. Savour this book. Its images, its insights, and its often-splendid turns of phrase are well worth pondering. Like "the eagles, ... the fawns at the river bend, / The storms, the sudden sun, the clouds sheered downwards" of Douglas LePan's famous poem evoking memories of a canoe trip through the north, they will help those who attend to them to "face again the complex task" of understanding this curious and fabulous country.[7]

Permissions

An earlier version of Chapter 1 appeared as "'Behold Me a Sojourner in the Wilderness': Early Encounters with the Georgian Bay," in the *Michigan Historical Review* 28 (Spring 2002): 33-62, copyright Central Michigan University. An earlier version of Chapter 5 appeared as "'Our Dear North Country': Regional Identity and National Meaning in Ontario's Georgian Bay" in the *Journal of Canadian Studies* 37, 4 (Winter 2002): 68-92.

Excerpts from the following books appear in this book and permission to use them is gratefully acknowledged:

Weathering It: Complete Poems 1948-87 by Douglas LePan, *The English Patient* by Michael Ondaatje, and *Crossing the Distance* by Evan Solomon, used by permission of McClelland & Stewart Ltd., *The Canadian Publishers*.

Angel Walk by Katherine Govier. Copyright © 1996 by Katherine Govier and reprinted by permission of Random House Canada.

The Immaculate Conception Photography Gallery and Other Stories by Katherine Govier. Copyright © 1994 by Katherine Govier and reprinted by permission of Random House Canada.

The Age of Longing by Richard B. Wright, published by HarperCollins Publishers Ltd. Copyright © 1995 by Richard B. Wright.

Credits for illustrations appear with the illustration.

Acknowledgments

In still waters, things reflect most clearly.
– Camp Hurontario motto

In the stillness that comes from finishing a work like this, what becomes clear are my debts to a great many people. I wish to thank all those who shared their memories of Georgian Bay with me, especially the alumni of Camp Hurontario. Their enthusiasm for my research and their love of the Bay assured me at the outset that what I was doing was worthwhile. Many others provided assistance and information: John Birnbaum and members of the Georgian Bay Association; officials at the Ministry of Natural Resources, Parks Canada, Department of Indian Affairs, and municipal offices in Muskoka and Parry Sound; Michael MacDonald and others at Library and Archives Canada, and the reference staff at the Archives of Ontario; and especially the reference librarians at the University of Western Ontario's D.B. Weldon Library, J.J. Talman Regional Collection, and Map Library. I am very grateful for the financial support provided by the University of Western Ontario, Ontario Graduate Scholarships, and the Social Sciences and Humanities Research Council of Canada during my doctoral studies, and the Killam Trust for postdoctoral funding. Reading and thinking about something that interests you without having to worry overmuch about money is truly a blessed existence. It has been a privilege to work with Graeme Wynn, Randy Schmidt, and Holly Keller as part of the Nature/History/Society series at UBC Press, and I want to thank them for the welcome encouragement and guidance that saw this manuscript to publication. I am indebted to the anonymous reviewers, whose thoughtful comments greatly improved the work; and to the many people who supplied illustrations, especially Ed Bartram.

At the University of Western Ontario, A.M.J. Hyatt always challenged me with his warm but no-nonsense demeanour – I may soon be able to call him Jack. I found an exuberant fellow Bay person in Jan Trimble, as generous with her cottage one rainy day at Go Home as with her laughter. Their combined efforts in the Public History program – and the friends and colleagues I found there – seduced me into graduate work and, ironically, a career in academia. Margaret Kellow was a valued mentor for a young woman scholar. Andrew Johnston offered helpful comments on the original dissertation, along with Bruce Hodgins, Douglas Leighton, and Michael Troughton. Alan MacEachern spiritedly answered endless questions, and is an informed guide in the wilds of academia. I would also like to thank Cecilia Danysk, whose interest in my honours thesis at King's College, Dalhousie, led me to consider graduate school in the first place. I am grateful to the Englesk Institut at Aarhus Universitet in Denmark for the marvellous opportunity to teach abroad as a visiting professor in Canadian Studies while revising the manuscript; my time there is a most cherished memory. My students constantly inspire me as a storyteller and make teaching Canadian history more rewarding than any other career I could imagine.

I was immensely fortunate to begin doctoral studies the year Jonathan Vance arrived at Western. I couldn't have asked for a better supervisor: thoughtful, funny, unflappable, and infinitely patient, his door was always open to offer encouragement, answer a silly question, and puncture my rose-coloured view of the Bay when need be. Imagine my delight when I arrived in Edmonton to find an equally supportive, funny, and knowledgeable advisor in Gerhard Ens. To both, I offer my deepest thanks.

The community of graduate students at Western, though now scattered far and wide, continue to be a source of support and friendship. A note especially to my fellow public historians – a long way from Point Pelee, and I still think we made the best argument for teamwork I have seen in academia; and to the supportive circle of alumna of 4413 and 4415. The Department of History and Classics at the University of Alberta warmly welcomed yet another Ontarian into its midst and by treating me as a young colleague, with humour and kindness, affirmed I was on the right path. Finally a heartfelt thanks to the friends I found there, who share my aspirations, bolster my resolve, and cheer my achievements. And especially to the companion seeking a north-west passage with me.

My family may have wondered what exactly I was doing, but always had faith in me even when *I* wasn't sure. My father insists he deserves co-author credit, but he will have to settle for the knowledge that this would never have been written if it weren't for him, and a promise to go canoeing in the Bay with him whenever he wants.

SHAPED BY THE WEST WIND

Writing a History of Place

The wide expanse of fresh water,
the granite rock, the insistent presence of the elements,
have steadied the perspective when at times
the whole undertaking threatened to become overwhelming.

– Kathleen Coburn, *In Pursuit of Coleridge* (1977)

I fell in love with *Night, Pine Island,* the first time I saw it (Plate 5). *Night,* now owned by the National Gallery of Canada, is a small oil painting by A.Y. Jackson from 1924. It shows an island of coral-coloured granite fractured by crevices, heaped with sworls of rock cradling a dark and still pool of water, the island guarded by a line of upright pines silhouetted darkly against an indigo sky pierced by pure white stars. Despite the incredibly vivid colouring there is a feeling of utter stillness and timelessness. In a heartbeat I was transported back to a summer night on Georgian Bay. I had a similar reaction when I first read Douglas LePan's poem "Islands of Summer"; I was stunned to discover someone had been able to articulate my impression of the Bay so perfectly:

Abrupt granite rising from the clearest
water in all the world. Crowned with a tangled diadem
of blue green foliage ...
And always beneath birdsong the sound of water.[1]

By the time I was five years old, my father and grandmother had each spent thirty years in Georgian Bay, working at a boys' camp called Camp Hurontario, located just north of Twelve Mile Bay. In the camp's dining hall is a series of mottoes nailed to the rafters, which say things like: "In still waters, things reflect most clearly," "To every island there is a lee," and "The west wind shapes the pine." Growing up, I saw how people can become truly passionate about a place: my father, our friends at Hurontario, and, it seemed, everyone I met who knew "the Bay." The west wind had shaped their lives just as it had shaped their beloved landscape.

If the Bay could evoke this reaction on a personal level and in the arts, then clearly landscape mattered – and has mattered more than we often think. From this emerged questions about how people have responded to the Bay in the past. I began to discover a treasure of cultural artifacts ranging from local ghost stories to some of the most famous paintings in Canadian art, which, like the dining hall at Hurontario, express how people have felt about the Bay. This response may be practical or imaginative, personal or collective, but it demonstrates the extent to which our culture shapes the environment or is shaped by it. Our reaction to Georgian Bay reveals our ideas about nature, but at the same time, the landscape itself shapes these ideas. Looking at the silent night on Pine Island, we are looking into the past, at how the Canadian experience has been inextricably intertwined with the natural environment.

By virtue of its size, location, and history, Georgian Bay deserves to be considered one of the Great Lakes. Though inextricably bound to the Great Lakes system in general and Huron in particular, its rich history and distinctive geography do seem to warrant a special status – certainly more than just an arm of Lake Huron. It is almost closed off from the main body of Lake Huron by the limestone ridge of the Bruce Peninsula and Manitoulin Island. At two hundred kilometres long, it is almost as large in surface area as Lake Ontario, and large enough to be among the world's twenty largest lakes. Not surprisingly, the Bay was identified as a separate lake, "Lake Manitoolin," when Capt. William Owen of the British Admiralty drew up the first modern charts of the Great Lakes in 1815.[2] James Barry thus expressed a popular sentiment when he subtitled his 1968 history of the Bay "The Sixth Great Lake."[3] Holding one-fifth of the world's surface freshwater, the Great Lakes bring maritime culture into the continental interior.

The Lakes basin as we know it dates from about 10,000 years ago, shaped by a succession of ice sheets and glacial lakes. Georgian Bay angles northeast alongside Lake Huron, with a shoreline that encompasses four very different geological sections. The southern shore around to the Severn River is the northern edge of the St. Lawrence lowlands; the limestone ridge of the Niagara Escarpment runs through the Bruce Peninsula and Manitoulin and adjacent islands; and the two sections of the Canadian Shield are divided by the French River, the older Huron province to the north (which includes the La Cloche hills), and the younger Grenville Province on the east shore. Incidentally, it is usually known as Georgian

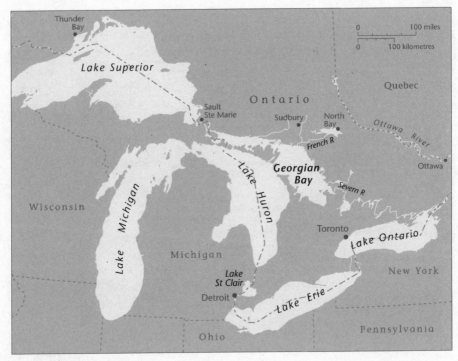

Map 1 The Great Lakes

Bay, but historically references to "the Georgian Bay" are also quite common. I grew up saying "the" Georgian Bay, but I use Georgian Bay here in a concession to modern usage.

This study focuses on the archipelago of the east shore, from the Severn to the North Channel, a territory sometimes referred to as the Thirty Thousand Islands. It is the least hospitable of the four shores: the gneiss bedrock is home to scrubby vegetation such as juniper, white pine, red oak, and cedar, but the acidic, shallow, poorly drained soils could never support the mixed agriculture and density of settlement found on the south or west shores. First Nations used the Bay seasonally for fishing and trading, and whites only began settling the southern islands at the end of the nineteenth century. In human terms, the archipelago is best defined as the vertical corridor of communities that depend on the shore and "derive their character and economic survival from the waters of the Bay." Communities on the islands and inland lakes might use a slightly more stringent definition of water-access only.[4] But it is not a narrow or isolated subject. The Bay is positioned at an intersection of historical influences that have converged from the rest of Ontario, the Great Lakes, and eastern North America.

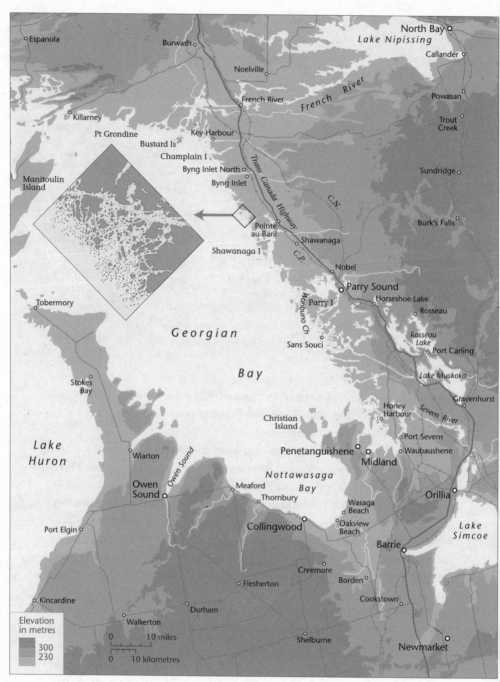

Map 2 The Georgian Bay area. The inset is a detail of the coastline showing some of the many thousands of islands that dot Georgian Bay.

This convergence is evident in the remarkable range of sources that refer to Georgian Bay, though the bulk of these dates only from the past century or so. Cartography, art, and photography offer different visual portraits of the landscape. Literature includes a vernacular record in voyageur, logging, and shipwreck ballads; personal memoirs; fiction and poetry, in which landscape acts as setting and symbol; and tourism material, which, by treating nature as a marketing tool, reveals what it was presumed people sought in the outdoors. Oral histories are a source of colourful stories but also a valuable measure of popular environmental attitudes, user conflicts, and public reactions to political policy. Boat or cottage designs, along with archaeological remnants of underwater cribs, sunken logs, and anchor rings, provide clues to ways in which people used the land. Thanks to the division of powers in Canada's federalist system, all levels of government have been concerned with natural resources. The most important records in this case come from the federal Departments of Indian Affairs and Marine and Fisheries, and the National Parks Branch, together with the provincial ministries responsible for natural resources, municipal affairs, and tourism. Working with such disparate sources – impressionist and documentary, textual and visual, high and popular culture – is challenging, but ultimately it rewards us with a much more complete image of the past. We might "read" the influence of the Canadian Shield, for instance, in geological and mining surveys; in patterns of travel on the rivers that formed in fault lines, or in failed agricultural settlements attempted in stony soils; in architectural adaptations to an intractable rocky foundation; or in paintings or literary descriptions of "miles of ancient rock twisted in lava layers, laid out in swirls, rolled into boulders, opened in crevices."[5] These clues fit together for a coherent telling of a place's history, particularly when the same messages about a place recur across time and in different media.

Studying landscape means taking a broad view of history. The very word "landscape" implies seeing things in context, a context that may be thematic, spatial, or functional. It draws on a wide range of subjects in the humanities and the social sciences: history, geography, political science, economics, literature, art, and design. Integrating different disciplines into a single story of the past is a better reflection of actual historical experience, and may keep academic scholarship accessible to the general public. As these different disciplines became more and more specialized, they became increasingly isolated from one another and less intelligible to the rest of us. This has been a major problem with cultural theory and art criticism, for instance. Canoeist, filmmaker, and painter Bill Mason probably spoke for many of us when he wrote, "Much of what I read about art is beyond my comprehension. In some cases I haven't the foggiest idea what they are writing about, and sometimes I think the writers, reviewers and critics don't know either."[6] Art or literary criticism normally has a very narrow focus: a close reading of a select group of works or the career of a single artist (or writer), emphasizing the formal or stylistic qualities of his or her work. Even then, interpretations vary widely depending on the critic's point of view. One looks at Tom Thomson's work and sees flattened foregrounds, another sees distant horizons and expansive sightlines. Frank Underhill decided the "tortured" rocks and trees in the paintings by the Group of Seven were an unconscious metaphor for the urban wasteland. Evidently he was unaware that the rocks and trees *are* twisted and ravaged, a physical reality that has nothing to do with Toronto's moral bankruptcy or the Group's psychological state and everything to do with the effects of the glaciers and the northwest wind.[7] And the enduring popularity of the Group is a classic example of how the opinion of the general public – who believe they know "what they like about art" – is rarely mirrored in the writing of professional art critics. But for the historian, art (and art criticism) can be a valuable piece of evidence when seen as an illustration of attitudes toward nature at a certain time and place.

In the latter part of the twentieth century, though, nature fell out of favour among cultural scholars. It has been decades since Northrop Frye stated that "everything that is central in Canadian writing seems to be marked by the imminence of the natural world" – a conclusion that many would now consider dated in its fidelity to geography.[8] (Curiously, arts criticism turned away from the environment just as other disciplines, such as folklore, landscape studies, or geography, began to consider it in new ways.) Equally surprising is the dearth of regional studies about cultural production in Ontario. Fortunately, the interest in regional cultures elsewhere, specifically the Maritimes and the Prairie West, maintained the interest in landscape and identity. Indeed, scholars of Ontario and Quebec would benefit from conceptualizing central Canada as a region more often.[9] Ontario has always assumed it speaks for Canada; the Group of Seven, who made a name for themselves by painting northern Ontario, somehow became Canada's national artists, and the Canadian Shield is held up, by the Group and others, as the landscape most representative of national experience. Yet, in a cultural mosaic of Canada, Ontario is likely to be represented either by the Toronto skyline or by small-town life in the southern part of the province, by writers such as Stephen Leacock, Robertson Davies, and Alice Munro. Whether their Ontario is rural or urban, social rather than environmental factors are the main concern. "Wilderness" is reserved for the agricultural frontier of the early nineteenth century (for instance, Catharine Parr Traill and Susanna Moodie) or the far north. We have yet to fully catalogue the cultures of Ontario's diverse landscapes, including its resource and recreational frontiers. It seems odd that no one has asked why and how an iconic landscape was created from the Shield and pine of the near north, especially given the major figures in the arts associated with Georgian Bay, from Paul Kane to the Group of Seven, from Margaret Atwood to Douglas LePan.

As one historian has observed, "Lake Huron and Georgian Bay remains one of the best kept secrets of Canadian history."[10] The region has been a significant cultural, economic, and political resource in modern Canada but has never really been studied, apart from a few popular histories. James Barry was the first to devote a full study to it, with *Georgian Bay: The Sixth Great Lake*. In 1994 Lynx Images published *Ghosts of the Bay: The Forgotten History of Georgian Bay*, followed in 1996 by *Alone in the Night: Lighthouses of Georgian Bay, Manitoulin Island and the North Channel*. As part of a series of popular books and videos about the Great Lakes, these effectively combined a dramatic, even lurid, writing style, with detailed directions to shipwreck sites around the Bay. General histories of the Great Lakes rarely mention the Bay, a remarkable oversight in light of the history and geography it shares with the rest of the Lakes. It is given a small part as either seventeenth-century Huronia with explorers and missionaries at the dawn of a saga of nation building, or as the site of rising industrial centres at Midland, Owen Sound, and Collingwood in the mid-nineteenth century. This is because historical writing on the Great Lakes has traditionally focused on shipping – the romance of the age of sail followed by the energetic progress of the age of steam – and the golden age of industry with its sense of optimism and prosperity. The natural environment is simply a resource incidental to "progress" or a backdrop to human heroism in "the eternal battle of men against the elements, of courage shining in a dark hour, of sailors going down with their ships ... fighting to the last."[11] Local histories also tend to be anecdotal: biographies of ships, lumber companies, or cottage communities told in isolation from their cultural or geographical setting. The story of the transformation into an overwhelmingly recreational landscape in the twentieth century has not yet been explored, yet the complexities and ramifications of this industry are as important and as intriguing as any aspect of the region's history.

That said, there are excellent examples of regional writing in the Great Lakes area, many of which emphasize the role of the natural environment. Though Victoria Brehm savages the existing historiography of the Great Lakes, she makes an impassioned argument for regional history when written as part of a continental history as well as a place-based history.[12] Patterns of exploration, transportation, industry, political evolution – all these pull us outward from a locality. Bruce Hodgins and Jamie Benidickson have chronicled how competing actors converged on a single landscape in northern Ontario in *The Temagami Experience: Recreation, Resources, and Aboriginal Rights in the Northern Ontario Wilderness* (1987). The collection of essays in *The Great Lakes Forest: An Environmental and Social History* (1983) uses a regional ecosystem as a starting point for historical analysis. An increasingly common term in the literature is "bioregionalism," which essentially means place-based behaviour: using natural systems as a reference for human agency or activity, and the idea that human societies should exist within patterns set by the natural environment; incorporating indigenous knowledge and community participation into environmental management to respect local economic needs; and what Doug Aberley calls a "defiant decentralism." John Wadland explains the benefits of this approach to the historian:

> This bioregional perspective, far from being parochial and inward, is essential to an understanding of national history apprehended as the relationship between its parts. It becomes quite literally a grassroots study, seen from the ground up. It is a history meant to be guided by a felt sense of place, attuned to the subtleties and nuances of the particular ... On the ground, and at the root, is the landscape of the place. No cultural activity occurs exclusive of the environment required to sustain it.[13]

Taking a thematic approach, asking questions about the relationship between culture and environment, gives regional history a theoretical sophistication. Yet, at the same time, regional experience and local sources give academic concepts a much-needed relevance by anchoring them to a specific place and reality.

This has been the trend in environmental history outside the Great Lakes as well. Many of the best-known works in the field are in fact histories of environmental thought. Roderick Nash's *Wilderness and the American Mind* (1967), Donald Worster's *Nature's Economy: A History of Ecological Ideas* (1985), and Max Oelschlaeger's *The Idea of Wilderness* (1991) trace the evolution of ideas about nature in western culture from biblical times, through the science of the Enlightenment and its answer in Romanticism to their descendants in the utilitarian and preservationist wings of modern environmentalism. But we need to know what people were thinking of and looking at when they expounded on "nature" – in other words, their physical as well as their societal environment. Increasingly, historians have sought to situate this kind of intellectual history in a regional and experiential context, to show how these ideas worked out in practice in specific places. Some studies have been more oriented toward functional relationships between humans and nature, as in William Cronon's *Nature's Metropolis: Chicago and the Great West* (1991), or the ecological consequences of human interaction, as in Worster's *Dust Bowl* (1979) or Barry Potyondi's *In Palliser's Triangle: Living in the Grasslands, 1850-1930* (1995). Others have concentrated on policy, particularly park policy, as applied to local communities, as in Alan MacEachern's analysis of Parks Canada and its agenda in *Natural Selections: Natural Parks in Atlantic Canada, 1935-1970* (2001). Still others have explored how a place is constructed differently over time through its representation in popular culture, scientific thought, and the arts. Doug Owram applied this approach to the Canadian West in *Promise of Eden: the Canadian Expansionist Movement and the Idea of the West, 1856-1900* (1980), while Stephen Pyne focused on the creation of meanings for a single physical phenomenon in *How the Canyon Became Grand* (1998). And there is new interest in regional communities that evolve as social and environmental, if not political, entities, like that in Beth LaDow's *The Medicine Line: Life and Death on a North American Borderland* (2001).

My study draws on all these different approaches to some extent. Chapters 1 and 2 discuss the attempts to establish some kind of useful or profitable role for the Bay. Chapter 3 notes the historical image of the Bay (and its Aboriginal inhabitants) in popular culture, and Chapter 5 traces its changing representation in science, literature, and art, as well as the emergence of community identity in a process of regional definition. By focusing on the record of encounters with the two defining elements of the Bay – rock and water – Chapter 4 presents a different perspective on daily life on a Great Lake and the evolution of a "maritime" identity. Chapter 6 explores the experiments with conservation and parks, and the emergence of community activism. Like these other historians, I am interested in the interplay between historical context and local conditions. In this case, it means the reactions to a place for which people were frequently unprepared. Often Georgian Bay challenged the usual attitudes toward nature, but sometimes it reinforced them: as Chapter 3 demonstrates, its physical and racial features (notably the First Nations) could illustrate and affirm popular stereotypes about wilderness in Canadian culture. Writing environmental history from a regional perspective and from the ground up shows how ideas shaped, and were shaped by, encounters with specific places.

This was not always the case. For a century, North American historians wrote from the outside looking in, envisaging landscape in grand, national, almost abstract terms. Frederick Jackson Turner, for example, pronounced the American frontier as the formative influence of a national character – a thesis which itself became a formative influence in American historiography. For Canadian scholars, continental geography provided a natural basis for a national narrative, an explanation – and justification – of a transcontinental dominion of the North. The expansive foundation of the Canadian (or Laurentian) Shield was particularly important to demonstrating a natural coherence, and so lent its name to much of the writing (and art) that emerged from central Canada in the early and mid-twentieth century. Their territorial imaginings were shaped by different emphases – for Harold Innis, it was the technologies used to penetrate the continent and to export natural resources or "staples"; for Donald Creighton, it was the dynamics of a commercial empire based in the St. Lawrence valley. But they shared a fascination with the vastness of the continent and, in particular, its waterways as transportation routes, as well as a belief that political union emerged naturally from economic patterns that in turn were derived from geography. These theories rested not only on the Canadian Shield but on rather imperial assumptions about the importance of centralizing forces and the leadership of Central Canada's urban centres – assumptions that implicitly fashioned other parts of the country as hinterlands and supporting players to the national story. The preoccupation with Shield country was closely related to another fundamental aspect of historiography and nationalist thought: expansive, heroic ideas about the North. Scholars identified the Canadian Shield as northern terrain and, as such, the defining landscape of the national character.[14]

By the 1970s, however, the landscape of historical writing looked quite different. First, the homogeneity of a national framework, geographic or historical, had come under attack. Led by William Morton's early criticism of its dismissive attitude toward hinterlands, and energized by J.M.S. Careless's call for writing history by tracing the evolution of "limited" rather than national identities, historians embraced a variety of regional landscapes. To borrow Cole Harris's wonderful metaphor, the illusion of Canada as a single unit bound by continental networks dissolved into an archipelago of disconnected islands.[15] Second, with scholars across the humanities and social scientists, historians increasingly approached historical records and landscapes as texts – an artifact of human thought and action – rather than as an objective reality. Reading historical documents as a subjective account, one which expressed a point of view, undermined the concept of a single narrative of Canadian history, as did the interest in writing a region's history as a product of a distinctive combination of historical factors apart from a single central actor or infrastructure. It also led to some profound questions about what we mean by nature and wilderness, and the recognition that such concepts are both historically and culturally specific:

> Not only is nature affected pervasively by human action, but our very conception of nature has emerged historically and differs widely from one cultural tradition to another. What we mean by nature, our beliefs about wilderness, the recognition of landscape, our very sense of environment have all made a historical appearance and been understood differently at different times and places.[16]

Influenced by postmodernism and its concern with power relations, competing voices, and symbolism, scholars also became interested in the ways in which landscapes are constructed and contested: constructed according to factors such as class, gender and ethnicity, cultural preferences and political agendas; and contested between different groups.[17] Daniel Clayton, for example, explored the competing geographies and spatial politics of First Nations and Europeans in *Islands of Truth: The Imperial Fashioning of Vancouver Island* (1999). Whites and First Nations have had economic dealings in Georgian Bay since the early seventeenth century, and non-Natives have selectively borrowed ideas and practices about the "wilderness" of the Thirty Thousand Islands from Native residents. I, however, deliberately limited my study to perceptions of Natives in Georgian Bay – in part out of respect for cultural difference; in part because Native history and pre-history draws so heavily on disciplines such as archaeology and ethnography, in which I am very much a layperson; and in part because other historians have profiled these Native societies in compre-

hensive and admirable research, as Bruce Trigger has in his work on the
Huron. There are profound differences between European and Aboriginal attitudes toward nature, and complex historical and pre-historical
relationships between the Wendat, Iroquois, and Anishinabe, which are
beyond the scope of this work.[18] Most of the records of Georgian Bay were
produced by a fairly homogenous group of generally white, generally male,
and generally middle-class tourists or seasonal residents, who delivered a
consistent message about the place. Those sent to investigate the Bay in
terms of official agendas for the hinterland reported its uselessness; those
who came for leisure described a blissfully recalcitrant wilderness. So it is
impossible to write a history of place that does not acknowledge cultural
processes of perspective and negotiation. But in the past decade some
historians have challenged the concept of nature as a purely cultural creation. Instead, they suggest thinking of our relationship with nature as
dialectical, and "a process of discovery and adaptation."[19] Place is a site of
exchange where influence flows in both directions. We need to understand the cultural assumptions that people brought to Georgian Bay –
and the physical setting in which they found themselves.

The relationship between nature and culture lies at the heart of landscape study. The landscape school emerged in the 1920s within the field
of historical geography, as a reaction to the environmental determinism
that characterized the discipline at the time. Instead of emphasizing the
effect of nature on society – such as the effect of the frontier on American history or the American character – the landscape school, led by
geographer Carl Sauer, inverted the relationship to examine the impact
of culture on the environment. As Sauer said, "Culture is the agent, the
natural area is the medium, the cultural landscape is the result." This is a
critical distinction for scholars: unlike the "environment" or "nature,"
landscape is a synthetic space. It is a product of human history, interests,
and ideas.[20] Accordingly, landscape geographers were concerned particularly with man-made, material features, artifacts (such as architectural
and settlement patterns) that demonstrated changing cultural impera-
tives over time. For example, Thomas McIlwraith's *Looking for Old Ontario*
(1997) is a visual reading of the built and natural landscape, reading architecture and patterns of settlement as a kind of archaeology of past use
and past ethnic influences (American, British, and so forth) on community designs. Landscape study is intrinsically geographical because it be-
gins with a particular place – its distinctive appearance, or collective
knowledge about it. But it is concerned with the history of that landscape's

construction, and the cultural forces applied to it. Not surprisingly, the emphasis on culture has made landscape study extraordinary fluid, increasingly interdisciplinary, and theoretically complex. In the past few decades, it has embraced a range of strongly humanistic approaches that have also appeared in related disciplines, such as historical geography and cultural history: psychological and emotional perceptions of nature and the bases of place attachment; reading ideas about nature in artistic representations of place; and the use of physical places as a means to secure social power, community identity, or group cohesion. Landscape study is about the making of places, not only through functional relationships with the natural environment, as was the case before, but as the "subjective and intersubjective construction of places through imagination and discourse." As in the field of history, there has been a shift from identifying landscapes of national concern to studying more localized ethnic, class, and regional landscapes.[21]

Research about landscape can often be applied only tangentially, because the study of landscape is by definition essentially place-specific. Whereas environmental history has been concerned with discussing ideas about wilderness, culture, and nature in a broad sense, landscape study is rooted in the study of particular (and especially vernacular) communities and places. Yet, increasingly, scholars have sought to draw from both in order to understand the evolution of landscapes. Simon Schama's *Landscape and Memory* (1995) is arguably best described as an intellectual environmental history, for it explores the symbolism and myths of nature in Western culture. But by locating these myths in particular times and places – from classical Rome to Elizabethan England to Nazi Germany – he demonstrates that they are not universal.[22] The meaning of nature depends largely on its relationship with the local culture. A concept like wilderness means different things in different communities and in different national myths. But this raises another problem. How much information can be gleaned from Renaissance Tuscany, early modern England, or the American South and put toward understanding twentieth-century Ontario? Such place biographies favour the study of distinct, long-established, coherent settlements in rural areas (or to a lesser extent, urban neighbourhoods), where there is a visible ethnic or class imprint. The seigneurial system in rural Quebec would be an ideal choice; Georgian Bay, a resource and recreational space with a history of transient populations, is not.[23] And yet Canada, a postcolonial nation-state of enormously diverse geography, may be uniquely poised to contribute to the field of landscape study precisely for this reason: endless case studies of encounter and adaptation, as ideas of nature inherited from European colonizers are acted out in distinctly North American environments to produce new forms.

That said, one of the most common, and most useful for our purposes, ways of thinking about landscapes is the classification of vernacular, designed, and associative landscapes. Vernacular landscapes are usually working landscapes that evolve organically over time, often in rural communities. Designed landscapes are ordered arrangements such as the formal gardens created for aristocratic Baroque estates. Ethnographic or associative landscapes possess cultural or heritage value for a particular group.[24] Any place, though, can exist as more than one type simultaneously. Georgian Bay and the Canadian Shield are easily associative landscapes ("Oh, like the Group of Seven" is the most common response I get when I tell someone I study Georgian Bay) but, after a century or more of settlement, they have elements of the vernacular as well. Vernacular landscapes in particular are indicative of the adaptation of imported agendas to local conditions; early cottages in Georgian Bay were shaped by both the pragmatic concerns of local builders and cottagers' enthusiasm for rustic "wilderness" escapes. Not only are the boundaries blurred between these classifications but so are the lines between "here" and "there." I agree that any cultural landscape is the product of human imagination – how we think of a place, and what we learn to expect from it, determines how we act there and how we alter that place. But these ideas and expectations are best understood not only through in situ artifacts such as architecture and patterns of settlement but by artifacts such as art and literature that record encounters with the natural environment and then are exported from the place to shape a wider collective imagination. How Canadians think about Georgian Bay, in other words, is shaped by the early cottages of Go Home Bay and Cognashene, but also by the National Gallery of Canada.

Indeed, scholars in a variety of disciplines have been addressing the subjects of environment, region, and identity in different ways. Much of the environmental history in Canada has come from historical geographers. This is particularly evident in Atlantic Canada, and the work of scholars from Andrew Hill Clark to Graeme Wynn. Though they retained a geographer's interest in spatial patterns, concerning settlement and land use, they turned to archival research and a regional focus – in the case of Clark, at a time when Canadian historians were still preoccupied with national and political stories. Closer to home, Conrad Heidenreich's

study of Huronia announces that one of the principal themes of historical geography is the reconstruction of past landscapes; in this case, in
order to assess the Huron's functional relationships with their physical
environment. Like their compatriots in landscape study, however, historical geography is increasingly diverse in its methodologies, encompassing
the spatial and quantitative interests dominant in geography and social
history in the 1960s, the newer "humanism" and theories about cultural
analysis, and ground-level experience.[25] Here they share interests with
scholars of folklore, who study the concept of a sense of place, "homelands," and how experiential knowledge of a place is contained in such
diverse sources as maps, material culture, and personal narratives. The
folklorist literature, however, often selects a single ethnic community to
study how it has incorporated the landscape into its cultural life or bonded
to that place, not necessarily how the landscape set out boundaries between communities.[26] It is important to remember that in Georgian Bay
there is no single, homogenous population responsible for a local vernacular; First Nations, seasonal resource workers, and cottagers all contributed to the regional culture. Oral history is another indispensable
tool in local and environmental history because it unearths information
not found anywhere else, about unusual subjects such as youth camping,
ice boats called scoots, or cottage communities, and from the perspective
of people with a historical investment in the area. And it is a record of
lived experience, to complement the prescriptive sources often used in
environmental and social history. Reading only camp brochures or the
correspondence of camp directors, for example, would suggest children's
camping is a highly structured and manipulated social experience, unless it is balanced by campers' own thoughts about their exposure to wilderness. The study of tourism originated in literary criticism with a critical
reading of travelogues, though this tends to emphasize the judgmental
perspective and the preconceptions of visitors commenting on a new place.
Nothing in England's Lake District had prepared poet Rupert Brooke for
the Great Lakes: "I have a perpetual feeling," he writes, "that a lake ought
not to be this size."[27] More recently, scholars have focused on the consumption of nature through the construction and marketing of tourist
space. Patricia Jasen's *Wild Things: Nature, Culture, and Tourism in Ontario,
1790-1914* (1995) addresses the Ontario experience, but some of the most
interesting work in tourism, heritage construction, and contested landscapes has been done by scholars in the Maritimes.[28]

The emphasis on representation, imagery, and ideal or constructed landscapes combined with the critical textual readings of cultural studies has

infused landscape history with a kind of skepticism. Scholars assume that when people described a place, they were expressing learned prejudices or agendas typical of the age. A landscape painting or a travelogue is as much a portrait of contemporary aesthetic ideals – what people thought nature should look like – as of a particular landscape; a picture of what the artist expected or wanted to see (or thought his or her audience expected or wanted to see) rather than what the artist actually saw. A scene might be described negatively because it did not match expectation, some details omitted and others emphasized to conform as closely as possible to the prevailing ideal. Or it may be framed for visual consumption by dressing, for example, the landscape of North America in the softened pastels of contemporary British watercolours. This kind of scholarship has emphasized the cultural and mental arena into which nature is received and manipulated, the "perception *of* landscapes rather than experience *in* landscapes." Maps are read as an ordering of place, a projection of political agendas and imposition of cultural authority; art as a mechanical application of learned conventions and aesthetic preferences; architecture as a product of social exchange; place identity as a psychological, cognitive phenomenon.[29] The distinct physical features of a region are often pushed to the background. W.H. New's study of space in Canadian literature deals entirely with "the politics involved in the possession of the language of place"; Greg Halseth's interest in the "politics of turf" makes little mention of the nature of the turf itself.[30] What the place actually is often seems almost irrelevant.

This is an unfortunately anthropocentric and misleading way of understanding history. Nature "cannot be obliterated," Ronald Bordessa has written; "geography is beyond denial ... at the very heart of our understanding of what Canada is and who we are as moral and social beings."[31] Studying a map or painting or government report, we obviously need to consider the author and the context in which it was created, asking what was considered beautiful or useful or valuable at the time. But to assume that our response to nature is entirely predetermined by cultural circumstances – that someone describing Georgian Bay in 1850 is really only expressing the prejudices and opinions of 1850 – ignores the dynamic relationship between environment and history, and between individuals and their surroundings. We may experience nature subjectively according to our personal, social, and historical circumstances, but it also exists independently of us. "You can't create a landscape," A.Y. Jackson flatly told *Canadian Art* magazine in 1960; "I can't make them up and I don't know anyone who can."[32] Descriptions of Georgian Bay reflect mainstream attitudes toward the environment, but they are also the product of the specific circumstances of the Bay itself. Storms that blow in from "the Open," the open water to the west, appear in the Ojibwa propitiation rituals witnessed by Alexander Henry in 1761; the mysterious loss of the steamer *Waubuno* and all its hands in November 1879; the power in J.E.H. MacDonald's 1916 painting *The Elements,* which signalled a new direction in Canadian art (Plate 8); and Douglas LePan's thrilling experience in the poem "Red Rock Light," "cowering and exultant / in a November gale."[33] None of these responses could have been evoked in a vacuum. Looking at the vibrant coral and azure of Arthur Lismer's painting *The Happy Isles* (1925), critic Barker Fairley acknowledged that it reflected "the intense mood of the beholder," but added, "no other part of the world could have helped the artist to such a result."[34]

Perceptions of nature need to be situated in their historical and geographical contexts. Pinning history to a particular place emphasizes the influence that the physical environment has had over human history. This book is the story of the reactions people have when they encounter an unfamiliar landscape: the plans they concoct for it, and the changes it forces them to make. "Agency" is a term historians sometimes use to describe how people have influenced their surroundings and the historical process. As Ted Steinberg writes, we need to take a more humble view of human agency. Nature is not a passive object of human design, but something capable of upsetting human plans and directing historical events.[35] The environment is a factor, a force that has never been controlled or predicted, and a consideration in almost every decision made in Georgian Bay. Assigning it "agency," however, suggests a personification of nature, a consciousness or agenda that isn't there. Instead we might think of human agency working within the constraints and opportunities nature provides. People have to always contend with nature. Unlike the older kind of determinism, which took physical geography as a starting point, guiding principle, and permanent force directing human history, this approach sees nature and people as interrelated and dialectical. Each constantly affects and reshapes the other.

The same is true of that double helix of history, change and continuity. Here too, focusing on a single place reveals how supposedly opposing forces coexist and interact. On the one hand, Georgian Bay challenges the assumption that nature has been seen differently over time. It resisted manipulation, physical and imaginative, and people had to respond to the same basic elements. A surveyor in 1868 and a cottager in 1998 faced the same granite bedrock, just as a fisherman in 1850 and a yachter a century later shared concerns about dangerous shoal-filled waters (albeit with different technologies). At the same time, the archipelago was constantly buffeted by the ideas, events, and changes of modern North America. The "wilderness" of popular myth has been criss-crossed by Native peoples, invaded and harvested by industries, and settled by cottagers – a succession of users with different expectations of the Great Lakes and the interior. To distill the story into one of environmental resistance or steady progress would make the history of both Georgian Bay and attitudes toward nature too simple. Sometimes people adopted a new type of transportation or modified a park policy because inherited technologies or methods simply didn't work well, and it was easier to adapt to the physical reality of the archipelago than to alter it. But in many ways, the archipelago has been transformed by human presence. If Samuel de Champlain and his Huron allies were to paddle through the Bay today, their canoes rocked by motorboat wakes, they would see a largely foreign landscape: a shoreline divided into cottage properties and Indian reserves, waterways marked by Coast Guard buoys herding a steady stream of traffic along the main boat channel. We might think the Bay of a century ago equally alien: our outboards trapped by massive timber booms, the famous wind-bent pines surrounded by pine slash, wharves at Parry Sound or Killarney laden with huge nets of whitefish. Yet, the landscape would still be recognizable. Erase the boundaries of a provincial park, and the La Cloche Mountains still gleam white from the north. Remove the huge shore light at O'Donnell Point, and the exposed point is no less dramatic or dangerous. It took billions of years to weather the Canadian Shield into these contours; they have not changed in a matter of decades.

What defines Georgian Bay as a region, then, is its history of negotiation and interaction. Human activity and ideas were constrained by the natural environment even as they pushed against its limits. The archipelago is a littoral, a zone of transition between sea and shore; its history too is a littoral, between human and natural forces, and between competing and coexisting perspectives. No single vantage point encompasses the history of the Bay; rather, it has always been refracted through several perspectives.[36] We may experience a place in several ways simultaneously. This is why this book is structured as it is. Each chapter profiles a different type of encounter, a different way of seeing the Bay. Within each perspective we can follow the tension between change and continuity, and between nature and culture. Chapter 6, for example, focuses on the political arena. Competing political actors, driven by changing economic motives, vie to manage a "wilderness" that constantly challenged the usual

Figure 1 Scenery from Killarney.

strategies of conservation or preservation. Circling through these different perspectives – explorers and surveyors, loggers and fishermen, campers and cottagers, artists and historians, politicians and bureaucrats – gives us a three-dimensional view of the Bay. No one characterization of nature has been exclusively responsible for shaping its public identity.

The first chapter "surveys" four centuries of human history in Georgian Bay from the perspective of mapmakers and surveyors, from the period of contact between Aboriginal and European explorers to modern-day marine charting. Mapping was one way in which people sought to make sense of the apparent chaos of the archipelago, to orient themselves in a geographical location, and to order it in the ideological and political contexts of the day. This chapter provides a chronological outline of the past four hundred years by introducing most of the major players who have asked, where is here? and whose maps explain their purpose on the Bay: continental exploration, military campaigns, industrial traffic, and cottage sales. Nineteenth-century naval and government surveyors played a particularly prominent role in authoring the enduring characterization of the archipelago as a hostile wilderness. Different kinds of mapping and different generations of mapmakers also articulated the most prevalent attitudes toward nature in North America: pragmatic, utilitarian, romantic, artistic. These reports emphasize the labour and practical diffi-culties involved, but the process of mapping also draws attention to the representational or imaginative dilemmas, and the recent political reper-cussions, of the struggle to reconcile measurement with description, ab-stract reasoning and geometric lines with lived experience. Maps were one way of providing information about the Bay, and demands for such information inevitably arose from plans to use the region in some way. This leads naturally into an overview of changing land uses and the suc-cession of industries that attempted to gain a foothold in the archipelago. Chapter 2 opens with a discussion of the declining fur trade and moves through the disappointment of agricultural settlement and the mixed success of mining exploration; the technological ambitions of canal and railway construction, and the healthy shipping traffic on the Bay; and the fortunes of the best-known industries in the region, the commercial fish-ery and lumber trade. I spend some time showing how forests, in particu-lar, were reinterpreted from a physical to an aesthetic resource as the Bay was remade from a resource to a recreational hinterland with the arrival of the final "industry" and the colonization of the islands by cottagers. Each group of users sought different things from the Bay and evaluated it differently as "nature," but they all adapted in some way to the limits of the landscape. The industries that endured – shipping and recreation – were usually those that built on those very limits.

Despite this activity, the archipelago's resistance to development and a history populated by figures of classic Canadiana (such as Samuel de Champlain) fuelled an image of the northeast shore as a primeval wilder-ness, a relic of an earlier age, and an intractable frontier. The construc-tion of that image – and its implications – is the subject of Chapter 3. The obstreperous landscape that defied physical manipulation lent itself more easily to intellectual and imaginative uses because it suited so well popu-lar romantic, anti-modern, and nationalist ideas about wilderness. The First Nations were particularly convincing proof of its wilderness status, and were deeply affected by their association with a *terre sauvage*. Histori-cal references became an essential part of the region's public identity. Recreationists were drawn to the Bay because it promised a physical link to a heroic past, and supposedly a chance to experience the kind of wil-derness experienced by the explorers and voyageurs. For those living and working in the archipelago, however, the unforgiving realities of an in-land sea were a much more immediate and much less romantic way of participating in history. Shared experience with an inflexible environment has been an important source of continuity and tradition for the community "up the shore." Much of the historical record can be traced directly to the basic elements of water and rock in a maritime environment (exposed open water, a confined archipelago, storm weather). Chapter 4 explores some of the pragmatic adaptations and aesthetic expressions directly inspired by these physical features. A wide range of cultural artifacts, from paintings to architecture to boat design, represent the responses of various users to similar conditions. I look at the atmosphere of peril in a dangerous environment and reflect on aspects of the environmental and historical complexity of a *mer douce* and an inland sea.

Its maritime history differentiates the archipelago from the interior, a difference integral to the local sense of identity. The process of defining that locality is the subject of Chapter 5, which examines when and how people acknowledged its distinctiveness and set out its boundaries. In other words, it is about the gradual construction of a regional identity, in both high and vernacular culture, by defining a place from without and from within. I argue that even the standard colonial languages of science and art begin to incorporate place-specific experiences in their descriptions of nature when they confront the Bay. The evolution is most famously expressed in the extensive body of work by the Group of Seven. Next I trace the long history of distinguishing the archipelago from surrounding landscapes within Ontario and the Lakes basin based on a recognition of their physical differences. Residents developed a prized sense of difference – and distance – from the rest of "cottage country" by emphasizing the distinct ways of life on the islands. Yet, the irony is that once Georgian Bay was recognizable as a distinct locale, its leaning pine and rocky shores became a highly visible symbol of nationalism in Central Canada. Imagining the Bay in the political realm is discussed in Chapter 6, which situates Georgian Bay in the history of North American environmental policy and management. "Protective" designations followed the arc of changing use from resource to recreation and were usually designed to facilitate those uses. Measures directed toward fish, forests, and parkland in Georgian Bay reflected the larger shift from conservation to preservation, from protecting resources for industrial production to providing recreational space. I emphasize, though, that the unique landscape of the archipelago often challenged the usual understanding of environmental management and again produced regional variations in standard policy. The Massasauga Provincial Park is one example of how typical park planning (with its divided loyalties to recreational use and ecological integrity) took shape within the physical and political constraints of the archipelago. This chapter brings us to the present and to some of the issues facing the Bay today, as a landscape inhabited by many stakeholders and layered under multiple jurisdictions.

This might seem a disjointed way of writing history but, in fact, the kaleidoscope of perspectives is tied to a chronological arc. (Chronology is still comforting, and necessary, for historians as much as for anyone.) An unknown frontier is exploited as a resource hinterland and then settled as a recreational space. Cultural expectations confront a particular physical reality and gradually define a distinct regional identity. The book opens with a discussion of the First Nations who met the explorers of the seventeenth century and ends with a look at the dilemmas of managing an international marine environment in the twenty-first century. The work as a whole is animated by the spirit of public history. I wanted to establish the significance of Georgian Bay in Canadian historiography but also to make the research relevant and accessible to a wider audience. I wanted to study Georgian Bay as an academically trained historian but also as someone who has loved the Bay since childhood. This is not hard to do – because in my mind's eye I am still simply looking out to the Open, listening to the water and the wind in the pine.

ONE

What Word of This Curious Country?
Surveying the Historical Landscape

Now what shall be our word as we return,
What word of this curious country?

– Douglas LePan, "Canoe-Trip" (1948)

By the time Douglas LePan wrote his famous poem "Canoe-Trip," Europeans had been trying to describe, draw, and chart Georgian Bay for three and a half centuries. Explorers, surveyors, and artists mapped the landscape in different ways, using everything from hydrographic charts to travel journals to poetry. Mapping encompasses science and art, documentation and interpretation, the pragmatic and the imaginative. Consequently, maps – in any form – contain a great deal of information about an explorer's personal response to the landscape, and about contemporary ideas of nature and the physical world. Tracing this tradition of mapping allows us to survey changing attitudes and agendas toward the environment of Georgian Bay. Asking, where is here? – a familiar question to Canadians – is a logical starting point for exploring new territory.[1]

Exploration of the New World was a product, in both spirit and technique, of the scientific revolution of the sixteenth and seventeenth centuries. New theories of astronomy, for example, became integral tools in maritime exploration and navigational technology. The Enlightenment placed unlimited faith in human reason and generated a new confidence about the human ability to understand and order the natural world. But despite its scientific pedigree, scholars now emphasize the subjectivity and selectivity of cartography. Maps are cultural texts, products of the intellectual and social order of a specific time and place, as well as of the practical and technological constraints of the day.[2] Seventeenth- and eighteenth-century mapmakers had to rely heavily on existing maps, and in 1821 Lt. Henry Bayfield drew on "good information of the nature of the North Coast of the Lake from Traders" even as he sounded the Bay over the side of an open boat.[3] At the same time, a map is a functional document, commissioned for political or economic purposes, which must represent the world in ways that are comprehensible and meaningful to its audience. Continental maps of *la Nouvelle France* or British North America asserted different territorial claims to the New World; Bayfield's charts spoke to the British Admiralty in the precise hydrographic language of the scientific revolution (see Map 3). In Georgian Bay, exploration and surveying were motivated first by military strategy and boundary definition; then by shipping and, half-heartedly, settlement; and finally by cottage sales. The maps, surveys, and field notes chronicle attempts to make sense of a place that did not always conform to their expectations of "nature," and to order that space as best they could.

Maps are thus integrally tied to prevailing ideas of nature. But they also record the reactions that these men (and in this situation they are all men) had to the physical environment of the Bay. Although shaped by cultural convention, mapping is the product of an experiential involvement with a particular place, an "art of the immediate."[4] Surveyors sought to be systematic but were never detached; nature for them was not easily reduced to the standard, abstract symbols of conventional mapping, especially in a place as complex as the archipelago. Men forced to dangle a plum line with one hand while beating off mosquitoes with the other or struggling across swamps in the rocky mainland constantly complained that the region "introduced practical difficulties of no light kind."[5] First-hand experience and physical proximity, combined with a self-conscious obligation to document the landscape accurately, produced descriptions of the Bay that are strikingly consistent from the seventeenth century onward. Differences of perspective exist, but within the parameters established by the specifics of a place.

The earliest known depiction of Georgian Bay comes from one of the most famous maps in Canadian history: Samuel de Champlain's 1632 *La Carte de la Nouvelle France* (see Map 4). Based in part on his 1615-16 voyage to Huronia, it is the first time that the Great Lakes take shape in European cartography.[6] *La Carte de la Nouvelle France* is a wonderful insight into the early modern European mind. As France's representative in the New World, Champlain was a politician as well as an explorer, and his map reflects not only his personal experience in the interior but his political agenda and the intellectual climate of the day. His map makes a grand imperial claim for France, reaching deep into the interior. It conforms to the expectations of an empirically minded age by observing different types of vegetation and topographical relief. And it fills a useful

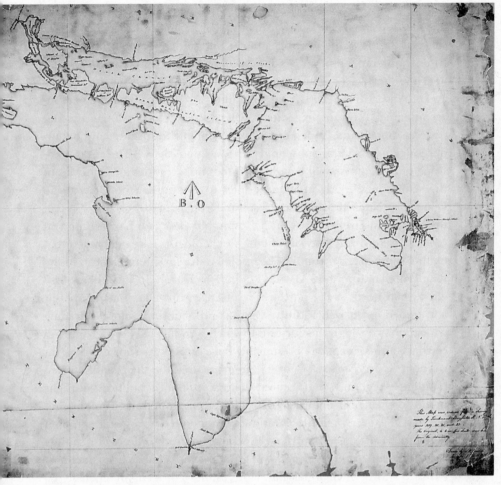

Map 3   Lake Huron, 1828, Henry W. Bayfield. Library and Archives Canada, NMC-21705.

strategic function by identifying the positions of France's new allies: the Huron, several different Algonquin bands, and *la nation neutre*. Champlain had sent Étienne Brûlé to live with the Huron, who called themselves the Wendat, in 1610 to further this alliance and to learn more about the interior. Five years later Champlain himself travelled to Huronia, or Wendake, for exactly the same reasons. His map shows the New World as both a new and strange environment and, to a lesser degree, an inhabited space.

The historical record indicates that First Nations to the north and south visited Georgian Bay on a seasonal basis for fishing, berrying, and trading. Champlain annotated the north shore on his 1632 map as *Lieu ou les*

Sauvages font Secheries de framboise, et bleus tous les anes. Fr. Gabriel Sagard, a Récollet missionary sent to Huronia in 1623, observed that the Huron camped on the islands during summer and fall. En route from Quebec he encountered a group of Ojibwa "who had come to station themselves near the Freshwater sea with the purpose of bartering with the Hurons and others on their return from the trading at Quebec." That October he accompanied a fishing expedition, where after "a long sail out into the sea, we stopped and landed on an island suitable for fishing, and put up our lodge near several households already established."[7] The rocky landscape of the Canadian Shield demarcated the territories and lifestyles of the semi-nomadic Algonquian peoples to the north (the Odawa, the Ojibwa, the Nipissing) and the agricultural Iroquoian nations (the Wendat, the League of Five Nations) of the south, but Georgian Bay provided an efficient water route between them. Archaeologists have concluded that Ojibwa camped as far south as Beausoleil to trade with the Huron in the Late Woodland period (AD 800-1700). The Huron, in turn, likely travelled as far north as the north shore of Lake Huron and Lake Superior and traded extensively with the Ottawa and Nipissing, who had trading connections through the northwest (which is probably where Champlain learned of the *grand lac* – presumably Lake Superior – shown on the far west of the 1632 map).[8] Georges Sioui has credited the Wendat with a "commercial genius" that secured them a position of political centrality in the northeast and in turn ensured their survival amid and against larger nations. However, their success as traders depended in large part on their agricultural surplus – a surplus that could be produced only in cultivatable soils. Hence, their villages were located some distance from the rocky, exposed shores of Georgian Bay. (Interestingly, though, archipelago terminology comes into play here, too, as scholars characterize the area of Huronia as an "island": Sioui translates Wendake as the peninsula country or "the island apart," in reference to the Wendat's status as a demographic minority and the shape of the Penetanguishene peninsula; Heidenreich sees the area as an "island" of well-drained soils surrounded by water, low country, and swamps, with a settlement history distinct from that of the area to the south.)[9] The Bay was as valuable as a transportation route in times of war as in times of peace: after the destruction of Huronia in 1649, it would carry Iroquois warriors north into the Hudson Bay watershed and Ojibwa territory. When, by the turn of the eighteenth century, the Ojibwa pushed southward as far as Lake Ontario, they used Lake Simcoe and the Severn River to Georgian Bay to return to their ancestral homelands north of Lake Superior.[10] Europeans would adopt similar

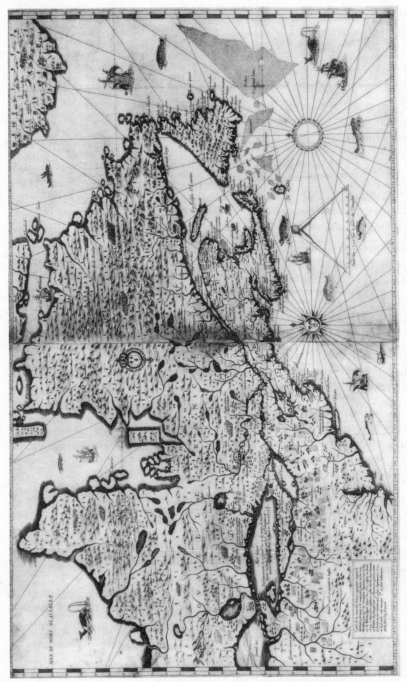

Map 4 *La Carte de la Nouvelle France*, 1632, Samuel de Champlain. Library and Archives Canada, NMC-51970. Originally printed in Samuel de Champlain, *Les voyages de la Nouvelle France occidentale, dicte Canada ...* (Paris: 1632).

patterns of mobility and transience, especially those who hoped to find a passage through the New World.

Champlain's first encounter with the Lakes was fairly typical. In 1615 he accompanied a party of Huron from Quebec to their home on the south shore of Georgian Bay. They travelled up the Ottawa River to Lake Nipissing, down the French to the Bay, then south through the archipelago. The next year he returned to Quebec by heading east and south through the Trent-Severn river system to Lake Ontario and the St. Lawrence. The "explorer," and the missionaries and fur traders who came after him, was following water routes that had been used by Natives for centuries. The vast expanse of Georgian Bay, accessible by water from the north and the south, clearly raised his hopes for a continental water route. This is one reason why he makes the *Mer Douce* so prominent and disproportionately large in his *Carte de la Nouvelle France*: literally and figuratively, the inland sea was central to his understanding of and aspirations for the continental interior. The islands of the east shore, however, are drawn simply as a handful of small circles – a convention that persisted into the middle of the nineteenth century. Looking at these few round shapes, it is easy to imagine a baffled Champlain scratching his seventeenth-century head. How on earth would he describe what he had seen? How was he going to represent such a shoreline in any detail? For the time being, the islands were little more than a passing curiosity. But as the physical obstacles of the tempestuous, shoal-rich *Mer Douce* became better known, and Europeans learned of alternate routes through the Lakes, the enthusiasm of the seventeenth-century explorers began to evaporate. Until the industrial boom of the late nineteenth century, the Bay would become more an obstacle to avoid.

Seventeenth- and eighteenth-century records of the Great Lakes are functional and transportational, noting primarily waterways, shorelines, and rapids. These representations of rivers and lakes were "the embodiment of intention."[11] Water routes flow where people took them, but also where explorers hoped they would lead. Champlain puts the *Mer Douce* on an east-west axis, betraying his hope of having it carry him across the continent to the Orient. Subsequent mapmakers continued to flatten the Bay into an east-west corridor north of Lake Huron until the end of the next century. John Mitchell's 1755 *Map of the British and French Dominions in North America* conveniently arranges a series of lakes between the Bay of Quinte and Georgian Bay to form a direct line through British territory

Map 5 *Partie Occidentale de la Nouvelle France*, 1755, Nicolas Bellin. Library and Archives Canada, NMC-113509.

to the upper lakes. Other maps, such as Vincenzo Coronelli's *Partie Occidentale du Canada ou de la Nouvelle France* (1688), show the Ottawa, French, and St. Lawrence rivers as roughly equal in width, a measure of their importance as transportation routes rather than their actual size.[12] With travel almost entirely dependent on water, the detail of shorelines is in stark contrast to the emptiness of the interior. The land between Georgian Bay and the Ottawa River and north of the North Channel remains largely blank, with only speculation about the major rivers, until the arrival of land surveyors in the 1860s. Even the north shore west of the French River, running alongside the most important fur trade route, was annotated on Nicolas Bellin's *Partie Occidentale de la Nouvelle France, ou Canada* (1755) as *"Toute cette coste n'est pas connue."*

The collapse of the Wendat confederacy in 1649 and the westward expansion of the fur trade ensured most traffic would bypass the archipelago altogether. Following the North Channel from the French River directly west to Lake Superior kept European and *Canadien* traders out of Iroquois territory. This was a major artery for voyageur canoes to and from the northwest until well into the 1830s. The importance of this route may explain why early maps often show the Bay split nearly at the halfway mark by the French, exaggerating the length of the better-known North Channel and minimizing the less-used east shore. But with the retreat and pacification of the Iroquois by the 1670s, Europeans began to penetrate the lower lakes. The St. Lawrence into Lake Ontario route had clear advantages: fewer rapids than the Ottawa-French route, with arable land for settlement along the way. By the end of the eighteenth century, these settlements along the north shore of Lake Ontario would be substantial enough to generate interest in an overland passage to Georgian Bay. But for now, just like the French River-North Channel, the preferred route west neatly bypassed the archipelago. It was far easier to reach the major destinations such as Detroit, Mackinac, and Sault Ste. Marie by sailing on the open lakes, along Lake Erie to Lake St. Clair, and then straight up the main body of Lake Huron.

Still, it is interesting that exploration material before 1850 makes so little mention of the islands. Explorers and missionaries were preoccupied with the destination, the exotic peoples encountered en route, or their chartered mission, whether imperial or spiritual. As a result, the islands get short shrift. Champlain says simply that the *Mer Douce* "is very large" and "there are a great number of islands." Father Sagard is equally brief: "Two days before our arrival among the Hurons we came upon the Freshwater sea, over which we passed from island to island, and landed in the country so greatly longed for." The Jesuits produced detailed maps of Huronia but like Sagard preferred to look southward, at the arable mainland settled by the Huron confederacy from the Penetang Peninsula to *Lac Ouentaran* (Lake Simcoe).[13] By the late eighteenth century, the word-maps of travel journals begin to provide descriptions of the northeast shore. In 1761 Alexander Henry arrived at the mouth of the French, where "the billows of Lake Huron ... stretched across our horizon, like an ocean," before continuing on to Michilimackinac. He was clearly impressed with the "small islands, or rather rocks, of more or less extent, either wholly bare, or very scantily covered with scrub pine-trees."[14] Yet, the eastern shore remained largely unknown and undocumented; the standard handful of round shapes, meant to symbolize the impossibly complex archipelago, are paired with dismissive comments such as that by Gother Mann of the British Royal Engineers, in his *Sketch of Lake Huron 1788.* "The whole of this Coast from Lake George to Matcha-dosh Bay," he writes, "is Rocky and Barren."[15]

However unenthusiastic Mann might have been, he was part of a new wave of interest in charting the Great Lakes and the newly acquired western territories of British North America. Mapping expeditions by British military parties accelerated toward the end of the eighteenth century in a concerted effort to secure this strategic region. Georgian Bay had been peripheral to the imperial wars preceding the Conquest but was not unaffected by the chronic tension between the French and English. After reaching the Bay in 1761, Henry had to disguise himself as a Canadian voyageur to avoid arousing suspicion among Ojibwa along the north shore, who had allied with the French; New France might have fallen at Quebec two years before but was still a reality in the northwest. For the most part, however, it was not until after the American Revolution that the British began to look seriously at the Bay. With the south shore of the Lakes lost to the new United States, they needed to find a new and protected route to the upper lakes of Huron and Superior. Georgian Bay lay conveniently midway between Kingston and York and the western posts, and angled northwest to boot.

The first Lieutenant-Governor of Upper Canada, John Graves Simcoe, travelled to Matchedash Bay in September and October of 1793, hoping to establish a naval base in the sheltered harbour. Alexander Macdonnell, who accompanied Simcoe, complained the rapids and carrying places on the Severn were numerous, reputedly infested with rattlesnakes, and little more than solid rock.[16] Nevertheless, Simcoe settled on Penetanguishene and set about establishing a direct route from York to the nearest of the upper lakes. By 1785 treaties with the resident Ojibwa bands had obtained parcels of land linking Lake Ontario, Lake Simcoe, and Severn Sound, and by 1800 the Crown had secured Penetang Peninsula as well as Mackinac Island and St. Joseph's Island.[17] Like their predecessors in the fur trade, though, the British adopted traditional Native travelways. The Toronto route ran up the Humber River – or later, the new Yonge Street – to the Holland River, which emptied into Lake Simcoe. From there one could exit through the western arm of the lake at Kempfeldt Bay and portage to the Nottawasaga River, which ran north into the Bay. Or one could head north and portage to the Severn, following it west to Matchedash Bay (now Severn Sound). Surveyors, however, would repeatedly warn that the Severn's many rapids, while "picturesque," presented a daunting obstacle to navigation. Watercolour sketches by Lt. Robert Pilkington, such as *Near Gloucester Bay on the Severn River*, show the river clogged with dauntingly large, angular boulders.[18] Other military artists sent back similar impressions of the French River, with imposing barriers of solid rock dwarfing survey parties in J.E. Woolford's *Rapid of La Duc, French River* (1824; Plate 1), and C.R. Forrest's *Grand Campement, French River, Lake Huron* (1822).

What forced the British hand was the impending conflict with the new United States. In June of 1812, the United States under James Madison declared war on Britain, but the war (such as it was) would be fought largely in Britain's colonies in North America – and on the waters of the Great Lakes. In the first major action of the war, British major-general Isaac Brock ordered an attack on the American-held Michilimackinac, the strategic post guarding the juncture of the upper Lakes (Superior, Huron, and Michigan) and, by extension, entry into the Mississippi Valley and the continental interior. With Mackinac would go the northwest, the confidence of the Amerindians, and the fur trade. Capt. Charles Roberts and British troops stationed at St. Joseph's Island swiftly and easily captured the American fort, but it remained the main prize for both sides in the western theatre. In January of 1813, the American secretary of the navy wrote to Commodore Isaac Chauncey that should Detroit fall into American hands, Chauncey should

> detach a part of your force to Lake Huron, to take the post at the mouth of the French River on the N.E. side of Lake Huron, were you will intercept the supplies for the western Indians [which are] distributed through the waters of Huron & Michigan to the tribes, even beyond the Mississippi. It is this commanding position, which gives to the Enemy the absolute controul of

the Indians. This force would also enable you to take Michilimackinack & command the waters of Lake Michigan.[19]

As things turned out, this was almost exactly what happened.

After the American naval victory on Lake Erie in September 1813, the British were unable to reach Lake Huron by the usual route of Lake St. Clair. By the following spring, the Americans were preparing to sail from Detroit to Mackinac, now the lone British post in the west. This set the scene for the one major incident of the war that occurred on Georgian Bay. In an attempt to starve the isolated garrison into surrender, an American naval force captured one supply ship, forced the burning of another, and blew up the third, the *Nancy*, at the mouth of the Nottawasaga River. But the *Nancy*'s resourceful captain, Lt. Miller Worsley, and its crew had hidden supplies and two bateaux upriver. (Worsley later claimed that *he* had blown up the *Nancy* to prevent it falling into American hands.) After the *Nancy* was destroyed, Worsley "and his gallant little band of Seamen," as Mackinac's commanding officer later reported, "traversed this extensive Lake in two boats laden with provisions for the Garrison," rowing along the northeast shore and reaching the beleaguered fort undetected on 31 August.[20] The indefatigable Worsley promptly led the capture of the two American schooners patrolling the North Channel, thereby in one fell swoop reclaiming the upper lakes for Britain.

Worsley's heroics are the most exciting aspect of this story, but his experience encapsulates thirty years of observations about Georgian Bay. The onerous task of mapping the Bay was a military responsibility, based on a perceived need for strategic information. Col. George Head arrived at Penetang in February 1815 to find the commanding officer "generally absent all day, employed in surveying the shores and taking the soundings of the bay."[21] But "this extensive Lake" was seen solely as a through route. The skirmishes occurred at the Nottawasaga depot and at Mackinac, with the *Nancy* ferrying supplies between the two. Of his stealthy slip up the shore, Worsley wrote only that "I left that place [Nottawasaga], on the 18th of August, with two Batteaux laden with flour for the Garrison of Michilimackinac and had the good fortune to arrive on the sixth day, within 8 miles of the Island of St. Joseph's, without any accident."[22] The need to provision western posts forced the British to construct overland routes to the Bay from Lake Ontario, set up boat-building facilities at the forks of the Nottawasaga River and later at Penetang, and to contemplate building a canal from Quinte (the future Trent-Severn) to move troops to the upper lakes in the event of another war with the Americans. (Fifty years later, T.C. Keefer suggested that a port on Georgian Bay "could maintain a fleet to contend for the superiority of Lake Huron [and] in the defense of the western peninsula from invasion by way of Michigan.")[23] The Bay was viewed not as a destination in itself but as part of a journey, which one travelled only out of necessity and only en route to somewhere else; and a poorer, rather out-of-the-way alternative when the usual avenues were unavailable, at that.

Most of the commentary on the landscape dwells on just how out of the way Georgian Bay was. The soldiers sent out to Lake Huron saw themselves as venturing deep into the wilds of North America. "Behold me a sojourner in the wilderness," wrote one man travelling with Worsley in 1814. The commander of the American squadron sent to destroy the depot at Nottawasaga described the river as "dangerous and difficult, and so obscured by rocks and bushes that no stranger could ever find it," and learned from a British informant that the Matchedash Bay route was equally "impracticable, from all the portages being a morass."[24] Of course, that was partly the point: supplies might be transported without harassment through this obscure back door. Yet, even the British were not sure it was worth the trouble. William Dunlop, supervising the construction of a road between Lake Simcoe and Penetang in 1814 "at mid-winter, in one of the northern-most points of Canada," tried to follow the blaze marked by the surveyors, got lost, and almost froze to death.[25] When George Head struggled into Penetang a few months later, he found more of a campsite than a naval base, "far removed in the woods ... a parcel of small huts, made up of a few poles thatched over with spruce boughs." "My future life and occupations," he observed wryly, "seemed likely to be sufficiently rural." It was, to be sure, sufficiently isolated: he did not receive news of war's end until 9 March 1815, at which point Head and most of the troops promptly withdrew from the fledgling post.[26] This was hardly a triumphant vanguard of empire.

The naval presence was equally tenuous. With a grand total of three British ships on the upper lakes in 1813 – and only one by August of 1814 – it is not surprising that Mackinac was in so much trouble. British and American ships alike were plagued by rapids and "carrying places" along the rivers, rocky anchorages and coastlines, and dangerous open waters. Such was the experience of the relief expedition rushed to Mackinac from Kingston in April 1814,

conducting open and deeply-laden batteaux across so great an extent of water as Lake Huron, covered with immense fields of ice and agitated by violent gales of wind ... For nineteen days it was nearly one continued struggle with the elements, during which the dangers, hardships and privations to which the men were exposed were sufficient to discourage the boldest among them, and at times threatened the destruckion [sic] of the flotilla.[27]

After the attack on the *Nancy*, the USS *Scorpion* and *Tigress* were ordered to patrol the south shore, but they stayed only long enough to weather their first Georgian Bay gale before hightailing it back to St. Joseph's. In short, encounters with this portion of Lake Huron had not been encouraging. It was remote, hazardous, and largely unknown. The war prompted the Admiralty to undertake a comprehensive survey of the Great Lakes, but these early impressions of the Bay would die hard.

Although there was relatively little action on the Bay itself, this era represented a major step in bringing it into the public imagination. The need to define the international boundary after the war produced some of the earliest comprehensive accounts of travel to and through the Bay, including those by J.J. Bigsby of the Boundary Commission and Bayfield of the Admiralty. As Benedict Anderson has argued, European mapmakers "profoundly shaped the way in which the colonial state imagined its dominion" – by (and there is a certain irony here) drawing lines that correspond "to nothing visible on the ground" in order to demarcate sovereignty.[28] In this case, for instance, the natural unity of the Great Lakes would pull a strong American presence into Georgian Bay by the end of the century, even though it fell entirely within British-Canadian jurisdiction. But neither opinions nor boundaries were firmly fixed. The British garrison at Mackinac was bounced first to St. Joseph's Island, when Jay's Treaty assigned Mackinac to the United States in 1796, displaced to Drummond Island after the War of 1812, and finally withdrawn to Penetang in 1828 after the Boundary Commission awarded Drummond to the United States. Nor is the detachment and political machination that Anderson attributes to imperial cartography present in the vivid response to the landscape these men record in their writings. British authorities would have a very different image of their so-called dominion from observers who concluded that the "part of the Coast of Lake Huron which belongs to His Majesty, from Penetangushene [sic] to the falls of St. Mary ... is a bleak, barren, and inhospitable coast, entirely uninhabited."[29]

But the war had made it clear that even this bleak and barren coast had to be thoroughly mapped. The single most important cartographic representation of Georgian Bay appeared in 1828, when the British

Admiralty published Lt. Henry Bayfield's maps of Lake Huron and Georgian Bay. Capt. William Owen had done a preliminary survey of Georgian Bay – which he called "Lake Manitoulin" – in 1815. Bayfield became Owen's assistant the next year, and over the next ten years would survey Lakes Erie, Huron, and Superior. The survey of Lake Huron, begun soon after his promotion to Admiralty surveyor in British North America, proved especially time-consuming and lasted four full years, from 1819 to 1822. This stands as an unmatched accomplishment, the basis for all subsequent hydrographic maps, and testament to a dogged faith in "the theoretical study and practical application of science," as he later wrote.[30] In addition to depth soundings in the Bay proper, there are unusually precise annotations identifying rock types such as mica, granite, and gneiss; beaches and shores; and species of tree. It appears uninhabited apart from one "Indian Winter Encampment" in Owen Sound and a trading post at Franklin Inlet; Bayfield was evaluating natural resources, not human ones.

Bayfield's depiction is really the first to be familiar to us today: the northeast orientation of the Bay, the details of the major bays, river mouths, and larger islands. For the first time, the entire detailed shoreline is visible at a glance; the Bay has become a single, knowable, and presumably malleable entity. Yet, his annual reports to the commander at Kingston betray incredulity at the scope and complexity of the archipelago – and a feeling that this environment is testing the limits of Enlightenment confidence in reason and scientific method. In 1820, the twenty-four year old tried to explain the magnitude of his task:

> I have no doubt Sir, but that you will think 45 Miles of coast in 10 weeks, very slow progress, but I trust you will think otherwise when I inform you, that, in that distance we have ascertained the Shape, size and situation of upwards of 6,000 Islands, Flats and Rocks: the main shore too is broken into deep Bays and Coves and together with the Islands is composed of barren Granite Rock ... The Survey of this coast and its numerous Islands, has been a work of great difficulty and labour and has put both our perseverance and patience to the test.

Completing this particular assignment, he confessed, was a "much desired object."[31] Although Bayfield's is the first modern survey, in another sense it marks the end of an era. Previously, mapping was driven by exploration, military activity, and imperial competition. Bayfield's successors in the mid-nineteenth century would concentrate on the interior rather than waterways, and were less interested in charting serviceable travel routes west than in scrutinizing the new frontier for possible use.

After the War of 1812, Upper Canada embarked on an enthusiastic mission of land acquisition. Pressure from white settlers coincided with a major shift in Indian policy by the 1830s. First Nations were no longer viewed as significant military allies but as wards of the state whose welfare would be best served by cultural assimilation. Their relocation onto permanent reserves would, it was hoped, encourage them to adopt an agricultural and Christian lifestyle, while at the same time freeing up land for the growing white population.[32] Treaty surrenders around the Georgian Bay initially targeted the arable land to the south and west, securing the Saugeen Territory south of Owen Sound in 1836, the Saugeen (Bruce) Peninsula in 1854, and most of Manitoulin Island in 1862. With the arable land of the Great Lakes lowland parcelled out, Canada West turned its attention to the Shield country on its northern border. The Robinson Treaties of 1850 acquired a vast territory north of Lake Superior and Lake Huron, including Georgian Bay, and there was a concerted effort by the 1860s to systematically survey the Ottawa-Huron Tract. Civil and military surveyors set out, by canoe and over land, seeking elusive pockets of arable land and a through route between the Ottawa River and Georgian Bay (a scheme which would reappear in numerous forms until the 1930s). The expansionist, proprietorial attitude that pervaded Central Canada in the mid-nineteenth century is captured perfectly in the instructions to surveyors of the Ottawa-Huron Tract in 1864:

First: As to the extent and general resources of the Territory lying between Lake Huron on the West, French River and the Ottawa River on the North and East; and the Townships on the South, previously surveyed; Second: The portions of such Territory suitable for settlement, their geographic position, extent and quality; Third: The portions producing Merchantible Timber, and the best method of dealing with therewith in order to combine the interests of colonization and the utilization and preservation of such timber.[33]

But this expectant optimism was short-lived. The Bay shoreline was soon relegated to the same role it had played in the imperial era: on the periphery of interest and more often a barrier to circumvent.

In part this was because the archipelago quickly became notorious for presenting no end of difficulties to surveyors. Ten years after his survey, Bayfield offered one of the most comprehensive litanies of complaints regarding "my labours," with very inadequate means, viz. two row boats and some young and inexperienced assistance. There has perhaps been no Survey made of greater labor of detail than that of the 20,000 Islands, Islets, and Rocks, of Lake Huron – nor any Survey ... in which greater privations were endured than in that of the Lakes, certainly none in which they were endured so long ...

We were frequently out in our Boats till the middle of November, and in two years till the middle of December ... [we] carried our whole stock of conveniences, of which we had fewer than the native Indians ... slept, in all weathers, in the Boat, or on the shore upon a Buffaloe robe under the Boat's mainsail thrown over a few branches placed in the ground. Many a night have we slept out, in this way, when the Thermometer was down to near zero, and sometimes even below it. Yet even this was not so wearing as trying to sleep, in vain, in the warm nights of summer (when the Thermometer was at 80) in the smoke of a fire to keep off the clouds of Moschettoes which literally darkened the air – fatigued as we generally were from getting from sun rise to sun set in an open Boat, no rest could be obtained under these circumstances ...

As the seas of Lakes Huron and Superior are as heavy as those of the Ocean, we had many very narrow escapes when caught in heavy gales late in the Season. We have frequently been obliged to run for some nameless island or shore where we have thought ourselves fortunate to be able to land, drenched to the skin, to find what shelter we could in the woods when the ground was covered with a foot or two of snow ...

Such is briefly, and literally, a picture of the Lake Surveying Service for nine years and I hope you will not think me above exaggerating it – in truth I do not.

(Bayfield also emphasized that "we were cut off entirely from Society, after we reached Lake Huron, particularly female society, for six years.")[34] David Thompson complained he could not even *find* the Muskoka River, an inauspicious beginning to his 1837 exploration of the interior:

We had to find the main land, as all before us were rocky, granite Islands, in order to be sure to find the River we were in search of. We had much difficulty in getting to the main Land, & in following it. In hopes of finding Rivers we were led into Channels & Bays like Rivers from which we had to return.[35]

Others grumbled about the rugged and broken topography, the multitude of portages, the confusing system of inland lakes, and the generally "chaotic character of the country."[36]

They also found translating it into the conventional language of maps nearly impossible. The surveyor of Foley Township admitted, "The difficulty of rendering anything like a correct representation of all the various characters which the country presents, is well known to those who have made the attempt, so that only the general subdivisions are attempted, as being all that is deemed necessary."[37] The islands themselves were apparently not "deemed necessary" until about 1900 amid a new interest in the archipelago for recreational properties. Even then, one surveyor remarked sourly in 1910 that "many divisions were made with the feeling that labour was being lost as no market would be obtainable for a long time."[38] For a long time, Crown land surveyors literally avoided this problematic territory by surveying around it: up the semi-arable Bruce Peninsula to Manitoulin Island, and along the north shore of Lake Huron, which at least offered decent timber stands and the possibility of copper reserves. Maps that proudly track the proliferation of township lines and road networks in southern Ontario abruptly become blank north of the Severn until almost the end of the nineteenth century.

By the 1860s, though, there were some efforts to attract settlement onto the southern edge of the Shield. Colonization roads penetrated the Muskoka interior to Parry Sound and the Magnetawan valley, townships ringing the port at Parry Sound (Foley, McDougall, Carling) were surveyed by 1873 and offered as free grant lands to homesteaders, and the 1879 *Guidebook and Atlas of Muskoka and Parry Sound Districts* supplied prospective settlers with generally positive descriptions of the terrain. But these inducements could hardly compete with the blunt reports from the surveyors themselves. The men who laid out the townships along Georgian Bay – no doubt wondering what they had done to deserve this fate rather than an assignment somewhere nice, flat, and arable – detailed the limitations of the landscape far more than its potential, and bequeathed Georgian Bay a singularly dreadful reputation. The agent who inspected the newly surrendered region for the Department of Indian Affairs in 1856 concluded the Department had in fact obtained very little: "innumerable small islands composed of gneiss & granite devoid of vegetation or like the main land sparsely covered with stunted Pine and of no value."[39] The commissioner of Crown lands for Ontario received letter after letter describing the shoreline townships as "almost a barren waste," with clumps of "inferior" pine, "wholly unfit for settlement, being composed of burnt land, rocks, swamps and lakes." J.W. Bridgland's opinion of the lower Muskoka River is not exactly a ringing endorsement:

a tedious monotony, of rocky barrens, swamps, marshes, and burnt regions; destitute of good water, good timber, in short of every-thing, necessary to make settlement desirable, or life supportable ... I beg leave again, to express my regret, as to the general result of the survey [that] so little should have been discovered, to promise advantage, or to induce settlement.[40]

Even publicity material that painted a rosy future for the Ottawa-Huron Tract, pointing to scattered pockets of decent land, a "noble class of set-tlers," and "an atmosphere both pure and bracing," conspicuously avoid referring to the west coast of the district. An 1878 pamphlet promoting the merits of *The Undeveloped Lands in Northern and Western Ontario* admit-ted frankly that "a strip along Georgian Bay is practically worthless." The Committee on the Ottawa and Georgian Bay Territory concluded there was little hope for "the harsh and forbidding coast of Georgian Bay ... The entire frontier, with a width of from twenty to thirty miles, may be described to the last degree sterile and desolate."[41] For industrial On-tario, this seemed to be the country of its defeat. Georgian Bay remained very much a *terra* both *incognita* and *non grata* for much of the nineteenth century.

If agricultural settlement seemed stillborn, there was mild interest from another source, the Geological Survey of Canada. The GSC played an important role in introducing the Canadian Shield to the public imagi-nation: as a scientific phenomenon, a continental geological formation (an appealing concept to imperialists dreaming of national expansion *ad mare usque mare*), and a potential treasure of mineral ore for a country on the brink of industrialization.[42] The Shield landscape was a significant political and cultural concern decades before the Group of Seven turned it into a visual icon. But once again, the GSC examined the eastern Geor-gian Bay quickly and perfunctorily before its attention was diverted by the larger copper and iron deposits on the north shores of Superior and Huron. Alexander Murray, who explored the Bay coastline between 1847 and 1858, was intrigued by the geological formations but could not see any use for them. His matter-of-fact evaluation of the area south of the French River by now sounds very familiar:

As an agricultural country a large portion of the region ... appears to be valueless, and the pine-timber, where it attains a size worthy of notice at all, is too much scattered, and besides usually too small to be of any commercial importance. The principal if not only recommendation which the coast at present possesses is as a fishing station ... [There are no] mineral indications worthy of especial notice, nor have any rumours through the Indians or others reached me of the presence of metalliferous lodes nearer than Lake Nipissing.[43]

According to almost every standard Victorian Canada held dear, the Bay had absolutely no value. Agriculture was inconceivable, timber quality poor, mineral ore minimal at best. Fishing was not yet a major industry in the remote area, but the Indians he met would have been increasingly dependent on it. The factor at La Cloche was reporting widespread starvation among Native hunters and their families by the late 1820s, as well as the decline of game in the interior, forcing the Ojibwa into the unenviable position of accepting reserve lands along the inhospitable shore.[44]

From a maritime perspective, the north arm of Lake Huron, as Georgian Bay was commonly known, remained well off the beaten track. The traffic of the early nineteenth century, with the fur trade in decline, was primarily political or military: Indian Agents travelling to Manitoulin Island for the annual present ceremony, or troops en route to the Sault area from Penetang. But most of the boat traffic stayed on Lake Huron proper. Georgian Bay had to be reached through unpleasant overland travel, and its waters were as yet largely unknown to navigators. En route to the Sault in 1830, Calvin Colton hears about the Thirty Thousand Islands from a fellow passenger, but he dismisses them as simply too fantastic to imagine: "In my opinion, thirty-two *hundred* was quite enough; and that there must be a mistake ... I could hardly believe there were *thirty-two thousand* islands, in all the waters of the continent of America." A few years later, David Thompson squelched the possibility of steam travel on the French River, noting the "difficult and dangerous access" in narrow channels, the "almost incredible sterility of the country," and the massive expense of excavating the river and constructing navigational aids.[45] Until the railroad reached Collingwood in the 1850s, there was little reason to travel to or through the Bay.

But with railroads providing a direct overland route to the upper lakes, and with the dramatic invasion of the Bay watershed by loggers in the 1860s, the small ports ringing the south shore shot to new importance. Increased shipping, though, meant increased shipwrecks. Railways provided easier access *to* Georgian Bay but crossing it remained as difficult as ever. After the 1882 *Asia* disaster (when over one hundred lives were lost), the federal government finally commissioned a second maritime survey.[46]

This time the lucky fellow in charge was Commander J.G. Boulton, who took the next ten years to produce the *Georgian Bay and North Channel Pilot*. The *Pilot* is a methodical boating guide, most entries noting shoals, reefs, and islands. But it is also a useful record of activity on the Bay at the end of the century, with a thorough inventory of built structures (wharfs, summer houses, fishing huts, anchorages, mills, fishing stations, villages); navigational aids (spar buoys, iron cage bell buoys, range lights, beacons, lighthouses); and place names. By this point people were obviously making use of the Bay – but they were still ambivalent about it. That the Georgian Bay Survey was the first hydrographic survey of the new Dominion and the debut of what would become the Canadian Hydrographic Service indicates that the Bay was well established on the official radar. Yet, it could not shake its longstanding role as something to go through or around. Most ships went directly from the south shore (Owen Sound, Midland, or Collingwood) to other Great Lakes ports (the Sault, Port Arthur, or Chicago). Despite his own careful calculations, Boulton constantly warns the reader about "the notorious north-east shore of Georgian bay."[47] Business interests enamoured of a canal route from Montreal encountered the same problems. The survey for the proposed Georgian Bay Ship Canal by the Department of Public Works between 1904 and 1909 produced what one historian has called "one of the greatest of all Canadian engineering documents": a series of elaborate engineering diagrams of the connecting rivers, rapids, and proposed canal sections. (It didn't produce a canal, but that is another story.)[48] But for all its visual precision and technological optimism, Public Works found the French River mouth just as Thompson had seventy years before: clogged with shoals and an excavation nightmare.

By this point, industrial traffic was only one reason for mapping the Bay. Since the 1860s, surveyors had been recommending tourism as the only viable use for such broken and rugged land. The same reports that bemoaned the inferior timber and barren land proclaimed the scenery "extremely wild and romantic," with "many attractions to the sportsman and tourist."[49] The yawning disparity between these pronouncements and the generally abysmal opinions of the place is a surprisingly accurate reflection of the conflicting attitudes toward nature at this time. An Arcadian view of wild nature, exemplified by the Romantic movement, had offered a critique of utilitarianism – nature as resources for use – since the eighteenth century. (These opposing schools of thought would underwrite the two streams of modern environmentalism, preservation and conservation.) So the very same qualities that made the Bay a dismal prospect

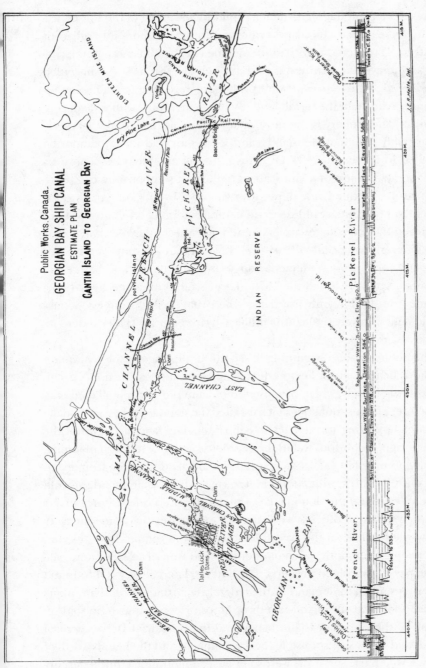

Map 6 Canal diagrams, French River. Reproduced from Canada, Department of Public Works, *Georgian Bay Ship Canal: Report upon Survey with Plans and Estimates of Cost, 1908,* 2 volumes (Ottawa: C.H. Parmelee, 1908).

for resource exploitation made it a promising example of the sort of aesthetic, recreational landscape beginning to draw visitors into adjacent Muskoka. Disordered and chaotic could be read as wild and natural; rugged and uneven as picturesque and intriguingly asymmetrical; unsettled as primeval. Thus, surveyors and promoters were able to find some redeeming feature in the terrain and discharge their duty with a clean conscience: "The scenery is the most picturesque imaginable, the lakes are dotted with innumerable islands, and for beauty and variety cannot be excelled in the Province."[50] Of course, fishing and viewing scenery require somewhat less of a landscape; granite and scrubby pine can be attractive if one need not depend on it for a livelihood. Areas once dependent on primary industry and natural resources often turn to tourism when these resources are exhausted, for a second career based on a very different perspective of "nature." In Georgian Bay, however, neither industry nor orderly settlement had secured the landscape before the tourists began arriving in the 1870s. Large-scale fishing and forestry descended on the Bay only just before the cottagers and campers, so aesthetic values actively rivalled the utilitarian viewpoint, as evidenced in the surveyors' reports.

The public was also beginning to see the archipelago as the realization of the Romantic ideal. For the first time, attention was focused specifically on the islands, which promised to redeem the much maligned eastern shore. The very qualities that frustrated the standard grid system made the area appealing to the back-to-nature set. The irregular lines of the archipelago, an obvious foil to the geometric order of modern planning, seemed to manifest values such as individuality and freedom from external control. "This country imposes its own law, these islands and this cold dark water refuse the dominance of man," declared one visitor.[51] In the 1880s government officials were amazed to discover that someone actually wanted to live on the Bay, at least during the summer months, and that people from both local communities and American cities were willing to pay to own island property. The Ontario Land Survey (OLS) found its services much in demand by the federal Department of Indian Affairs (which managed the islands south of Moose Deer Point, to be sold on behalf of the Christian Island band) and the provincial Department of Lands and Forests (responsible for the mainland and the islands north of Twelve Mile Bay). Surveyors were kept busy submitting technical descriptions for patents, estimating sale prices, and doing their best to depict the archipelago in accordance with popular expectations of a recreational landscape.

Sometimes they seem to be talking about a different place altogether. Barren? Worthless? Not at all. J.G. Sing, working for the OLS, sounds more like a real estate agent than a surveyor: "It is a pretty and desirable little island ... A cute little island ... A beauty indeed ... A pretty little dot ... A pretty little clump ... A perfect little gem ...[!]" North of Moose Deer Point, he promised, the scenery "reaches a state of magnificence not to be excelled anywhere."[52] Suddenly pieces of a garden were being found amid the scraps of bare rock in this land of Cain. The back-to-nature movement among urban North Americans had made wilderness a very appealing locale – and the Bay, as any nineteenth-century surveyor would testify, was certainly wilderness. Industrial interests still had little interest in "what is, at present, useless land." (Some surveyors remained unconvinced as well; one quietly informed Indian Affairs in 1900 that 905 of the 1,286 islands surveyed to date were "very small or mere rocks of little or no value.")[53] But surveyors were no longer looking for industrial resources; they were appraising the aesthetic character of the islands. Before 1890 no one had cared in the least whether the islands were pretty. Now that they were marketable as recreational properties, they were almost always described as pretty, picturesque, or beautiful. Surveyors also took care to emphasize features of particular interest to campers: canoe routes; tree cover; places for cottages, for camping, bathing, boating; quality of fishing; proximity to boat channels. Who could resist?

That the islands were now desirable did not make them any easier to survey. They might appear to be an ideal wilderness, but the ideal proved to be inconveniently uncooperative. Georgian Bay was still one massive cartographic and bureaucratic headache, albeit increasingly a lucrative one. After opening the islands south of Moose Deer Point for tender in 1905, the Department of Indian Affairs (DIA) began fielding requests for maps from people seeking to select an island or locate a newly purchased property. (This was hardly surprising since the lists of islands for sale published by the DIA consisted of entries noting only acreage and price, giving no indication of an island's location or appearance.)[54] This new demand for information about the archipelago contrasted starkly with earlier reports that had concluded either implicitly or explicitly that the area was valueless. But almost immediately, the DIA was deluged with questions from the interested but confused. Where *is* the island I bought? Which is 96A and which 96D? If the water drops far enough, the adjacent island/rock/mainland is joined to my island, is it also mine, and if not, why not? Is there a subdivision on the island? Why does the original survey say one thing and a more recent survey, another? If the water level was high/low at time of the survey, does that mean the island is available/not available? Why does my neighbour think the island is his? And so on. One woman named Ella Carmichael wrote politely in 1922,

> Dear Sir: –
> Re. Island No. 106 in Georgian Bay recently purchased from your Department. The writer took a guide and tried to locate this island and was unable to find any identification marks whatever on any of the islands as shown on the map ... any information you could furnish would be appreciated ... [Also,] we find that this island is joined to the mainland at the North end. Apparently this was surveyed when the water was very high.[55]

Other "helpful" correspondents sent in hand-drawn maps with the desired islands circled or with recommended corrections to the official blueprints. In retrospect it is all quite comic, though the harried (and equally confused) officials in Ottawa and Toronto probably did not see it this way at the time.

Surveyors in the field had their own problems to worry about. Some complications were political. A.G. Ardagh, working for the Department of Lands and Forests, complained that private patents and privately commissioned surveys complicated *his* surveying. Overlapping and adjacent political jurisdictions – which would dog environmental policy in the Bay

Figure 2 Surveyor's stake, south of Killarney.

after Confederation – further confused the picture, as the National Parks Branch discovered in 1937:

> The difficulty arises over the fact that the plan of the survey of the islands as issued by the Department of Indian Affairs does not correspond with the plan of the survey of the mainland which shows the shore line as issued by the Province of Ontario and from which later plan the titles to the properties on the mainland were issued. It is now claimed that some of the original grants by the Province include areas which the Indian Department's plan shows as islands and which have been sold and title issued to same by the Department ... the exceptionally low water in Georgian Bay complicates the situation.[56]

So too did the introduction of new mapping technologies. The most important of these in the interwar period was aerial photography, adopted by both the provincial and federal governments by the 1920s. Air photographs provided a fundamentally different way of seeing the archipelago – and "discovered" the existence of even more islands. Here again cottagers were quick to take advantage of the situation; one Ottawa resident had his lawyers contact Indian Affairs to point out an island in the popular Go Home Bay area that must be available, "as its presence was only discovered by aerial camera and it has no official number," and therefore could never have been sold.[57] With different types of mapping and different systems of classification, the islands were numbered and renumbered in ways that seemed more random than ordered. Often the numbers did not appear to correspond to mainland townships, nor was there any overall system for the archipelago as a whole. By the middle of the twentieth century, so many names and numbers had been issued by different surveys that the government and its surveyors were clearly having trouble keeping track.

For the surveyors though, other problems were more immediate, and not unlike those faced by their nineteenth-century predecessors. Many reported that logging and fires had destroyed the original blazes on the mainland. J.T. Coltham, who surveyed the townships of McDougall in 1934 and Carling in 1953, complained that too much of his time and public money was spent searching for the first survey lines set over seventy years before. He tried interviewing local residents to find out where the markers were located, but since the original settlers had long since emigrated for greener pastures, no one knew the whereabouts of the lines. J.H. Burd could not find the older markers in Shawanaga either, but since water levels had dropped six feet, his map bore no relation to the actual landscape anyway. Rough weather later forced him to abandon the survey.[58] By the 1930s the Department of Lands and Forests was constantly revising its Georgian Bay charts in an attempt to keep up with fluctuating water levels as well as with growing demand for "summer resort lands," factors that seemed to produce new islands annually. (In 1832 Bayfield referred to Lake Huron's 20,000 islands; shortly thereafter the count rose to the Thirty Thousand Islands by which the archipelago is still known; but current estimates place the total at approximately 100,000 islands. After all, it depends in part on where you draw the line between a rock and an island.) The difficulties of mapping the islands seemed as numerous as the islands themselves.

There was pressure too for small-craft charts. Most recreational boaters still followed the old steamer route up and down Lake Huron, but small-craft boating in the archipelago was on the rise as part of the booming interest in outdoor recreation. In 1957 a regional business lobby known as the Georgian Bay Development Association asked the federal government to chart a protected route along the east shore, hoping to lure more, specifically American, boat traffic.[59] Cottagers were abandoning public steamers for private outboards by the 1940s and demanded easy-to-read charts and clearly marked boating routes. A 1958 cruising guide complained that Pointe au Baril cottagers were blackmailing passing boaters by setting up private markers which "create confusion in a body of water, and then charg[ing] a dollar for the map that explains it."[60] John Hughes must have made quite an impression on his new bride when they encountered an even more enigmatic system of markers during their honeymoon at the mouth of the Moon River:

> I guess I'd never been in those particular waters with a canoe, but it all looked pretty good to me ... I noticed a pile of rocks on the tip of this island there and a pile of rocks on the tip of that island there, and I thought, now what the hell is that all about. I cut the engine, the boat flattened down, and we went over a shoal that was maybe [six inches] underwater. This was in late fall, *nobody* was there. We would have taken the bottom and the engine right out of the boat, and we'd have been floundering around in the middle of Georgian Bay.[61]

The Canadian Coast Guard issued the first strip charts in 1964 and finished its soundings of the Bay in 1986 – over a hundred years after Georgian Bay Survey was created.[62] But the torturous process of mapping the Bay was far from over. The issue promptly surfaced in the political arena, and the inability of cartographic science to fully document the terrain of the archipelago continued to inspire other forms of mapping.

During the past few decades, mapping has been deeply embroiled in the politics of central Ontario. Local government reviews in Muskoka and Parry Sound in the 1960s and 1970s brought the entire system of township grids under scrutiny. The grid was an artifact of the nineteenth-century survey and its "bald application of geometry" that had systematically carved up the frontier across North America.[63] But as the landscape became more densely inhabited, it drew attention to the disparity between the abstract zones assigned by surveyors and planners, and the communities taking root up the shore, which measured and divided the landscape according to patterns of use and daily experience. Bay residents began to question the merit of a system that served the interior by using the Bay as a convenient end point for boundary lines, thereby leaving the islands as simply the tail end of whatever township in which they happened to fall. District governments seemed equally arbitrary "creatures of the province": the line between Parry Sound and Muskoka cut straight through the Muskoka Lakes.[64] Maps had legitimized this approach. By abstracting the environment into universal cartographic symbols, they mute the effects of distinctive physical features, and by looking down on a place, they adopt an imperious gaze in which landscape is easily manipulated. Township and district maps oversimplified physical idiosyncrasies in the interests of political efficiency. As residents questioned the logic of district and township boundaries, they brought the limits of conventional mapping under new scrutiny – and began pushing at the boundaries of conventional political wisdom as well.

By the 1960s it was apparent that the township grid had had profound political and ecological repercussions. With shore communities facing outward to the Open and the boating channels, there had evolved an enduring divide between west (water-access) and east (land-access), which, as one study reported, resulted in "little interaction and many social and economic differences."[65] The township unit became an increasingly strained marriage between the overwhelmingly seasonal communities on the islands (represented by the Georgian Bay Association and its member associations) and the permanent or year-round municipalities inland. To further confuse the issue, some postwar planning studies envisioned an overextended "Georgian Bay Region," which ballooned out to a 120-kilometre radius reaching from Lake Nipissing to Metropolitan Toronto. This was an attempt to incorporate a wide enough variety of industries

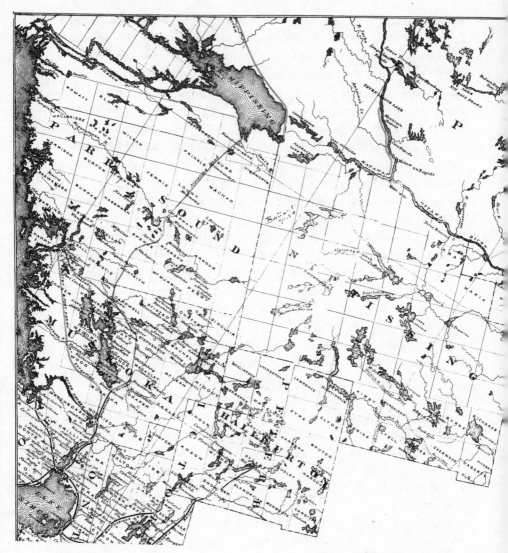

Map 7 Map of Muskoka and Parry Sound districts. From Ross Cumming, *Guide Book & Atlas of Muskoka and Parry Sound Districts,* with maps by John Rogers (Toronto, Port Elgin: H.R. Page, 1879, 1971).

to ensure a balanced regional economy, but did little more than lump together disparate communities and dissimilar physical regions. (One provincial forestry study admitted its "Georgian Bay" had to be divided into fourteen physiographic subregions.)[66] By the 1950s new subdivisions appeared annually on the mainland as the province tried to meet the burgeoning demand for summer properties. Proposals for municipal consolidation or large-scale development in towns such as Britt or Parry Sound saw these nascent urban centres as the saving grace of the regional economy. (The Parry Sound District Review, for example, was struck after the Ontario Municipal Board rejected two proposals to annex adjacent townships to the town of Parry Sound.) But this approach devalued outlying rural areas, and the natural environment. None of these configurations acknowledged the physical unity of the archipelago, or facilitated environmental policies tailored to the area. So calls for stricter planning controls and pollution legislation implicitly questioned the rationale of municipal boundaries.

This increasingly untenable situation became the focus of local government reviews in both the Muskoka (1967-9) and Parry Sound (1973-6) districts. The shore communities led the battle for redrawing township lines into "ecologically sensible units of local government."[67] A north-south municipality that followed the shore would be better positioned to enact legislation appropriate to the archipelago, such as more stringent controls on severances and building density to reflect its limited carrying capacity. As the Parry Sound study observed, "the most consistent demand for change has been for the establishment of local capability to undertake planning responsibilities."[68] The study group's final report included a series of thematic maps which provided graphic evidence of the biophysical differences between the islands and the interior. Severing the western portion of existing townships was even endorsed by Intergovernmental Affairs Minister Darcy McKeough in 1978.[69] At the same time, there was a concerted effort to envisage the districts as a single watershed. The reviews called for a greater role for district-level governments in coherent water-body planning and in coordinating inspection and enforcement, especially in sanitation and water quality. By 1980 two significant revisions to the provincial map had been made – revisions that didn't revolutionize local government in Ontario but at least reflected the questions being raised about political representation and ecological reality.

In 1970 the Act to Establish the District Municipality of Muskoka created the Township of Georgian Bay. This was a conservative first step in municipal reorganization, simply amalgamating the three townships in the Muskoka District that shared Bay shoreline: Baxter, Gibson, and Freeman. The province was wary of substantially revising jurisdictional lines; for example, it left Lakes Joseph and Rosseau inexplicably dissected between two townships in two different districts, despite petitions from the townships involved.[70] But the new framework did grant municipal

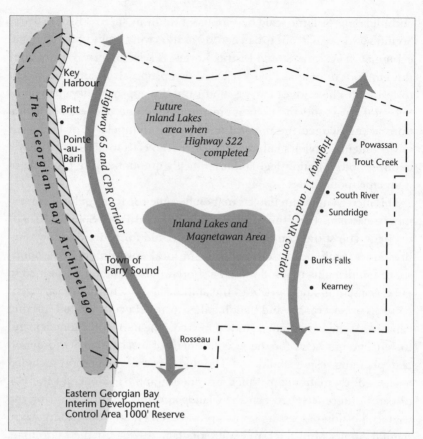

The Georgian Bay Archipelago

Key Harbour

Britt

Pointe -au- Baril

Highway 65 and CPR corridor

Future Inland Lakes area when Highway 522 completed

Powassan

Trout Creek

Highway 11 and CNR corridor

South River

Sundridge

Inland Lakes and Magnetawan Area

Town of Parry Sound

Burks Falls

Kearney

Rosseau

Eastern Georgian Bay Interim Development Control Area 1000' Reserve

Map 8 Norman Pearson, "Significant Entities for Local Government Units." Reproduced from *Environment Control, Planning and Local Government in the Georgian Bay Archipelago,* Sans Souci and Copperhead Association (1975).

government to two formerly unorganized townships (Gibson and Baxter), thus enabling them to formulate official plans and environmental guidelines and to participate more formally in district planning. The township's political centre also shifted westward, by giving the well-organized seasonal majority greater political clout. The idea of a shoreline municipality was further realized ten years later when An Act Respecting Local Government in the District of Parry Sound (1980) created the Township of the Archipelago. This included four unorganized townships – Harrison, Shawanaga, Cowper, and Conger – and hived off Blackstone and Healey lakes from Foley Township. Touted as Canada's first archipelago municipality, it also recognized patterns of settlement by basing its internal ward structure on the existing "colonies" along the shore. Interestingly, the

Archipelago was originally created as two townships – a North and South Archipelago, separated by Parry Sound and Carling – but their councils immediately petitioned for a ministerial order to incorporate, demonstrating both the new degree of local control and the desire for a single archipelago municipality.

Redrawing the century-old township lines opened a can of political worms which belied the neat and orderly character of the simple grid. The problem was that nothing like this had been done before; the provincial government at Queen's Park was not used to revising patterns of governance based on local conditions. Politicians haggled over the balance of power between regional and local governments, the justification for a municipality with such a sparse permanent population, and a formula for electoral representation of the overwhelming seasonal majority. (Cottagers associations later complained that the wrangling over administrative functions "overstressed economics at the expense of geography, ecology and interrelated planning problems."[71]) The process also threatened to polarize opinion between seasonal residents, strongly in favour of an archipelago municipality, and year-round residents, more reluctant to join a municipality in which they would be wildly outnumbered by the "summer people" who favoured stringent environmental regulations. Severing the islands was an unusually contentious proposal because municipalities were loath to surrender the lucrative assessment base of cottage properties, especially given the cost savings to be had when fewer municipal services were extended to water-access communities. On the other hand, others wanted a much larger archipelago municipality, possibly stretching from Honey Harbour all the way to the Key River or the French River.[72]

The archipelago, though, has never been formally identified as a single political unit. Non-governmental organizations, led by the Georgian Bay Association, have pursued this idea of a shoreline designation in various forms, most recently as the "Littoral." The Great Lakes Heritage Coast set out in the province's 1999 Crown land strategy indicates some awareness of the shoreline's distinctive physical character. The old township grid has been modified on all the other shores of Georgian Bay, with more fluid and more "natural" boundary lines for townships along the Bruce Peninsula, the French River, and the North Channel. In the past, explorers descended on the Bay, travelled through it as quickly as possible, and evaluated it according to external agendas, whether military, trade, industry, or recreation. By the late twentieth century, the environmental and political consequences of such drive-by mapping led to a questioning of conventional cartography and its supposedly objective representation. The controversy over municipal reorganization in Georgian Bay revealed how politically charged a map can become. More to the point, it became evident that politics (and political history) needs an environmental reference – to take into account the natural environment to which a political system is applied.

Yet, after nearly four centuries of being surveyed and sounded, we often have the impression that Georgian Bay remains unknown and unknowable. On a modern provincial road map, the area west of Highway 400/69 stands out from the rest of densely marked southern Ontario as strikingly empty. Apart from a few of the largest islands – Parry Island, the Christians, the Manitoulin group – the archipelago is invisible. The broken shoreline does not neatly define the Bay, so it is never fixed, delimited, or contained. It changes hourly as the countless shoals are exposed or covered by changes in wind, currents, or water level. Even the frequently revised boating charts, detailed to a scale of 1:20,000, are guides at best. The all too apparent deficiencies of conventional maps never inspire a confidence of knowing the area, undermining any real sense of territorial possession. Historically, few maps attempted to claim any sort of authoritative knowledge of the area. Boulton's *Pilot*, ostensibly the official record of the national hydrographer, frequently cautioned that, "whilst every possible care has been taken to chart the shoals, some *may* and in all probability *have* been left out."[73]

A lack of cartographic information had a profound, usually dampening, influence on activity in the Bay. During the War of 1812, an American squadron sailed up Lake Huron planning to destroy the supply depot at Penetang. They could not find it:

- [We] imagined that we should arrive in a few days at Malshadash Bay. At the end of a week, however, the commodore from the want of pilots acquainted with that unfrequented part of the lake, despaired of being able to find a passage through the island[s] into the bay, and made for St. Joseph's.

As the captain himself later explained, it was only by sailing directly to Mackinac that he avoided the "great danger of total loss, in this extremely dangerous navigation."[74] The unknown quality of the Bay later discouraged industrial activity as well. As late as 1908, the Department of Public Works complained that all the previous investigations of the French, Mattawa, and Ottawa rivers had collected such little data as to be no help at all to the proposed Georgian Bay Ship Canal project.[75] For the first thirty years of its existence, Georgian Bay Islands National Park reported extremely low visitor counts because of a lack of maps of the park and surrounding waters. Writing to oppose the creation of Blackstone Harbour Provincial Park, one resident warned in 1974 that the archipelago was no place to invite neophyte boaters: "The waters are basically uncharted to date ... I have no doubt that the local residents will have a hayday [sic] salvaging boats and motors because of multiple uncharted or unchartable shoals which make these waters exceedingly dangerous."[76]

This elevated the importance of experiential mapping, or navigating by memory, visual landmarks, or trial and error. It has ensured steady employment for local residents as guides since Champlain's foray into Huronia. Amerindians often seemed like an indifferent coast guard, rescuing hapless travellers such as J.J. Bigsby, whose party got lost in a storm in Franklin Inlet. En route to Manitoulin in 1839, Lt. Andrew Agnew observed that

> the thousands of passages among these clusters of Islands are of course to an unpracticed navigator as bad as the ancient labarinth [sic], but the voyaguers [sic] and Indians kept the proper course with a wonderful correctness. They have certain marks, for instance, at one place an impression resembling a caraboo's [sic] foot on a rock though under water they look for in passing.[77]

Boulton warned that his "directions are only given to supplement the local knowledge of vessel captains, as it would be impossible to write detailed directions, that would be of use to a stranger."[78] A hundred years later, cottage communities still depend on water taxis, whose drivers are familiar with local hazards. Newcomers "learn where the rocks are and it's all changed next year," one told *Cottage Life* magazine. "Me, I've been through all the cycles, high and low, and I drive as if it's low water all the time."[79] Since maps and charts show only part of the picture, necessity generated new types of knowledge.

For cottagers and campers, on the other hand, the opportunity to play explorer in an uncharted frontier was greatly appealing. An American trio sailing in from Lake Michigan in 1898 found that their map extended only to the North Channel, but happily "sailed off the edge of our chart" into the "foreign territory" of a *"terra incognita."*[80] The earliest cottage communities were located in townships such as Baxter, Gibson, Cowper, and Shawanaga, which were of little interest to surveyors and consequently left unorganized and frankly blank on most maps before 1900. Cottagers were often responsible for having their property surveyed. In Sans Souci, there was a standard twenty-five dollar survey fee by 1914, at a time when the land itself might cost only ten dollars an acre.[81] It was easy to imagine the islands, scattered chaotically off shore, as eluding the reach of civilization even in the latter half of the twentieth century. In this "Rough Sweet Land," LePan wrote, there is always "something still to find."[82] The indispensable guides, especially if Amerindian, added nicely to this image of permanent wilderness.

The limits of science also meant richer opportunities for the arts. The graphic shorthand of modern maps may summarize a great deal of technical information but it lacks a descriptive and narrative element. The cartographer and the storyteller tell us different things about the landscape.[83] This may partly explain the strong documentary tradition in Canada's art and literature, as if artists and writers, aware of the shortcomings of standard maps, sought to provide as thorough a description as possible. Paul Kane's *Indian Encampment on Lake Huron* (1845; see Plate 2) and Anna Jameson's pencil sketches of her "summer rambles" in 1837 were undoubtedly influenced by conventions of the picturesque, but they were also conscientious efforts to depict a little-known landscape to a larger audience, just as Bigsby insisted that "my views were selected less for the extremely picturesque than for the characteristic."[85] Sketchers who accompanied survey parties saw themselves as topographers and reporters rather than artists. The misshapen trees, fast-flowing rapids, and imposing boulders depicted by Pilkington, Woolford, or Forrest, even when rendered in watercolour, are not idealized rural settings. William Armstrong, a civil engineer working in Ontario's northwest at the time of Confederation, sought to document landscape and atmospheric conditions in a style that would approximate photographic records. Constrained by the demand for information about a new land, they had to produce images that were essentially factual, realistic, and largely unaffected by artistic licence.[85]

This extended beyond the colonial period. In the early twentieth century, the Group of Seven entrenched the figure of the artist-explorer in Canadian modernism. Maria Tippett has observed that the Group's version of modernism was relatively conservative, maintaining a close relationship with the recognizable physical world, in order to remain accessible to the public.[86] Adopting the role of explorers in the wilderness also suited the Group's public rejection of Old World conventions. Like surveyors twenty, fifty, or a hundred years earlier, twentieth-century artists travelled by canoe, literally immersed in the landscape, making on-the-spot sketches. Today, John Hartman "maps" an area by sketching it from several directions, and then hangs the viewer above the final canvas to view it as a whole.[87] Although coloured in a wildly unreal palette and peopled with characters from Hartman's own past, the shoreline and surface elevations are anchored by recognizable physical landmarks. The documentary tradition, or the motif of exploration, seems peculiarly Canadian, perhaps because it is intertwined with our long history of struggling to map a vast and complex landscape.

But art also allows visitors to express a more subjective response to the landscape. The archipelago, never fully explained by reason, has left much room for the imagination and alternative forms of representation. Often in literature there is a feeling of amazement or bewilderment at travelling through a place without the usual tools of measurement or direction. LePan wonders what words to use to describe "this curious country." The modern voyageur sees the stream he paddles, but as for the "whole wild dialectic of lakes and rivers[,] Angels alone would see it whole and one."[88] Steven Tudor, cruising through the Great Lakes, finally comes to accept the limitations of navigational technology and channel markings, asking, "How do we navigate if not by faith? / Staring down the night this exuberant / doubt." His own perception of shape, colour, and depth is more useful than the detached abstractions of his charts. His depth sounder fails him, his rudder is torn off, and he thinks wryly,

Look, a reef dead ahead:
you might have seen green shapes,
gray knees of granite ...

And later,

How can the pure world of numbers
correspond to the green turbulence
and rock smash of waters in their
gull-sung and accidental distances?[89]

Such ground (or water) level encounters offer impressions of place on a human scale, with an intimacy and immediacy often lacking in more objective representations. But LePan in his canoe and Tudor in his yacht are simply commenting in different ways on the same physical realities that left mapmakers in such a quandary. Whether documentary or subjective, the arts have played a central role in shaping our mental map of Georgian Bay.

Day to day, though, the collective memory of the community is kept alive through the place names crowded onto maps and charts of the Bay. These names are a colourful record of past encounters with the Bay, accumulated over centuries: a unique mix of the imperial, the prosaic, and the imaginative. Histories of cottage communities are full of interesting stories about the origins of local and cottage names, particularly since many cottages in Georgian Bay – more so than in other parts of Ontario's cottage country – have remained in a family for generations, even from the original purchase. The vocabulary of place names has become a kind of regional vernacular. Only long-time residents familiar with the dizzying number of names in the Bay would understand the comedy in Evan Solomon's 1999 novel *Crossing the Distance*. An obnoxious Toronto reporter thinks she has discovered the hideout of the protagonist, Jake, wanted for attempted murder, after a reluctant source utters the word "point." She turns to her cameraman and says urgently, "We need to get a map of Georgian Bay area and look for places beginning with the word Point."[90]

Many names in the archipelago date to the long period of exploration and colonization. The Bay is a classic example of what W.H. New calls "naming from without," presuming authority over a place, just like the survey grid.[91] Often European explorers and surveyors either did not learn or simply disregarded any existing designations used by Natives, fur traders, or other local residents. Bayfield and Boulton together are a perfect illustration of imperialism writ small. They apparently realized that however laborious the assignment, it was a golden opportunity to claim the area for the Empire or the new Dominion and make their own mark in history at the same time. Bayfield's choices are a far better reflection of him than of the archipelago: liberally sprinkled about are tributes to royalty (Georgian Bay for George IV), government officials, his heroes of the Admiralty, famous explorers (Parry Sound, Byng Inlet, Franklin Inlet), other naval officers (including his own midshipman, who is remembered at both Philip Edward Island and Collins Inlet), together with friends, family, and even "young ladies of his acquaintance."[92] Why not, he must have thought; there are more than enough places to go around.

Other illustrations of the imperial impulse include the Jesuits, who preferred saints, as at Isle St.-Joseph (now Christian Island) or Isle Ste.-Marie (Manitoulin). When the departments of Indian Affairs and Crown Lands began selling off islands, they resorted to numbering them, in a stab at imposing some system of orderly identification. While this did not resolve the confusion, it left its mark on the landscape nonetheless:

> In Trudeau's time, the Bilingual and Bicultural Commission changed all the English language national park signs right across the country into bilingual signs. So all these signs were taken down and left, chopped up or broken up, for area users as firewood. I can remember the one I got was still hanging; no, it was taken down and leaning against two posts on Hatch Island, south of O'Donnell's Point ... Our island number is 358.[93]

Purchasing an island known only by a number allowed cottagers to name their own private domains. Unlike much of southern Ontario, though, there are relatively few spots named for the Old World (Killarney being a rare exception), because they were named more whimsically and not named by settlers thinking of home.

The names grandly bestowed by the Admiralty were not always adopted into everyday use. People still referred to Georgian Bay as the north shore of Lake Huron late in the nineteenth century. The cumbersome Prince William Henry Island never dislodged the French Beausoleil, and Franklin Inlet has reverted to the native Shawanaga. As shipping and industrial traffic increased, a second, more indigenous category of place names developed. While Boulton added his roster of historical and political figures to those of Bayfield's, he also recorded many more prosaic designations in use when he arrived. These usually feature an aspect of local history or a reaction to the environment. Most common are those that describe navigational routes or act as a means of orientation: Passage Island, the Watchers (guarding Penetang Harbour), Barrier Island, Devil Channel, Snug Harbour. Some commemorate boats and ships, local residents such as ship captains, and shipwreck sites. Others speak to activities that took place in certain spots: Depot Harbour, Picnic Island, American Camp.

In addition, a great many names are derived from observations of natural phenomena. This group of names demonstrates an awareness of wildlife and topography, and even an aesthetic appraisal of the landscape. Some are simple and straightforward – there are a lot of Green, Stony, Round, and Little islands in Georgian Bay. Or they are variations on a theme: just south of the western edge of Manitoulin lies Great Duck Island, along with its lesser compatriots Middle Duck, Outer Duck, and Western Duck. But other names highlight details of shape or outline

(Double Island, Bear Head); colours (Black Rock or Green islands); species of birds, fish, and other wildlife (Gull, Serpent, Bass); trees, or lack thereof (Pine, Bald); or geology (Agate, Quartz). These could be another illustration of the documentary impulse, or a way of noting landmarks by which to navigate; either way, they indicate that people were studying the world around them. Even Bayfield found room between clerks of the Admiralty for the Limestone Islands and Mouse (Moose)-Deer Point. Certainly these descriptive names paint a more realistic and more attractive picture of the archipelago than surveyors' reports of the same period.

But names that have survived from the age of exploration also tend to rely on elements of natural and human history to identify places. Here a map becomes an archaeological record of past cultures.[94] Amerindian names are unusually prominent in Georgian Bay, in part because of their own history in the area and in part because Europeans adopted Native routes, such as the Musquosh and Moon rivers, to access the interior. At the same time, they failed to graft conventional European patterns of settlement onto, and consequently erase, the older cultural landscape. Cottagers and campers borrowed or revived Native words to suggest that, in the supposedly primeval wilderness of the islands, they were in some way the spiritual descendants of the "braves of the past" who had bequeathed "names soft sounding in the Indian tongue and exquisitely appropriate ... to add glamor and romance to the district."[95] Names still in use usually translate as a description of a particular site, such as Minnicognashene (place of many blueberries) or Massasauga (mouth of river). More generic names are based on observations of Native inhabitants or activities, such as the group that Bayfield calls "Indian Islands" east of the Limestones. The French presence is remembered along the old trading routes, aided by the influential role of French mapmakers to the end of the eighteenth century. Jean-Baptiste Perrault's work with the North West Company took him out west in 1815. In his memoirs he explained the origins of names along the *rivière des Français*, including such cheerful events as drownings *(le rapide de Parisien)*, Amerindian battles *(Pointe aux Yroquois)*, or a child stolen by animals *(l'Enfant Perdu)*.[96] Among those still in use are Point Grondine, where water thunders around an exposed point; La Cloche, where the rock rings like a bell when struck; and Pointe au Baril, where the original lantern in a barrel of local folklore has been replaced with a Coast Guard lighthouse. These names preserved a connection with the past and lent the Bay its anti-modern character, so prized by those seeking to rediscover a fragment of pre-contact wilderness. They have also left us a landscape littered with clues as to earlier views of the natural world.

I can pore over a small-craft chart of the Bay for hours, studying the names, depths, and navigational markings. Part of the fascination comes from knowing that the information crowded onto the map is more than a two-dimensional rendering of the topography: it is a trove of clues about past encounters with the Bay. But even these detailed charts only graze the surface of an extensive mapping tradition of four centuries. Explorers and naval officers, land surveyors, tourists, and campers have all asked, what shall be our word to describe Georgian Bay? *La mer douce ... pas connue ...* harsh and forbidding ... pretty and desirable. Some were bewildered and frustrated, others intrigued and excited. Few were entirely prepared, physically or intellectually. Their tools of measurement and description ranged from surveyors' chains and field notes to personal journals and paintings. When wilderness became a destination rather than an obstacle, what British observer Smythe described as the "bleak, barren, and inhospitable coast" became prime recreation country. But all these attempts to map the Bay share a quality of frontline reporting. These visitors saw themselves as translators, evaluating and interpreting a new and strange subject for their contemporary audience. The next question, then, is how modern Canada would make use of this information – and attempt to make use of this curious country.

TWO

A Region of Importance:
Industry and Land Use

By the middle of the nineteenth century, people began to look twice at the "practically worthless" coast that explorers and sailors had been in such a hurry to cross. Ironically, it was a surveyor, William Gibson of Willowdale, who marked the start of a real and prolonged interest in the east shore when he built a sawmill at the mouth of the Seguin River (present-day Parry Sound) in 1856. The vast wealth promised by its fish and trees was more than enough incentive for industrial North America once the railway reached the south shore, and by the 1870s the Bay was a continental leader in shipping, fishing, and logging. Here the furious pace of expansion and ecological effects of nineteenth-century industry were dramatically exposed. Soon cottagers and campers arrived with an alternative agenda, to claim the islands as recreational space and to create the landscape of cottages and yachts we know today. People have sought to cast the Bay in a variety of roles: sometimes the archipelago obliged and even encouraged them, but sometimes it obstructed these plans and forced people to adapt their technologies, work rhythms, and patterns of settlement. This is the story of how different plans for wilderness, whether as a resource or recreational hinterland, worked out in practice in a specific place. It is a story of regional opportunities and constraints, and of the differences between expectation and actuality.

Although prospectors, fishermen, and foresters sought different things from the Bay, they viewed nature in essentially the same way. Their activities were seasonal, export-oriented, and driven by a "cut-out-and-get-out" mentality.[1] Like the surveyors that preceded or accompanied them, their invasion of the North American wilderness was propelled by a utilitarian ideology that distilled two centuries of Anglo-American thought – two centuries that had seen the scientific revolution, the rise of capitalism, the Industrial Revolution, imperial expansion, and life on the frontier. Most North Americans shared certain ideas about their environment: that humanity, since Adam in Eden, was responsible for governing the rest of creation; that progress resulted from the "improvement" of nature, by transforming raw wild nature into productive land; and that this process was key not only to the economic health of society but to its moral and social integrity as well. The scientific character of exploration and industrialization had transformed attitudes toward the North American landscape, once seen as hostile and incomprehensible. The modern scientific method reduced and distilled the tangle of wilderness to a series of systems and component parts that could be identified, classified, and analyzed in turn. Identifying the potential use of each part, and the technology needed to realize that potential, made nature malleable and profitable (a shift in perspective succinctly captured in the popularization of the term "natural resources" by the middle of the nineteenth century). That is why surveyors so doggedly plotted and sounded Georgian Bay, noting species, soils, and minerals, cataloguing the chaotic country as best they could in hopes of finding a use for it. This line of thinking was particularly compelling in Canada, where resources seemed inexhaustible and nation building was the pre-eminent political concern. Positioned on the closest edge of Ontario's "frontier," in the rich Great Lakes Basin, Georgian Bay appeared to be a great example of potential wealth.

But it was never malleable, and it constantly challenged those who proposed making a fortune from it. The Bay proved at once lucrative and uncooperative. People soon discovered that to mine it effectively – or indeed, at all – they had to adapt to its physical conditions. As a result, the industrial age took on a peculiarly regional character along the northeast shore. How nineteenth-century economic agendas played out locally is captured in the range of photographs from the period. There are the familiar sepia portraits of bustling harbours, mills, and railway sidings, all strewn with machinery, cargo, or lumber, suggesting a tamed and productive landscape. But there are also images that capture the idiosyncrasies of the archipelago: steamers run aground on shoals, fishing stations on the outer islands, log booms passing canopied touring boats, moonscapes of granite laid painfully bare by any sort of clearing. Sometimes people had to adapt or rethink the standard means of development when confronted with environmental limitations and opportunities. Timber and minerals were limited or inaccessible, and the Shield resisted construction of canals and railways, but the archipelago offered a seemingly inexhaustible fishing ground, an ideal site for lumber mills, and a valuable transportation corridor for small steamers. Profiting from the Bay required adjusting to it. But the greatest adaptation was imaginative. The transition from primary industry to recreation occurred not because the environment had changed but because expectations of wilderness had.

∿

The nineteenth century began with one resource industry well established on the Bay. Furs from hunting grounds north of Lake Huron and Lake Superior were brought down to trading posts on the shore as early as the 1690s. The French River and the North Channel became the major highway for the *canots de maître* carrying trade goods between Montreal and the northwest. The Bay's importance as a trade route was evident in its small role in the War of 1812. The *Nancy* was commandeered from the North West Company, which had sailed it between Detroit and Sault Ste. Marie; the American schooners tried to intercept a party of loaded canoes coming down the French River; and both sides concentrated on Mackinac, the key to the northwest and the still-lucrative fur trade. Clearly the interest in Georgian Bay was as much mercantile as military. Although dominated by the Nor'westers after 1784, independent traders backed by merchants in Penetanguishene and York appeared on the north shore in greater numbers in the 1820s and 1830s, reflecting a new Upper Canadian influence in the area. In 1830 there were trading posts at La Cloche, Whitefish Lake, Mississagi River, French River, and Shawanaga.[2] That year, Hudson's Bay Company governor George Simpson and his wife Frances were nearly dumped in the French River when, "on going from the foot of the Recollet Fall, [we] were very nearly drawn under it, by a strong Eddy; indeed so near, that the Spray from the fall showered over, & gave us another drenching." But as the Simpsons departed the Bay for points farther west, the fur trade was also vanishing from the Bay watershed. By the late 1820s, the situation in the La Cloche district, which included small posts at La Cloche, French River, and Shawanaga, looked grim. Presiding over a district that he described as "a complete mass of rocks," Chief Factor John McBean struggled with a sharp decline of fur-bearing animals, an erratic fishery, and levels of starvation that forced the "inland Indians ... to abandon hunting grounds and come out to the establishments to save their lives and those of their children." By 1835 he concluded the country to be exhausted.[3]

To the nineteenth-century mind, what should have happened next was fairly straightforward. Once the fur trade is exhausted, orderly pioneer settlement can proceed to transform wilderness into productive land.

Canada West embarked on a campaign to settle the Ottawa-Huron Tract, the land between the Ottawa River and Georgian Bay, in the 1850s and 1860s. Legislation in 1853 and 1868 provided for a system of colonization roads and free grants for homesteaders, and there was a burst of surveying and settlement activity in the Muskoka and Parry Sound districts after 1866 through the 1870s. Townships were surveyed first around existing communities such as Parry Sound and Killarney: Foley and McDougall in 1866 (included as free grant lands under the 1868 Free Grant and Homestead Act), Killarney in 1868. Three colonization roads eventually cut through the western part of the Tract. The Parry Sound road branched off the Muskoka Road about 8.2 kilometres north of Bracebridge and ran northwest to Parry Sound (the current route of Highway 141) by 1868; a Northern Road ran through McKellar to the Magnetawan River by 1872; and farther east, the Rosseau-Nipissing Road passed Ahmic Lake en route to Lake Nipissing. The roads did provide potential settlers with access to the interior; many along the Nipissing road in McKellar and Hagerman townships, for example, were claimed as free grant lands in 1870 and 1871. Despite very mixed reports from the surveyors, most believed there was good land in spots, marked by rich stands of pine and hardwood, and that the "industrious" settler could make a go of it.[4]

But the Ottawa-Huron Tract would be a nasty shock for Ontario, a society whose land ethic celebrated agricultural settlement as its destiny and the foundation of progress, and whose settlement system was designed to establish and service a rural agricultural community.[5] The drawbacks of Shield country reached their most extreme in the western part of the Tract: thin and acidic soils, low-lying swamps and marshes, exposed granite, and a broken maze of inland lakes and rivers. The islands and shore were obviously not arable and were not viewed in the same way as the districts of Muskoka, Haliburton, and Renfrew. The archipelago was peripheral to the efforts to colonize the Tract and is rarely mentioned in either the literature of the day or histories of Ontario settlement. Parry Sound was a useful entry point, serviced by steamers from towns on the south shore, but settlers were cautioned not to be discouraged by the "frowning barriers of inhospitable rocks" walling in the Tract. The territory north of Lake Rosseau was "forbidding in its aspect and unfitted for cultivation," wrote the superintendent of colonization roads, "but must necessarily be crossed."[6] The Parry Sound road led directly *away* from the Bay, cutting through the adjacent "worthless" land to, it was hoped, slightly greener pastures inland; great hopes were pinned especially on the Magnetawan valley.

The infrastructure and enticements provided by the state, and the entrenched faith in the powers of the pioneer, proved no match for the Canadian Shield. Underneath today's recreational landscape, the near north is a palimpsest of disappointment. The neat regularity of Simcoe and the southern counties, criss-crossed by rural villages and road allowances, vanishes. The initial interest in the Tract was quickly followed by disillusionment and outmigration. In 1892, for instance, Thomas Pearce led 298 settlers from Parry Sound District to Strathcona County east of Edmonton. All along the southern edge of the Shield, abandoned homesteads quickly began reverting to bush. Roads were notoriously unpleasant, difficult to maintain amid rocks and swamps, and continually chewed by logging traffic. Surveyors' reports held few illusions about the east shore, and their characterization of the region as barren and inhospitable was not likely to inspire much confidence. Promoters of Canada West concentrated on pockets of soil in Muskoka and the Ottawa River valley, tactfully ignoring the western part of the district. A small cluster of farms along the Magnetawan and Still rivers, where some clay pockets were located, was the best that could be hoped for. Yet, even here, "agriculture is primitive; of roads there are none," reported one visitor in 1899. "If a settler can find five acres of soil in a forty of rocks he has a good farm."[7] The handful of communities on the Bay, such as Killarney, Byng Inlet/ Britt, and Parry Sound, developed not as service centres for surrounding farmland, as in southern Ontario, but as isolated ports, railway stops, or company mill towns. They relied on Georgian Bay (and later the railway), not the colonization roads, for communication and transportation. Most declined when the mills closed; French River Village disappeared altogether. Those that survived have remained islands of settlement, in sharp contrast to the ribbon urbanization along Nottawasaga Bay and the south shore – graphic evidence of the differences between Shield and lowland.

Instead of creating a prosperous, permanent agricultural landscape, homesteaders found themselves in a fight for the Shield's most saleable resource: timber. Timber sales promised immediate profit, when even promoters admitted that cultivating a homestead in the Ottawa-Huron Tract would take years. As Neil Forkey has argued, the need for cash wove settlers into the lumbering economy, rather than supplanting it. (Thomas Roach has shown that environmental limitations later created a similar dependency on the pulpwood trade among settlers in the claybelt of "New Ontario.") To the settlers, logging companies offered both paid employment and a buyer for the timber that had to be cleared from their homesteads. But not surprisingly, what was meant to be a complementary relationship soon became a competitive and at times antagonistic one. Loggers, for example, denounced "pretending" settlers, those who claimed marginal lands only to harvest its pine. Colonization roads – designed to invite settlement but used more often by logging traffic – became focal points for the conflict.[8] Yet, one-time sales of timber from homesteads were not enough to support the inhabitance of subpar land. In short, timber limits in the Tract were desirable; ownership of poor quality land was not.

Thus, even with free grants, the district was most profitable as a hinterland from which to export raw materials. Logging companies preferred to secure a licence to harvest the timber but leave ownership of the land itself with the Crown. The nature of the Shield country and official policy together kept the vast majority of the Great Lakes Basin in public hands through the twentieth century. This turned out to have some benefits. By the 1970s, for example, the Ontario Ministry of Natural Resources had neither the funds nor the inclination to undertake substantial acquisitions of private property for the purposes of park creation. In Georgian Bay it was possible to design a park (Blackstone Harbour) in which 96 percent was already Crown land as late as 1976 – over a century after the age of the free grant and the homestead. In a sense, cottagers would be the first real colonizing population.[9] As for the islands, the confusion in the documentary record mirrors the geographical, legislative, and jurisdictional confusion in the archipelago itself. Here it would be cottagers, not settlers, who sought to secure control over the pine, and their goal would be to *retain* the tree cover, not harvest it. Generally, sale of an island by the government extinguished any licence to harvest its timber. That did not, however, stop cottage communities from intervening to buy out timber limits on or near their islands in order to protect the trees and the scenery.[10]

For political and geographical reasons, then, the east shore could not share in the backwoods-to-civilization narrative of progress typical of frontier mythology. (Policy studies characterized it as a frontier until the 1980s.)[11] This presented a problem for Indian Agents, who encouraged farming as a method of "civilizing" a traditionally semi-nomadic people. Poor soil forced the Beausoleil Island band to move to Christian Island after only fifteen years, but here too they had to supplement agriculture with traditional subsistence activities such as fishing and berrying, in addition to wage work. This was typical of Algonquian farmers in southern Ontario, as on the Parry Island Reserve.[12] Chiselling a homestead in this part of Ontario was a Sisyphean task well into the twentieth century. In paintings such as Carl Schaefer's *Before Rain, Parry Sound* (1935) and Yvonne McKague Housser's *Little Clearing, Georgian Bay* (1952; Plate 12), houses teeter precariously on jumbled knolls of bare granite, in lonely clearings encircled by a solid barrier of hills. Settlements rest uneasily here; the human presence is oddly *un*natural, impermanent, superimposed on the physical environment. It is, as Thomas McIlwraith writes, a shallow cultural landscape.[13] Instead of marking the twilight of the wilderness, attempts at agriculture served only to emphasize its intransigence. Homesteaders were forced to supplement subsistence farming with other types of seasonal employment. Opportunities to earn money from logging during the winter may have encouraged settlers in northern Ontario to remain on marginal farms.[14] For those near Georgian Bay, servicing tourists quickly became the more lucrative alternative to farming rock. The 1879 district atlas for Muskoka and Parry Sound pointed out that tourists and sportsmen already rivalled the numbers of the "emigrating class." "True, there are other parts where they have less rock and can boast of better farming country," admitted Thomas McMurray in 1871, "but there is no spot more healthy or romantic than this. Here the sportsman and the pleasure-seeker can enjoy the richest possible treat."[15]

Disappointment with the unarable surface of the Shield was tempered by enthusiasm for what might lie beneath. The possibility of iron, nickel, and copper reserves partly redeemed this difficult landscape for a Central Canada eager to fuel its industrialization. Growing interest in its mineral resources led directly to the 1850 Robinson treaties, which arranged for the surrender of the north shores of Lake Superior and Lake Huron.[16] Bruce Mines was an active site on the north shore of Lake Huron after 1855, but was hampered by the difficulty of supplying camps in such a remote area with necessities such as food and fuel. Bigger finds on Superior then overshadowed Huron's north shore, though interest shifted eastward again when the construction of the CPR in 1883 uncovered nickel and copper pyrites in the Sudbury Basin.[17] Sudbury and Algoma became the epicentre of Ontario's mining industry, feeding smelters and steel mills in Sault Ste. Marie and other cities around the Great Lakes. Closer to home, Canadian Explosives Limited built an explosives plant just north of Parry Sound in 1913-14, and a new village of Nobel, to supply Shield mines farther north and the possible construction of a Georgian Bay canal (about which more is said below). But wresting minerals from the Shield proved to be almost as difficult as farming it. Franklin Carmichael captured the struggle in paintings of Sudbury-area mining towns. In *The*

Nickel Belt (1928) and *A Northern Silver Mine* (1930), dark, powerful hills
threaten to engulf the narrow rows of houses and mining shafts.

Useful ore proved elusive along the Bay, but despite erratic findings
the eastern shore was revisited periodically for close to a century. It was as
though people's expectations for Shield country convinced them that
minerals *must* exist there, and only after years of digging in the archipelago would they concede otherwise. William Norton calls this an "in-
ferred environment," when knowledge and aspirations for a place are
projected onto an unknown landscape.[18] Three centuries after Champlain
exaggerated the *Mer Douce* into the desired transcontinental water route,
hopes for New Ontario and the gleam of Sudbury-area deposits masked a
lack of solid evidence about accessible ore. This appears to have been the
case around Parry Sound. By 1894 two mines had opened in pursuit of
copper in Cowper and Foley townships, south and west of the Sound.
Exploration for garnet, feldspar, and anorthosite took place on Parry
Island. Sporadic test drilling south of the Sound occurred well into the
1940s and 1950s, and, as late as the mid-1970s, the Division of Mines refused to extinguish staking for copper and zinc in the Blackstone Harbour Park Reserve.[19]

This would be understandable if there had ever been some indication
of worthwhile deposits, but reports had always been lukewarm at best. In
1848 Alexander Murray reported to the Geological Survey of Canada that
"the north shore of Lake Huron is destined sooner or later to become a
mineral region of importance," but the rest of the shore consisted almost
exclusively of contorted granite or gneiss. A representative of the Bureau
of Mines visiting Foley and Cowper in 1899 was most unimpressed: "the
large majority of the prospects do not look particularly promising. The
owners of these prospects seem to have quite exaggerated ideas of the
value of their properties." Surveyors for the Georgian Bay Ship Canal did
not bother with test bores along the French River, "as practically only
granite is encountered there." Even with the contingencies of war in 1943,
the Department of Mines remained cool to the "erratic and negligible"
findings, and recommended reforestation for "what is, at present, useless
land." Although the Department fought to keep Cowper Township open
to staking, officials privately admitted that "the known information about
these copper showings indicate they are very insignificant and unimportant." Research in the 1970s determined there were definite deposits, but
exploration was not pursued.[20] By this point it is difficult to imagine re-
introducing extractive industry into the heart of cottage country. Those
looking for minerals would have to search elsewhere.

It would be the railways, and to a lesser degree canals, that would pull Georgian Bay firmly into the industrial age and the orbit of urban North America. Before the 1850s the Bay was too inconvenient to warrant much attention, since it could be reached only by canoe, poor dirt and corduroy roads, or the few steamers that travelled to Owen Sound. But railways provided a fast, direct, all-weather link between the upper lakes and the St. Lawrence, and from there to the Atlantic Ocean. Collingwood was the first town on the Bay to welcome the train, in 1855; by 1880 other lines had reached Waubaushene, Penetang, and Midland. Suddenly the Bay found itself at the centre of proposed transcontinental routes. From the south shore ports, ships could head north and west directly for Lake Superior, making, as one observer wrote in 1854, the upper lakes "innumerable canals in one, a cheap, and universal railway" to the northwest. Rail lines could also cut through the Shield granite, so it was now possible to contemplate a route across the Ottawa-Huron Tract, to make "this present wilderness a thoroughfare for a great portion of the continent."[21] A railway or canal from Lake Ontario to the Bay's south shore, or an east-west route from Montreal to a port on its east shore, would be shorter than the principal all-water route down Lake Huron, across Lake Erie, and up Lake Ontario to the St. Lawrence. In the frenzied enthusiasm for railways, provincial maps of the 1860s and 1870s are criss-crossed with dozens of rail lines shown as "chartered or constructed" that never materialized – what Alan Morantz has called an "anticipatory geography."[22]

The east-west route *was* realized, however, and by one of Canada's most famous railway magnates and lumber barons. John Rudolphus (J.R.) Booth built the Ottawa, Arnprior and Parry Sound (OA&PS) Railway as a feeder line for his Canada Atlantic Railway to New England. The OA&PS was running its full length from Ottawa to Depot Harbour by 1897 and enjoyed a brief heyday shipping lumber, grain, and coal until the Depression. In 1905 the CPR abandoned its port at Owen Sound in favour of one at Victoria Harbour, named Port McNicoll, for a route to Montreal as short as the Canada Atlantic Railway. Soon thereafter, both the Canadian Pacific and Canadian Northern railways were pushing north to Sudbury, and by 1908 were laid along the full length of the eastern shore. This was an important catalyst for development in the isolated area. Railway stops along the east shore formed the nucleus for industrial ports such as Britt and tourist stations such as Pointe au Baril Station and Rose Point at Parry Sound. On the south shore, port towns such as Owen Sound, Collingwood, Midland, Penetanguishene, Port McNicoll, and Waubaushene boomed, competing for local and Lakes business in tourists and bulk cargo.[23] Around the Bay, these communities existed as transshipment points, conveying cargo from boat to rail and vice versa. Britt and Key Harbour, for example, survived as Canadian National Railway (CNR) ports to unload coal and oil shipments off tankers coming from Lakes Superior and Huron.

The environmental effect of railways was not confined to a line of track a few feet wide. Railway (and canal) construction was arguably *the* definitive industry of the nineteenth century, and permanently reshaped perceptions and use of the natural environment. The arrival of the railway on Georgian Bay marked a turning point in the history of the Great Lakes. It transformed the scale of all other industries and catapulted shipping, timber, fishing, and tourism to a new intensity. In particular, fish and lumber could now be sent in vast quantities to southern markets, a powerful incentive for accelerated harvesting. Railways also built a landscape that endures as one of the most visible legacies of the industrial age, a landscape of grain and coal elevators, bridges, abandoned spurs, and railside towns or ports such as MacTier or Britt. At the same time, they could be a destructive force, as sparks from engines lit forest fires in the lumber slash left by loggers. And they introduced people to a formerly remote area in unprecedented numbers. The very technology that exported natural resources and symbolized the Victorian veneration of progress was soon importing wilderness seekers attempting to escape modern life by visiting the "primeval" archipelago. Railway and steamship companies played a key role in building an infrastructure for commercial tourism in nineteenth-century North America, providing everything from advertising to hotels and resorts. As *Canadian Magazine* commented in 1900, the camping boom was "the inevitable accompaniment of the growth of railroad and steamship lines and of the perfecting of travelling comforts," able to serve the increasing numbers of people "desirous of getting 'back to nature.'"[24] Railways moulded the Bay as a resource and a recreational hinterland almost simultaneously.

Canals, on the other hand, appeared in the realm of imagination far more often than in the granite reality of the near north. For decades the favoured technology of the early nineteenth century was thrown relentlessly and ineffectively against a landscape that, despite its numerous natural waterways, resisted man-made ones. In fact, neither of the major canal projects in the region had much to do with Georgian Bay itself, except as a destination; both generated far more paper than construction; both were surpassed by the Welland Canal-St. Lawrence route, which did not require digging in the Canadian Shield and which did incorporate flat, open stretches of lake and much more urban settlement along the way. However, the convoluted histories of the Trent-Severn Waterway and Georgian Bay Ship Canal are useful mirrors of the ambition of the industrial age and the politics of natural resources through to the mid-twentieth century. They are also evidence of how uncooperative the Georgian Bay region could be.

The idea of a navigable water route between Lake Ontario and Lake Huron was not new. Natives, explorers, fur traders, and military men from Samuel de Champlain to John Graves Simcoe had used portions of the connecting waterways for centuries, though finding a route through British territory to the upper lakes was particularly important after two wars with the Americans. However, it was the growth of the western grain trade that prompted serious consideration of a canal between Lake Ontario and the Bay. Different routes were proposed, including the old Humber River-Holland-Nottawasaga route and a canal via Lake Scugog, but these were never carried out because they required extensive excavation.[25] Instead, a canal system was begun along the Trent River system, propelled by the timber trade and colonization of the Kawartha Lakes. The Trent and Severn rivers meant a winding, indirect route to Georgian Bay, and it took almost ninety years before a through route was completed from the Bay of Quinte to Port Severn. In his history of the Trent-Severn Waterway, James Angus argues that the entire project was completed flukishly and piecemeal, according to local business pressures and sporadic government funding. Georgian Bay was on the periphery of Ontario politics in the mid-nineteenth century, so it is not surprising that the last section to be completed was that next to the Bay, along the Severn between Sparrow Lake and Gloucester Pool, in 1920.[26]

Construction here was particularly arduous, for the Severn empties into the Bay in a string of dangerous rapids framed by steep granite walls. Marine railways were built in 1918 and 1919 at Big Chute and Swift Rapids to haul small boats over the rapids, sort of a poor man's lift-lock. (The Port Severn dam and lock, completed in 1915, cost $137,000, whereas the marine railways at Big Chute and Swift Rapids cost only $11,000 and $10,000, respectively.) The marine railways were a tacit acknowledgement that the Waterway would never see the major commercial traffic for which it was originally intended. By the time it was finished, there was no need for covert military movement, and lake ships and barges were far too big for the canal. The Department of Canals, trying to put a good face on things, pointed out that motor launches, at least, could make the trip and "this possibility is rapidly attracting the attention of tourists."[27] Rather than adapting the landscape for human use, humans had had to selectively adapt their activities to the landscape. A project justified alternately for defence, lumbering, colonization, and bulk grain transport finally found permanent use as a recreational waterway, which makes the Waterway a rather fitting symbol for the near north as a whole.

The story of the mysterious Georgian Bay Ship Canal is a convoluted mess of transportation and political history, most of which centred on the Ottawa River rather than on the Bay for which it was named. The idea of a road or waterway between Ottawa River and Lake Huron had been bandied about for most of the century. The Ottawa-Opeongo colonization road was meant to extend from the Ottawa right across to the Magnetawan or the Muskoka River, though this was abandoned by the early 1860s.[28] A canal project linking the Ottawa, Lake Nipissing, and French River had several other things to recommend it. It was a shorter distance from Lake Superior to Montreal (the same logic behind the OA&PS Railway); most of it could be pieced together from existing natural waterways; it fell entirely within Canadian territory; and it passed through resource-rich northern Ontario. Construction on a canal finally appeared imminent in 1894 when the Montreal, Ottawa and Georgian Bay Company was chartered and the route surveyed by the Department of Public Works between 1904 and 1909. But thirty years later, as one furious MP told the House of Commons, the proposal had "survived under nine prime ministers, under two kings and three queens of this great empire, under eight or nine parliaments, and is again today before this House during which not a spadeful of earth has been dug on this canal."[29] Finally, the House put its collective foot down and refused to renew the company's charter – for what would have been the fourteenth time – in 1927. The federal government was smarting from massive overspending in railways and war production and was already committed to the Welland and St. Lawrence route; it was not about to undertake another hundred-million-dollar megaproject. Much of the controversy stemmed from a section of the charter that granted the company rights to hydro power from the Ottawa River. The clause had seemed inconsequential in 1894, but twenty years later the profits to be realized from hydroelectric plants on the Ottawa had overshadowed the value of the actual canal, further minimizing the significance of Georgian Bay in the scheme. MPs were reluctant to become embroiled in a political and constitutional quagmire with the governments of Ontario and Quebec by granting the rights to hydroelectric power on an interprovincial river to a private company. The canal proposal that had dominated discussions of northern Ontario for decades vanished into obscurity.

But it remains a useful gauge of the technological and political dimensions of environmental thought at the turn of the nineteenth century. First, the project proposed to completely reshape the physical landscape, with a breathtaking confidence in the capability of modern engineering. The most extensive work was planned for the mouth of the French River, where colossal excavation to a depth of 7.3 metres would remove entire islands to permit navigation. According to the prevailing philosophy of "wise use" development was a form of public service that transformed a redundant landscape into a productive one; hence, public works projects were referred to as "improvements." If earlier reports had dismissed the French River as an unnavigable, barren waste, it was because its potential had not been unlocked. Modern technology could harness the landscape "to the benefit of humanity."[30] But this rationale presented a moral dilemma that fuelled the political controversy over the ship canal. How could Parliament permit "the hand-out of these magnificent resources," of Canada's "natural heritage," to corporations from St. James Street or Bay Street, or worse, the United States? How was the public good served by alienating waterways and water power to private investors? "The interests of the whole people come before the avaricious ambitions of a few," declared the Toronto *Globe*. In the end, however, the debate was moot. The difficulties posed by the granite-choked French River and the expense of excavation were indeed "sufficient to preclude the practicability of the contemplated water communication," as David Thompson had gloomily predicted eighty years before.[31] The phantom Georgian Bay Ship Canal demonstrated that popular faith in engineering and the ideology of progress was insufficient to overcome fundamental political and environmental obstacles in nineteenth-century Ontario.

If the interior was an obstacle to industrial expansion, the open Bay was an invitation. "For indeed in this Freshwater sea," an astonished Fr. Gabriel Sagard wrote in 1623, "there are sturgeon, Assihendos [whitefish], trout, and pike, of such monstrous size that nowhere else are they to be found bigger."[32] Europeans were especially intrigued by the Native technique of spearing fish from canoes by torchlight. George Head, the British military officer stationed at Penetanguishene in 1815, tried it and promptly set himself on fire with the birch-bark torch.[33] Local fishermen and small-scale logging were operating on the south shore by mid-century. By the 1870s, however, commercial fishermen were descending on the Bay in increasing numbers, forced northward by the depletion of the lower lakes to Canada's richest Great Lakes fishing ground. Although

Georgian Bay was still considered remote, fishers were propelled by demand from southern markets (especially American cities such as Buffalo and Chicago) and new technologies such as rail links to those markets, new net and hauling gear, and fishing tugs. Within a decade Georgian Bay consistently led the Great Lakes in catches of Canada's most important commercial species, lake whitefish and lake trout, and a variety of lesser species, notably chub and herring. In 1888-9, for example, the Georgian Bay catch of 3,229,000 pounds of whitefish represented twice the catch of the North Channel, four times that of Lake Huron proper, and 63 percent of the total catch of whitefish in the Lake Huron district.[34] The Royal Commission studying the Bay fishery in 1908 stated bluntly: "The fisheries of Georgian bay and the north channel are, in many respects, the most valuable fresh water fishing grounds in the world." Catches remained fairly steady until an explosion between 1950 and 1956, followed, somewhat predictably, by the collapse of the entire industry. Whitefish catches fell from a peak of 6,242,000 pounds in 1953 – 90 percent of the total Bay catch – to 49,000 pounds in 1959. Lake trout plummeted from 1,066,000 pounds in 1945 to 67,000 pounds ten years later, and the trout fishery was closed in 1962.[35] A century of overfishing, compounded by the arrival of a natural predator (the sea lamprey) and a man-made one (toxins, especially DDT), had exhausted the final holdout of the Lakes fishery in a crash that in some ways prefigures the recent fate of coastal stocks.

In fact, people had been watching Georgian Bay with a worried eye for years. Royal Commissions were appointed in 1892 and 1906 to investigate the state of the fisheries (with characteristically Canadian bureaucratic redundancy: struck within fifteen years of each other on essentially the same subject and reaching many of the same conclusions). At hearings around the Lakes, commissioners grilled fishermen and overseers about their methods, nets and boats, poaching, sales, and the future of the industry. Their reports reveal that industry sources were well aware of the specific environmental conditions that made the archipelago such an aquatic gold mine. The two core species, whitefish and lake (salmon) trout, migrated in the fall months to the sheltered spawning and feeding grounds provided by the "honeycombed bottom rocks" and shoals, then retreated out to the cooler, deeper waters of the Bay in summer.[36] In pursuit, fishermen had to adapt to these same conditions. They operated seasonal camps on outer islands such as the Bass, the Mink, and the Bustards, near the spawning grounds, but retreated in winter to their homes on the south shore. Fishers also depended primarily on the gill net, which, unlike a pound, trap, or seine net, did not have to be anchored to the lake bottom. By hanging a wall of net vertically in the water, fishermen could set nets both far out from shore and over a rocky granite lakebed. (By the 1930s Georgian Bay fishermen were using a suspended form of trap net that worked on the same principle.) The gill net could be managed by the smaller, one- or two-man crews using small sailboats, known as Mackinaw or Collingwood skiffs.

These boats also had evolved to service the Bay fishery. They had lapstrake or clinker-built hulls for sturdiness in open water, and were shallow in draft (some versions had centreboards rather than keels) so they could be rowed or even run ashore. The skiffs were fore-and-aft rigged with unstayed masts, so their booms swung out of the way when it was time to haul in the nets. They dominated the fishing fleet until the First World War, probably because they were so well suited to the Bay, especially with the addition of gasoline engines by the turn of the nineteenth century, which enabled fishermen to cover long distances. (They were also very popular as recreational boats and would remain on the Bay for that reason.)[37] In 1913 Raymond Spears observed Canadian net fishermen still favoured the skiff – "it rides the waves like a duck, pulls easily, considering the size and weight, and he can haul his net up over the side of it without upsetting it" – even though trappers and other "lake men" were beginning to use motorboats of about six horsepower among the islands, which provided shelter for the boats. The larger fishing tug appeared by the mid-1870s to assist sailing vessels in navigating busy harbours or unfavourable currents; later, they enabled fishermen to reach their remote island stations even during the storm-filled autumn months.[38] What fishermen could not change was the reproduction rate of overfished species threatened by increasingly polluted waters. In this sense, they were forced to comply with the limits inherent in the ecosystem.

The intense overfishing that characterized the industry from the 1870s sparked a great deal of comment over the Bay's carrying capacity, water quality, spawning seasons, and migration patterns. Testimony before the two Royal Commissions provides a barometer of a mature industry wrestling with the economic and political implications of resource exhaustion. Fishermen agreed stocks had suffered an unprecedented decline over the past thirty years and were extremely vulnerable. While they admitted to exhaustive fishing practices, they tried to redirect blame onto the sawlog industry. At issue were the log booms towed from the French or other northern rivers to mills on the south shore or the American side of the Lakes. Booms and deadhead logs destroyed nets and blocked fishing vessels. Bark and mill waste polluted the water, clogging nets and settling on spawning grounds. "They [the log booms] are sometimes delayed by winds and shelter in bay or behind islands," complained one fisherman, "all this time the rubbing and grinding is going on, and the bark and soft parts [settle] to the bottom." The commissioners were inclined to blame overfishing for declining catches, and at times their attitude made the witnesses rather testy:

Q. Do they tow them [log rafts] past the island?

A. Yes; they cannot very well tow them across it

...

A. It is the fine bark and is stringy and doubles across the nets and you can't get it off the nets and they are no more use.

Q. It is of great benefit to this fish I should say?

A. Well, you won't catch them then.[39]

And yet commercial fishermen were conceding Georgian Bay to the tourist industry. They acknowledged the growing importance of the sport fishery, the value of game species such as black bass and pickerel, and the revenue generated by sports fishermen. By 1908 the commission could report that "very few commercial fish are caught in the limits" of the archipelago.[40] (Recreational fishermen, it was true, usually found muskellunge, pike, and largemouth bass – some of Canada's most popular sport fish – in the weedy bays along the shore.) In short, fishermen showed an informed understanding of species exhaustion and habitat destruction, and an awareness of changing land use. The physical features of the archipelago received unusual attention because they played an essential role in sustaining one of the last frontiers of the inland fishery. But like every major group on the Bay, fishers' interpretation of the landscape was coloured by self-interest. Shoals were important because they sheltered spawning fish, November storms were problematic because they swept away nets, and fish populations declined because of industrial pollution or overfishing by other groups. Despite professing concern for collapsing stocks, the industry resisted regulation in pursuit of a retreating resource, and few of the commissions' impassioned recommendations for protecting these "most valuable grounds" were ever implemented.

Fishing skiffs and tugs were not the only boat traffic on Georgian Bay. The Bay was a crowded place in the late nineteenth century, for it offered a mobility and a means of transportation unavailable on the mainland. Although not an extractive industry like mining or fishing, shipping was a major commercial activity on Georgian Bay and a key part of its distinct history. Here again the patterns of traffic and types of boats used evolved in response to the natural conditions and constraints of both an open lake and a jagged shoreline. A long tradition of supply boats linked the isolated communities along the northeast shore until recently; the town of Killarney was serviced by boat until 1963. Independent traders first coasted the north shore in wooden schooners.[41] By the 1850s, the expansionist optimism of the railway era, the increased traffic on the other Great Lakes, and the explosion of logging and fishing on the Bay itself brought steam vessels to Georgian Bay. Here they would be a characteristic and indispensable feature of life for the next century. But most traffic was oriented westward, with the new trade in grain and emigrants travelling between old Ontario and the American Midwest and later the Canadian northwest. Steamers ran between south shore ports such as Midland and Owen Sound for Chicago, the Lakehead, and other destinations around the Great Lakes, continuing the old eighteenth-century pattern of by-passing the "empty" northeast shore.

The steamer fleet stratified according to destination and cargo. There were oceanic steamers for the Lakehead run, like those operated by the CPR after 1884; lake freighters for bulk grain or fuel; and smaller steamers still carrying passenger and freight for the local trade. A series of rival shipping companies, beginning with Georgian Bay Navigation Company in 1876, built and raced steamers to fishing and logging camps on Manitoulin and the North Channel. (Most of these companies were eventually consolidated into Canada Steamship Lines in 1913.) They soon discovered a lucrative trade in providing tours of the inside passage, and by the turn of the nineteenth century their bread-and-butter traffic came from ferrying cottagers and campers to a string of public docks between Midland and Parry Sound. A standard run from Midland stopped at Honey Harbour, Minnicognashene, Whalen, Go Home Bay, Wawataysee, Manitou, Copperhead, Sans Souci, and Rose Point. The trip from Midland to Parry Sound was an eight-hour journey for the *Maxwell* in 1888; fifty years later, it still took the *Midland City* six and a half.[42] Other supply ships or "floating stores" brought groceries and supplies. The awkward side- and paddle-wheels gave way to screw propellers, which offered more power and manoeuvrability, but these local steamers stayed relatively small in order to navigate the narrow channels. From stem to stern, they ranged from 74 feet (the *City of Dover*) to 152 feet (the *Midland City*).[43]

Figure 3 Midland City (steamer) docking at Camp Hurontario, c. 1955.

The casualty rate was high. Most steamers came to a sad end through engine-room fires, collisions, storm wrecks, running aground, or any combination thereof. The growing number of wrecks was a predictable consequence of increased shipping in such hazardous waters and earned Georgian Bay the ominous title of "graveyard of the Lakes," despite the best efforts of the federal government, which commissioned Boulton's *Georgian Bay Pilot* while it constructed two dozen lighthouses around the northeast shore by 1915.[44] Summer visitors, however, looked with great affection on these ships – especially the last, *Midland City* – which they regarded as a great adventure and a quaint anachronism. Although steamers were used elsewhere in cottage country (such as the H.M.S. *Segwun*, relaunched on Lake Muskoka in 1981), they were a prominent reminder of the island communities' unique dependence on water communication. Also worth mentioning is the *Penetang 88*, a converted Fairmile (wooden-hulled minesweeper) which ran briefly after the *Midland City*. The *Penetang 88*'s primary business was offering day trips to tourists, but evidently there was still a demand for a service boat up the shore into the 1960s. By now, however, public transportation was no longer indispensable, for cottagers preferred the convenience of highways and private motorboats.

Logged over, and then burnt over. Till all that's left
are scrub pin-oaks and jackpine on bare rock
a scruffy landscape ...[45]

The stark effects of logging belie the complex history of nineteenth-
century forestry. More than any other natural resource, forests were the
subject of competing views of nature, and became the focal point of
reinventing this "scruffy landscape" in the public mind. Forestry also reveals most clearly the two-pronged dynamic of industry in Georgian Bay:
how it appropriated and moulded the landscape as a resource hinterland, yet to some degree conformed to its constraints. There was some
logging on the south shore before the 1850s, but the transformation into
a large-scale, highly mechanized, capital-intensive industry occurred just
as loggers were beginning to exploit the north and east shores. The tim-
ing was perfect – or perfectly devastating. The Department of Crown Lands
was a vigorous promoter of lumber sales, as timber (as well as mining and
fishing) licences constituted a major source of government revenue. Loggers ploughed into the Upper Great Lakes watershed with full force, and
by the 1870s Georgian Bay was at the heart of what historian Arthur Lower
called the "North American assault on the Canadian forest."[46] Eastern
white pine, the most valuable species, grew well where little else did, in
the sandy, glacial outwash soils typical of Shield country. The timber trade
continued to dominate the economic landscape of Ontario and Quebec
thanks in large part to the rise of the pulp and paper industry in the early
twentieth century and its use of other Shield species – notably spruce –
after the pine was exhausted.

Typically, forestry studies of Ontario include the archipelago in a region of mixed deciduous forest stretching across the width of the province's "near north," with its Shield landforms, and bounded by the edge
of the Shield on the south, the French River and Lake Nipissing on the
north, and the Ottawa River on the east: the old Ottawa-Huron Tract. Yet,
the forests on Georgian Bay have a slightly different history.[47] Trees on
the rocky shoreline had to contend with both a heightened degree of
exposure to the elements and a rooting restricted by bedrock and especially shallow, low-nutrient soils. As a result, these trees paled beside the
rich stands of the interior. Loggers looked *past* Georgian Bay: they secured licences in Algoma along the north shore of Lake Huron as early
as 1852, and the major auction for the north shore was held in October
1872, only a year after the first public auction of timber limits in Muskoka
and Parry Sound.[48] In the words of one lumbering song, "Away to the

woods we thought we'd strike. / We hired with Riggin on Georgian Bay / To go to work at Wanipitei." Frank King recalls his older brother leaving Penetang in the winters for logging camps at Sunridge and Burk's Falls in the 1920s and 1930s.[49] Even the pulpwood industry, which might have readily utilized the smaller trees on the islands, seems to have been more interested in the territory north and west of Lake Huron, where it could acquire stands large enough to sustain its capital-intensive development. By the 1910s the Spanish River Pulp and Paper Mills was the largest newsprint producer in Canada thanks to its massive holdings north of Sault Ste. Marie and Sudbury.[50] Here it would not have to field complaints from cottagers, who by now were colonizing the Thirty Thousand Islands.

Georgian Bay itself was primarily a transportation conduit and processing hub. Logs were driven down rivers to the Bay, collected in large booms, and towed to mills around the Great Lakes. Mills were built at the mouths of major rivers, at places such as Waubaushene, Muskosh Mills, Parry Sound, Byng Inlet, and French River, where the open Bay provided easy water access for the tugs and the rivers provided the necessary water power for the mills. By 1900 Midland rivalled Ottawa as the leading lumbering centre in Ontario, and Byng Inlet and French River Village were thriving settlements in an isolated region.[51] Milling on the Bay peaked by the end of the 1910s, propelled by demand from the American Midwest and 1898 legislation requiring all timber cut on Crown land be processed in Ontario. (This was An Act Regarding the Manufacture of Pine Cut on the Crown Domain, passed by the Ontario government to retaliate against an American tariff on processed lumber.) This level of activity could not be sustained, and the industry moved to greener pastures. In 1906 the government tacitly acknowledged the new state of affirs by stating a preference for sales of cottage lots over timber dues.[52]

If there is a villain in Canadian environmental history, it is the nineteenth-century logger. He personifies rapacious exploitation, rubbing his hands in gleeful anticipation of profits, with reckless disregard for ecological consequence. Such is James Angus's portrait of A.G.P. Dodge, founder of the Georgian Bay Lumber Company:

> Entrepreneurial greed allowed him to see the huge conifers only in terms of investment. Screening out the natural beauty of the forests, his calculating eyes measured only the potential of the trees for generating wealth ... These virgin forests contained hundreds of millions, if not billions, of board feet of prime-quality pine lumber ... The enormity of potential profits was staggering.[53]

Seeing trees as nothing more than marketable lumber explains why loggers mowed across the Great Lakes Basin at record speed. It also fits the classic image of lumbermen as the manly conquerors of the backwoods: "Come young men a-wanting of courage bold undaunted ... And we'll make the valleys ring at the falling of the pine."[54] But this picture of single-minded avarice, indiscriminate consumption, and environmental insensibility does not prepare us for what they actually thought about the archipelago.

They were distinctly underwhelmed. Even before the great logging boom, observers were dismayed at the "inferior, frequently forked, crooked, and punkey" pine. The islands and shore were barely worth the paper it took to mention them, "timbered with stunted Pine, cedar and poplar of no value whatever," "too much scattered, and besides usually too small to be of any commercial importance."[55] The knobby, wind-twisted pine was disappointing to men expecting towering old-growth giants, particularly in the first phase of the logging cut. The exposed islands offered only small clumps of pine trapped between rock shears. While it is easy to identify the effects of logging on a landscape, it is important to recognize that logging is also affected by the place in which it operates. Loggers measured the landscape in commercial terms – the pine was "inferior" if it looked too thin or scattered to be profitable – but they were not oblivious to it. The nature of the archipelago shaped their decisions about how to exploit the resource: the speed with which crews moved past the rocky shores to the interior; the use of the open water for transporting logs, and of rivers for water power in milling; and inventions such as the alligator tug, a flat-bottomed, paddle-wheel amphibian designed to winch logs across small lakes and boggy portages, which first operated on the French River in 1889.[56] The entire industry took on the seasonal rhythms of the Lakes basin, from the winter cut to the spring river drive.

Meanwhile, company owners and their families from Midland and Penetang found a new use for the archipelago, reinterpreting the landscape as a recreational retreat. The Dodges, Campbells', Becks, and McGibbons were summering in the islands by the 1880s even as logs were towed past to their mills. Middle-class Victorians were able to see the environment in more than one way, and in apparently contradictory ways at that. If the archipelago had limited use as a timber reserve, they would make the most of its aesthetic qualities. James Hamilton's 1893 account of the Bay's *Position, Inhabitants, Mineral Interests, Fish, Timber and Other Resources* is as pragmatic as its title, yet he concludes on a surprisingly sentimental note:

However interesting such themes may be, it is on other topics that our memories will most kindly dwell ... the majesty of forests and granite shores ... the beauty displayed in winding, glassy coves among the islands, in flowers and verdure in sunny nooks, the Aurora dancing each clear night in the north.[57]

Many of the first campers and cottagers came from industrial states such as Ohio and Pennsylvania, but they were quick to oppose logging, which might strip their scenery. We tend to think of the romantic wilderness seeker and the profit-minded logger as ideological opposites and political foes competing for the environment. But their attitudes are neither mutually exclusive nor unlike our own in Canada today. Our economy remains fundamentally dependent on primary industry; we are world leaders in resource consumption; our environmental policy is based almost entirely on the principle of conservation for sustainable use. Yet, Canadian myth still embraces the wilderness as our cultural core and our retreats to that wilderness as a defining element of our national character. The complex views of Georgian Bay held by local logging families prefigure the modern Canadian attitude toward nature.[58]

What did the archipelago look like to them? Certainly those hoping for a peaceful, ancient wilderness were in for a bit of a shock. The fresh northern air was filled with smoke spewing from the smokestacks of passing steamers and from sawmills "burning the sawdust and refuse, and from which night and day the smoke ascends." The quiet was punctured with the whine of saws. Forest fires started in resinous pine slash left an "air of desolation ... nothing but bleak, barren rock, with blackened sticks."[59] Riverbeds were littered with a grotesque combination of waterlogged timber, bark, discarded boat hulls, and scrap iron, all covered with a thick layer of black sawdust. Cottagers butted heads with loggers who held local timber limits; in 1913 the Madawaska Club confronted a Penetang logger who clearly recognized a blackmail opportunity when he saw one and threatened to "strip the limit" unless the club bought out his timber rights.[60] Accordingly, we should eye the references to "verdure" cautiously. Promotional literature had a vested interest in painting such an image, since "denuded" or "barren" would not sell well with tourists. Some accounts mention only hardwood forests, "beautifully wooded with groves of Oak, Cedar, Maple, etc." Other visitors may be describing second growth; this took longer in the shallow soils, but many of the species in Georgian Bay (and much of the Lakes basin) are considered "ecological pioneers" for their ability to colonize burnt- and cut-over areas. Jack pine, red pine, and aspen are all pioneer species in burnt-over areas, and all are prominent in Georgian Bay today.[61] Perhaps we also have the loggers to thank for cultivating Georgian Bay's characteristic leaning pine. They may have made these trees even more prominent on the landscape either by ignoring them in favour of straighter logs or by removing mature trees that acted as a windbreak and thereby exposing younger pine to the full force of the elements.

Nevertheless, it appears that the archipelago retained much of its tree cover, such as it was. Sometimes the shore is described as a border shielding the decimated interior, akin to today's corridor cutting. "Along the coast the forest seems illimitable," wrote one resident, "although should you go ashore you would find that the woods have all been culled of everything that would square four inches."[62] Limits on the islands may not have been worth buying or carrying out. One surveyor credited the policy of reserving timber rights to the Crown for preserving the "pine clad shores," "*for the pine was not sold* and the destruction of the pine would destroy much of the charm of the neighborhood."[63] Franklin Island retained its white pine even though Franklin is one of the largest islands in the archipelago, at 8.77 square kilometres, and is located just outside the lumbering capital of Parry Sound. A 1907 timber licence was never carried out, and the island was designated a park reserve in 1923, in part for its "virgin stands." Seventy-five years later, it was again designated a conservation reserve for its old-growth stands.[64] The islands also enjoyed a partial immunity to the forest fires started by logging operations or sparks from railway engines (though they would be more vulnerable to campfires. By the 1910s, annual reports from the Department of Lands and Forests state that campers accounted for over one-third of all fires in the district). According to a 1929 study of the future Georgian Bay Provincial Forest, an area stretching from Shawanaga to the French River, only 3 percent of the 1,752 square kilometres had been leased and a scant 0.2 percent cleared. Similarly, Bruce Hodgins and Jamie Benidickson found it "astonishing" how much of the Temagami forest had not been exploited as late as 1940.[65]

The archipelago was not immune to logging, but it seems likely that the topography of the northeast shores discouraged it. Crown land records indicate a much greater interest in Algoma, north of Lake Huron and east of Lake Superior. If the Thirty Thousand Islands were seen as a wilderness retreat, this too suggests that they had escaped mauling by logging crews. Most visitors describe seeing both cleared land and "unbroken verdure." A correspondent for *Canadian Magazine* wrote that the Moon River was filled with logs but the islands were "shaggy with the forest growth of ages," while a writer for *Mer Douce* magazine reported that, within a few miles of Sudbury, "we have lost sight of the rocks, for they are now covered with forests, and we are looking out upon Nature at its beauteous best, reaching a climax at McGregor's Bay and the islands." The photographic record is equally ambiguous, showing some islands well treed, others achingly bare.[66] Ultimately, the contradictions themselves are probably the most accurate image of what Georgian Bay looked like a hundred years ago: "islands of every conceivable shape and size, some are bare and rocky, others are clad in verdure"; "some densely wooded with Canadian timber, but many being bald, naked rocks without any vegetation"; "bare islands and wood-clad islands – islands comprising a patch of rock ... and islands whose hundreds of acres are indistinguishable from the wooded shores of the adjoining mainland."[67]

Mere islands of wilderness were not, however, sufficient for the growing number of recreationists interested in the Bay. They undertook a campaign of imaginative rehabilitation that erased the logging era and substituted a primeval ideal. For the Algonquin Historical Society, the study of forestry meant understanding the role of forests in "the romance of Canadian history" and "the genial companionship of trees." "For me always the backward view: the solemnity of a forest unviolated," declared Marlow Shaw, paddling through the islands in 1926. Lewis Freeman stressed that nature "had restored to the ravaged regions a semblance of the smiling face they had presented to the red Indians and the pioneers." Condemnation of the lumberman's attack on the forests reflected the ambivalence about the costs of industrial expansion that underlay the back to nature movement of the early twentieth century. The aspects of the archipelago that had discouraged settlement now made its reclamation from modernity that much easier, until eventually it was delivered entirely out of the older march of progress. "This beauty does not change from age to age for here Nature is left alone," began the 1931 novel *The Great Fresh Sea*. "Commerce has not touched these islands to mar their freshness."[68]

The Group of Seven was the single most influential force in re-establishing Georgian Bay as wilderness in the public mind. Their paintings did not so much erase the human presence as depict a landscape that never submitted to the industrial yoke. Dark, dense forests guard the shores in A.Y. Jackson's *March Storm* (1920). Arthur Lismer's pine wracks are mammoth, contorted things that yield no quarter of the rock they grip so fiercely. As we have seen, this was not exactly a nostalgic fiction; Georgian Bay resisted manipulation by all comers through the nineteenth century. The Group's landscapes, with their jumbled boulders, lurking shoals and narrow crevices, gnarled pine, and tumultuous skies, explained why. At the same time (and in part thanks to the Group's influence), references to a generic "verdure" give way to a preoccupation with the single leaning pine. The bent pine invited all sorts of anthropomorphic projections. "Bent by storm winds but defiant / Proud though shattered by the lightening," it symbolized steadfast courage and tenacity: "I stand undaunted still." The pine clinging desperately to the rock embodies resilience, but is also a daily reminder of the severity of the landscape. "I know I couldn't survive there," says a cottager at Go Home Bay, "so I have tremendous respect for the things that *do* survive there."[69]

Recreationists had found a new use for the Bay, mining it now for iconography and pleasure rather than ore or timber. Wishful thinking, dogged determination, and colonization roads could not change the Shield, as Virgil Martin has observed; instead, people had to change their expectations of it.[70] Within a few years, second-growth began to hide the damming evidence of mills and fishing nets, and the burnt and scruffy landscape was concealed by "a green palisade that quite blanks out the granite."[71] The enthusiasm for wilderness as recreational space and nationalist symbol found its perfect match in the wilds of Shield country. The story of how outdoor recreation competes with heavy industry for a natural landscape has been replayed across the country, making it one of the most relevant in Canadian environmental studies. From Cape Breton to northern Ontario to Vancouver Island, every region in Canada has experienced the cyclical fortunes of primary industry and the bumpy transition to tourism. By the 1920s the Bay was being absolved of its industrial past, slipping into the role of an unspoiled (or nearly so) wilderness waiting to be rediscovered by canoeists and campers.

∾

Thirty Thousand Islands call from good ol' Georgian Bay ...
All the wilds are waiting, boys, and here I cannot stay.[72]

Wilderness seekers had, in fact, been coming for close to forty years by the time the Group of Seven discovered Georgian Bay. Camping parties and boarding houses appeared in Muskoka in the 1860s and in the Bay a decade later.[73] Recreation would prove to be the most durable and lucrative industry of all (James Barry describes the collapse of the logging boom as a "bust from which the Bay has not fully recovered even to this day"[74]; in fact, loggers and fishermen never dreamed property in the Bay could approach the prices we see today). Cottages developed along a more sustainable trajectory than did fishing and logging, but introduced permanent settlement on a scale unimaginable in the industrial era. Still, the obstreperous nature of the archipelago continued to shape its use by campers and cottagers just as it had among farmers, loggers, and fishers. Its "frontier" character was now appealing, but continued to pose certain challenges that gave the region a sense of difference from the rest of cottage country.

Outdoor recreation had been popular in parts of Ontario since the early nineteenth century. The well-to-do summered on the Thousand Islands on the St. Lawrence, the middle class enjoyed excursions to natural wonders such as Niagara Falls, and sportsmen sought more remote wilderness for fishing and hunting. In the eastern United States, recreational communities were associated with distinct landscapes by 1860. Resorts in the Adirondacks and the Appalachians, for example, provided a rustic health-oriented retreat for industrial New England. Summer retreats ranged from aristocratic second homes to modest fishing camps. As *Harper's Weekly* commented in 1896, "it is better to play at camping than not to camp at all, since even the pretense ensures more or less of an out-door life." Camping, or "tenting," became popular with the growing urban middle class because it could be tailored to suit any income, any distance – just outside a city or farther afield – and a weekend or two weeks' holiday.[75]

The canoeist, camper, sportsman, and tourist all appeared early in Georgian Bay. Residents of the south shore visited the islands for picnicking and blueberrying, as in this description of an 1856 outing for a family from Penetang and their guests:

> A week's picnic to the Islands was planned and their mackinaw, "The Seagull," was got under weigh [sic] to take us out, with tents, cushions, blankets, etc., and plenty of provisions packed in ... We camped there for the best of the week, sailing up and down the islands and enjoying the capital bass fishing and bathing, with evenings spent round the camp fire with songs and stories ... The wonderful combination of wooded islands, winding channels and sunlit waters left an indelible impression of natural, unspoilt beauty on our youthful minds.[76]

These casual boating and camping parties were emulated by (frequently American) fishing clubs such as the Pennsylvania Club, the Iron City Fishing Club, and the Yankanuck Club. The Thirty Thousand Islands quickly gained a reputation for excellent sport fishing of muskie, pike, pickerel, catfish, and bass. Each club built a large-frame building to serve as a main clubhouse for social events, while families lived in white canvas tents erected on the islands. The first island sale was made in 1884, but most visitors continued to tent for years before applying to buy a favourite

IN THE INSIDE CHANNEL, GEORGIAN BAY.

Figure 4 "*In the Inside Channel, Georgian Bay.*" Reproduced from George Monro Grant, ed., *Picturesque Canada: The Country as It Was and Is,* 2 volumes, illustrations under the supervision of L.R. O'Brien (Toronto: Belden, 1882), 589.

Figure 5 Yankanuck clubhouse.

camping spot. The Iron City Club, for instance, arrived in 1882, bought its island in 1901, and tented until 1923, when a three-day blow demolished eight tents in one night and the club decided to allow small cottages.[77]

Steamship companies began to promote the inside passage from Penetang or Midland to Parry Sound as a touring route, and after 1890 a series of hotels catered to those who preferred to view nature from a gingerbread-lattice verandah. Although most of these hotels have long since vanished – victims of fire, interwar decline, or both – photographs show ornate, three-storey, two-verandah structures aspiring to the heights of Edwardian elegance. Apart from telltale patches of bare rock poking through their neat front lawns, they could be anywhere; there is nothing in their design to suggest Georgian Bay was different from any other fashionable resort area, where experiencing nature meant an afternoon jaunt in a canopied tour boat. In Richard Wright's novel *The Age of Longing*, passengers on the *Huron Queen* are preoccupied with socializing and entirely indifferent to their surroundings:

> The factory owners in their slacks and short-sleeved shirts leaned against the railing and smoked White Owls, staring out at the shoreline of Georgian Bay. After five minutes they descended to the lounge to play pinochle. On the upper deck, their wives, handsome and tanned in sundresses, talked about their children who raced about the ship eating egg salad sandwiches and drinking Coca-cola.[78]

Fishing clubs and cottage communities also respected the usual Victorian conventions of formal propriety and social station. Women wore white shirtwaists. Maids were brought from the city, guides and caretakers hired locally. Shipping magnate James Playfair and his family cruised the Bay "in the grand manner" in luxury yachts up to 226 feet long. Even the humble bungalow, adapted from British India, suggested imperial leisure.[79] For the status-conscious pleasure seeker, Georgian Bay might have seemed like a generically stylish destination for a comfortable back to "nature" experience.

This is misleading, because a kid-glove approach that kept the Bay at arm's length was not at all representative of the first summer communities. From the start, the Thirty Thousand Islands were touted as a more authentic wilderness experience. The hotels on the islands, like those at Minnicognashene, Franceville, and Copperhead, were smaller than the mainland hotels, simple four-square buildings more like boarding houses or the club houses. Inevitable comparisons with the Thousand Islands derided the decadence of the St. Lawrence archipelago and rejoiced that the Thirty Thousand Islands remained in an original state of nature. "The gimcrack palaces which so mar the natural beauty of 'The Thousand Islands' in the St. Lawrence have not, heaven be praised, overflowed to Georgian Bay," said Lewis Freeman in 1926. A correspondent for the *Globe* compared the Bay favourably not just to the Thousand Islands but to every major vacation spot in Ontario:

> Muskoka had been ruled out as harbouring too many people and too few fish; Nepigon as too far away and too expensive; Lake of Two Mountains as too steamboaty, too Frenchy, and too much surrounded by temptations to go ashore. To North Bay, the Nipissing, the Thousand Islands[,] Balsam Lake and a dozen other places every member of the company had been before.[80]

The image of an unspoiled wilderness was partly a product of niche marketing that skilfully exploited the isolation of the islands, their unsettled and rocky shores, and the inconvenience of water travel. As the most rugged destination of the near north, Georgian Bay was an easy foil to more popular, more accessible areas, whether the resorts of Muskoka and the Thousand Islands with their croquet lawns and tennis courts, or the amusement-park attractions and campgrounds of Wasaga Beach. Capt. W. Whartman of Waubaushene told the Ontario Fisheries Commission in 1893 that, "taken altogether, the tourists are not objectionable, the better class of them having some little money to expend – most of them bring their tents and live cheaply." The Minnicognashene hotel advertised itself to "lovers of nature and solitude." A guest at the Ojibway Hotel at Pointe au Baril found it a "more rugged, and two-fisted, and third-helping sort of thing," where he could go fishing with an English earl and a plumber and not know the difference between them. Another complained tongue-in-cheek about the comforts of the Belvedere Hotel in Parry Sound, adding, "I would have thought it *more in accordance with the fitness of things* had I been able to contemplate a few evergreen boughs thrown together behind a sheltering rock as a couch for the night."[81] This was probably what attracted so many Americans to the Bay from booming Midwestern cities such as Cincinnati and Pittsburgh. The Bay recalled the rustic spirit of the popular Adirondack region but was as yet less peopled and more "genuine" a wilderness retreat. And yet, for those in New England or the Great Lakes region, Ontario's near north was a more accessible and more convenient place to get "back to nature" than, for example, the wilderness parks of the American West.

The archipelago's frontier identity was supported by the fact that it remained a difficult place in which to build. Urbanites getting back to nature wanted a rustic wilderness retreat; handymen working in isolated areas on rock terrain could not provide much more than that. Local conditions, practical concerns, and vernacular building patterns meshed nicely with cottagers' aesthetic preferences. Cottages grew organically and haphazardly: wooden tent platforms were partitioned into rooms and verandahs; sleeping cabins and kitchen buildings were added on wherever flat patches of rock allowed. Smaller cottages and outbuildings were a practical choice in a place with few flat building surfaces, where building materials had to be boated in, and where keeping low to the ground meant better protection from wind and other elements of weather. Smaller buildings were less conspicuous, and using local rock and lumber helped camouflage them even on small islands.[82] Full verandahs offered cooling circulation and opened the living space to the outdoors. For Bay cottagers, unpretentious cabins were an important statement about community identity: proof of their "woodsiness," their willingness to rough it, their appreciation of nature. These would not be "gimcrack palaces."

But they were not, in fact, all that unusual. There was widespread support for "naturalism" in architecture across North America at the turn of the nineteenth century. The idea that a building should echo the landscape or be subordinate to it influenced such mainstream styles as the Shingle or Adirondack, Arts and Crafts, and the Prairie School. All advocated using local materials, handcrafted workmanship, unobtrusive siting, and borrowing from regional vernacular styles. As one text instructed, a building should be an expression of practical need and "fitness to local conditions": simple in form, screened from sight, "harmonious with the landscape in texture and color" and "incongruous with natural expression." Architects realized the back to nature movement's idea of "the simple life" in the new city suburbs as well as resort communities. National parks likewise followed a policy of naturalistic design in order to convey a feeling of undisturbed nature.[83] The more rustic Shingle and Adirondack styles were popular for camps in the northeast United States, and their influence is visible in early Bay cottages: simple lines, broad roofs, exposed rafters or interior framing, and decorative woodwork. The grand dame of Cognashene, Longuissa, is perhaps the best example. Built in 1887 by the owner of Muskoka Mills, its lofty octagonal hall is framed with pine laid in intricate herringbone patterns. The Ojibway Club at Pointe au Baril, a rare survivor of the age of Edwardian hotels, has also preserved its wooden framing and light pine interior. While its châteaux exemplify the Victorian grand hotel, on the French River the CPR opted for an oversized log cabin with a roughly hewn stone fireplace and broad verandah under a low roof.[84]

As cottages dotted the islands, the older industrial landscape faded to a largely archaeological presence. Growing up in the 1930s near Big Chute on the Severn River, James Angus remembers seeing the scars left by lumbering forty years before:

> Giant, fire-blackened pine stumps were scattered through the second-growth forest; iron rings, once used to anchor catch booms at the foot of the rapids, were still fastened to the shore ... water-logged sawlogs could be fished from the bottom of shallow bays ... further back from the river, on tributary streams, one could find the rotting remains of log dams and timber slides.[85]

The thriving ports of French River Mouth and Depot Harbour became overgrown ghost towns; others such as Parry Sound, Killarney, Byng Inlet, and Key River survived as supply centres for cottagers and boaters. Logs, cribs, wharves, and wrecks litter the lakebed. Abandoned buildings, such as fishing sheds, were either dismantled and recycled or converted for cottagers' use. Recreation is merely the latest industry to mine Georgian Bay, cottagers the largest and most visible group of users. The difference between it and other industries in the Bay is that it operates on the premise that there is money to be made in the place – by bringing people in rather than shipping resources out. During the industrial era, the place itself had little or no value. Resources that could be separated from the land, harvested, and exported, such as timber or fish – these meant revenue. Sales of Crown land to cottagers and sports clubs represented a sea change in attitudes toward the landscape. Ownership of property implies people find value, economic and emotional, in the place itself. A landscape once thought of as utterly without value is now home to some of the most expensive recreational properties in the country.

But attitudes toward Georgian Bay as a recreational landscape are still affected by its geography. Islands, more than other types of landscape, encourage a feeling of possession and sovereignty as well as solitude. "There is something so satisfying in owning a whole island," says Barney in L.M. Montgomery's novel *The Blue Castle*. As early as 1910 a surveyor reported that "most people desired to control what they termed rocks (viz.: islets of very small acreage with a few shrubs or treelets on them), when opposite their property" to protect their privacy and view. Artist Clive Powsey has encountered this proprietorial attitude:

> I've been advised time and time again to paint Georgian Bay ... there is no doubt a very real sense of affection for the place. But I also have to think that there is a sense of propriety as well. I would be painting their rocks, their trees, and their view. Possessing an image of what you possess is deeply gratifying.[86]

Paradoxically, the more crowded the Bay becomes, the more it is associated with exclusivity and privilege. Cottagers may insist that their elitism is one of attitude – a fierce love of the Bay – rather than income, but their impassioned statements of place attachment creates "a landscape heavily infected with class denial."[87] Glossy books about cottage life are designed to grace the coffee tables of an affluent readership. Newspapers and magazines have become fascinated with the fantasy lifestyle of the well-to-do in the near north. Periodicals such as *Toronto Life* and the *National Post* routinely portray Georgian Bay and Muskoka as the preserve of celebrities and the blue-blooded of Upper Canadian society. Like tourist brochures of a century ago, they serve up Georgian Bay as a fantasy locale by featuring the grandest homes in the choicest locations.

Its elite social landscape is a product of the archipelago's physical layout. When the automobile brought much of Ontario's outdoors within reach for the urban middle class, the islands were held apart from this democratization of recreation. Spatial exclusivity enhanced their social exclusivity. Wright's novel *The Age of Longing* uses the landscape to demarcate the worlds of the working class, confined to the town of Midland, and the leisured summers of the company owners on the islands. The natural aloofness of the archipelago has been compounded by the dramatic escalation in real estate prices since 1970. Islands of a few acres can now sell for hundreds of thousands of dollars. "Since they have gone from him, years ago, into the lap / of the rich who can buy anything," Douglas LePan writes wistfully, "he has to imagine them, the islands / of summer."[88] This has split the cottage population into old and new, with a unique juxtaposition of highly expensive properties with an unusual longevity of single-family ownership – easily fourth and fifth generations in the older communities. Families whose deeds date to the era of five-dollar islands blame newer arrivals for tainting their rustic self-image with ostentatious displays of wealth, and for not understanding or appreciating the unique landscape of the Bay. For "old" Bay families, privilege means something rather different: "This feeling that you're only there with all these rich people, because you got there early ... those things that were accessible to everyone aren't anymore."[89] With the prohibitive expenses of real estate or yachting, the "commons" seems increasingly off limits, adding to the nostalgia of a wilderness lost. The costs involved in recreation – a cottage or a boat – also accentuate the disparity between year-round and summer residents, between the historically underdeveloped local economy and the wealth that migrates to the region in the summer. An Englishman visiting Ilfracombe in 1878 noted that since "we never lacked money for our needs ... it seemed to form a line of demarcation, never unfriendly, but always present between them and ourselves." Cottage community histories hint that summer residents occasionally suspected locals of "shady" behaviour, including breaking into their cottages over the winter.[90] Each group is dependent on the other, but for cottagers, the islands represent relaxation and recreation, whereas for locals, they represent necessary employment.

It is impossible, however, to draw clear lines of demarcation and ownership in the archipelago itself. The sovereignty of an island is illusory, for there is no clear dividing line in an archipelago between "the commons" of the water and private land. This ignites conflicts between, for example, cottagers and boaters. As a favourite cruising ground since the 1870s, issues of traffic, grey water, wake disturbance, fishing, and anchorage are not new. But the volume of boat traffic has exploded in the past two decades, leaving boaters and cottagers "to carve out their own space," as one boater put it. Boaters argue cottagers have parcelled out a perfect cruising ground. Cottagers regard boaters as intruders who "have no more stake in Georgian Bay than a tank of gas."[91] These conflicts – cottager and yachter, summer and year-round resident – belong to a long tradition of different groups competing for access to the Bay. As a resident of Parry Sound explains, their agendas for the Bay may differ but they all seek to use it in some way:

> Cottagers tend to be more protectionist and *hands-off* the environment. For them the Bay is a playland that they don't want to see changed. For many locals, the Bay is a source of income or jobs which may have some deleterious side effects. Although some are more exploitative than others, they are all largely very selfish.

He does conclude hopefully that ownership of the Bay somehow manages to elude them all. "These people may be associated with the Bay – but nobody can claim it as 'their own,'" he writes. "It doesn't belong to anybody except God and time."[92]

Not for lack of trying. For over three hundred years people approached Georgian Bay with hopes of claiming parts of it for profit or recreation. Their plans for the Bay reflected how North Americans thought about nature generally: as an assortment of resources or an idealized wilderness retreat. The pursuit of furs, minerals, fish, and timber triggered unprecedented activity on the northeast shores. Resources were harvested with such intensity that the ecosystem was profoundly and permanently altered. Yet, industrial activity evolved as the uncooperative landscape permitted; successes were lucrative, but sporadic and localized. Primary industry thrived in certain areas and failed miserably in others. When reimagined as a recreational wilderness, the Bay conformed more easily to expectation, and cottagers became the homesteaders the Ottawa-Huron Tract never had. The transition presaged that of resource-based communities across the country throughout the twentieth century. The transition, though wrenching, merely substitutes one expectation of nature for another. The commodification of wilderness is as ingrained in twenty-first-century Canada as enthusiasm for natural resources was in the nineteenth: to borrow a phrase from modern environmentalism, we remain a nation of appreciative users. In the next chapter, we will explore how popular ideas about nature, First Nations, and Canadian history were blended to create an identity for Georgian Bay as the archetypal wilderness landscape.

THREE

A *Vivid Reminder of a Vanished Era:*
Imagining Natives and History in a
Terre Sauvage

In 1999, Adrienne Clarkson began her installation speech as Governor General by describing the stone cairn at Sans Souci that commemorates Champlain's journey through "our beloved Georgian Bay." She spoke of her family's immigration to Canada, where they inherited a legacy of explorers "imagining themselves spanning this astonishing space," and a direct relationship with the land. "We became addicted to the wilderness," she said.[1] For Clarkson, and for many others, Georgian Bay has come to symbolize the essence of Canadian heritage as the twinning of nature and history.

History – as conveyed in both academic writing and popular culture – affects how a place is imagined and how people imagine their relationship to it. Historical imagery was unusually powerful in Georgian Bay because the archipelago *looked* like a wilderness, and because of its associations with the explorers and frontiersmen of Canadian history. In this regard, Georgian Bay affirmed and reinforced popular ideas about wilderness rather than challenging them. The construction of its historic identity says a great deal about our expectations of wilderness, especially its fundamentally anti-modern character and its related assumptions about First Nations. This identity in turn frames our approach to and activity in this kind of landscape. It also illustrates the Ontario tradition of writing Ontario history as national history and foregrounding elements of "national" history in Ontario settings (discussed further in Chapter 5). In short, a sense of history is an integral part of the creation of a cultural landscape. But if history shapes how a landscape is perceived, the landscape shapes what kind of history is told. If Georgian Bay represented a wilderness past, this past meant a return to the age of exploration.

∾

No one was more affected by this than the First Nations. There was a strong and enduring correlation between Europeans' reactions to the *terre sauvage* of Georgian Bay and their thoughts about *les sauvages* who lived there. Explorers and missionaries believed that the inhospitable environment forced Natives into semi-nomadic patterns of trade and fishing, and that the place's "savage" character mirrored the violent relations between various First Nations. European exploration coincided with (and contributed to) the destruction of the Wendat Confederacy by the Iroquois in 1639-40, as well as the heightened warfare between the Iroquois and Ojibwa later in the seventeenth century. By 1650 Iroquois raids had destroyed the trading networks around the Bay, scattered the Wendat and Algonquian nations farther north, and left the region uninhabited. (In 1929 Diamond Jenness, the chief anthropologist at the National Museum, reported that the Ojibwa on Parry Island still referred to "the hated Mohawks" as *nodawe* or "People who pursue in canoes.")[2] Jesuits reported that they established the Mission of the Holy Ghost among Ojibwa of the east shore, who had

> no certain abode along the coasts of the Great Lake, where they dwell sometimes in one place, sometimes in another, conformable to the different seasons of the year; or according as fears of the Iroquois compel them to move farther away. This means that our Fathers ... have led a wandering life among this wandering people, and have lived almost always on the water, or on desolate rocks beaten by waves and storms.[3]

Such an environment and lifestyle contrasted sharply with the sort of agricultural settlement that the missionaries had tried to foster on the fertile south shore of Georgian Bay. Close to two hundred years after the fall of Huronia, dismayed surveyors, such as David Thompson, would observe that in this "very rude Country ... all is desolation, and very little frequented by the Indians."[4] The archipelago fitted perfectly with the classical European definition of wilderness as a barren, uninviting waste.

Europeans found the hostility of the landscape was amplified and even personified by Amerindians. Inhabitants of a barbaric wilderness naturally would be cruel and treacherous. Fr. Jean de Brébeuf's description of a Jesuit expedition from Quebec mingled physical hardships with the "very rough treatment" inflicted by their Native companions, who stole their possessions and abandoned at least four Frenchmen along the way, including one "deserted at the Island, among the Algonquains."[5] Piles of bones left from battles and above-ground burials (an Ojibwa practice) added an air of the macabre to a desolate landscape. Jean-Baptiste Perrault noted sites along the French River "où les Yroquois s'étoient autrefois embusqués pour défaire les François qui venoient en commerce," and where a party of Iroquois was surprised by les Sauteux, "avec des os que les vicissitudes des tems n'ont pu effacer."[6] In the seventeenth century, Iroquoian raids defined Georgian Bay as a "terra aliend, in loco horroris et vastae solitudinis," as Fr. Paul Ragueneau wrote in 1649.[7] But as the Ojibwa expanded their territory southward, it was they who were regarded as fierce warriors, because they inhabited the wilderness of the east shore. "The Chippewas," observed Anna Jameson, "have long been reckoned among the most warlike and numerous, but also among the wildest and more untamable nations of the north-west."[8] Landscape and inhabitant fed a mutual image of physical and cultural savagery. Both the Iroquois and Ojibwa were considered unusually "wild" by virtue of their association with the place; their reputation as savages in turn reinforced perceptions of Georgian Bay as untamed wilderness.

This identification between an intractable wilderness and an uncivilized people became even more entrenched over the course of the nineteenth century. Georgian Bay and its Native inhabitants came to symbolize the antithesis of civilization for Upper Canada. In 1836 the Lieutenant-Governor of the colony, Sir Francis Bond Head, decided that the Manitoulin Islands and the "23,000 islands of Lake Huron" were a natural Indian reserve. He had the annual present-giving ceremony transferred to Manitoulin in an attempt to draw Natives from southern Ontario and around the Great Lakes onto the island. But what is often overlooked is that his plan included *all* the islands of the east shore. As he explained to the Colonial Secretary in London,

> we should reap a very great benefit if we could persuade these Indians, who are now impeding the progress of civilization in U. Canada, to resort [to a region] possessing the double advantage of being desirably adapted to them (inasmuch as it affords fishing, hunting and bird shooting, and fruit), and yet in no way adapted for the white population.[9]

The Ojibwa had little enthusiasm for the plan since traditionally they used the islands only as a seasonal stopping point, and they argued that Bond Head greatly overestimated its ability to support a permanent population. This did not stop the British from inviting their American Potawatomi allies to settle in Canada. Facing a policy of forced removal by the United States government, a few thousand Potawatomi migrated into southern Ontario by 1842. They too resisted settling on Manitoulin, complaining it was dangerous to cross the open Lakes to reach it and essentially barren anyway. They gradually dispersed around the Bay, pushed on to lands considered undesirable by both white settlers and the resident Ojibwa.[10]

Bond Head never persuaded colonial officials or the First Nations to abandon the standard British policy of civilizing-through-settlement in favour of his wilderness reserve. But he proved to be a better judge of contemporary opinion than anyone has given him credit for. Because Georgian Bay was considered an anachronistic but enduring fragment of an earlier wilderness, it seemed a convenient place to relegate the uncivilized from advancing civilization. Bond Head's evaluation of the archipelago expressed the conventional wisdom of nineteenth-century Upper Canada: the Bay could never support agricultural settlement, it was essentially useless to white settlers, so why not put the Indians there, safely and conveniently out of the way? Forty years later, farming prospects were so poor that the district atlas admitted "the idea was seriously mooted in high quarters of throwing the whole district into one vast Indian reservation" – and this in a publication intended to promote settlement.[11] Even the model agricultural settlements designed to teach Natives a properly "civilized" lifestyle were located on Manitoulin because of its isolation from whites. The first such settlement at Coldwater, for example, was relocated to Manitoulin in 1836 for this reason.

As things turned out, Bond Head's original plan was largely enacted twenty years later, when the treaties arranging for the formal surrender of the shore created an unusually large number of Indian reserves around Georgian Bay. Seventeen reserves were established by the 1850 Robinson-Huron Treaty, which dealt with the islands and mainland reaching from Penetang to Sault Ste. Marie. A second treaty concluded with the Chippewa of Lakes Huron, Simcoe, and Couchiching in 1856 reserved the three islands "by now known as the Christian Islands." The more fertile western shores, on the other hand, were not reserved precisely because whites wanted them. The semi-arable Bruce Peninsula and Manitoulin Island were reserved to the Ojibwa in 1836. In less than twenty years (1854), the Bruce was surrendered and opened to white settlers. When the government tried to open Manitoulin to settlement and its waters to commercial fishing in 1862, however, the Wikwemikong band refused to sign the new treaty, leaving the eastern peninsula of the island an unceded reserve to this day.[12] (The issue of treaty boundaries, which became a bone of contention between Ontario and the federal government, is

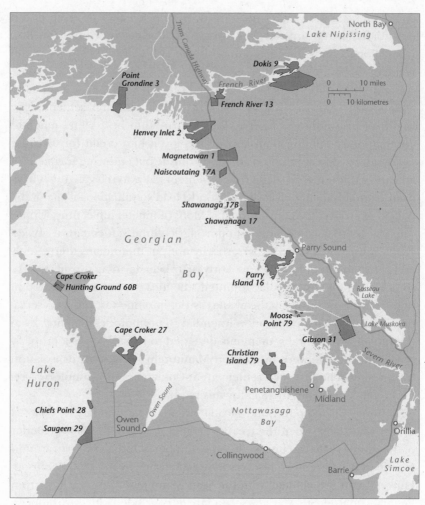

Map 9   Indian reserves

discussed further in Chapter 6.) That the conflict centred on Manitoulin, where permanent settlement *was* possible, served to only emphasize its differences from the granite islands of the east shore.

Yet, whites were ambivalent about what effect this wilderness environment might have on Natives' well-being. Récollets and Jesuits lamented the resilience of pagan beliefs among the nomadic Ojibwa north of Lake Huron. Paul Ragueneau thought their worship of manitou meant invoking the devil, while Gabriel Sagard continually referred to the Nipissing Algonquians as "sorcerers." Nineteenth-century missionary societies opposed Bond Head's plan on the grounds that a reserve on the islands was "calculated to render permanent their native superstitions, but perfectly useless ... for every purpose of civilized life" – the Natives would then never abandon their traditional ways of life. A government report in 1858 complained that the "heathen" Indians of Sandy Island "have hitherto resisted all the attempts made to civilize them, and cling with uncountable tenacity to the foolish superstitions imbibed from their fathers ... They support themselves principally by the Chase and by fishing. Their attempts at farming are of the rudest description."[13] Anthropologist Diamond Jenness visited Parry Island in 1929 and was fascinated by the range of traditional practices, potions, taboos, and rituals still in evidence. On the other hand, moral disapproval was moderated by Romantic ideas of the noble savage, which praised the uncivilized Native as the symbol of innate, primitive virtue untainted by the decadence of civilization. Lt. Andrew Agnew spent the present-giving ceremony on Manitoulin studying the Natives in attendance and looking for the "beauty supposed to belong to the majestic uncontaminated descendants of the ancient lords and ladies of the Lakes."[14] But for the "ancient lords and ladies" to remain "uncontaminated," it was essential they remain in their "natural" (and "uncontaminated") habitat. According to this line of reasoning, the remote wilderness of Georgian Bay protected the best remnants of a once admirable race. Here they could practice their traditional ways of life, away from the corrosive influence of whites. "Some Indians of the Gibbeway Tribe came from near Lake Huron," Elizabeth Simcoe reported in her diary in 1793. "They are extremely handsome & have a superior air to any I have seen, they have been little among Europeans therefore less accustomed to drink Rum."[15] Of course, the same Europeans who lauded the integrity of "the back Ojibwa" were also responsible for the degeneration they lamented in the southern tribes.[16]

The growth of tourist traffic in Georgian Bay built on the perception of Natives as de facto residents of a wilderness. Travel literature, that ubiquitous genre of the nineteenth century, included earnest observations about Aboriginal residents alongside other "natural" elements such as geological formations, climatic norms, lake conditions, and botanical life.[17] Unlike species of plants, however, First Nations endured endless commentary on their moral condition and concern for their moral character. The dichotomy of wicked/noble savage persisted, indicating that the question of whether wilderness boded well or ill was still unresolved. "Are they good Indians or bad Indians?" asked one character in a *Mer Douce* serial.[18] Yet, these same travellers were intrigued by traditional rituals such as the propitiation of manitou – leaving offerings at dangerous points, such as Shawanaga Bay and Moose Deer Point; or making offerings of tobacco before fishing – because they demonstrated the exotic, superstitious understanding of nature expected of the noble savage. (Alexander Henry had not found the ritual quite so quaint when his Ojibwa guides threatened to throw him overboard during a storm to appease the wrathful manitou.)[19] By the latter part of the nineteenth century, however, the public was inclined to look favourably on Ontario's remaining wild spaces. Rapid settlement and industrialization had dramatically eroded wilderness lands south of the Shield.[20] The mid-Victorian passion for progress gave way to a Romantic nostalgia for remnants of wilderness and any surviving "premodern" inhabitants. But admiration was reserved for the idealized Indian. This was a largely fictional creation that could embody whatever whites considered appropriate to a primeval wilderness: a primitive accessory, a cultural ancestor, or an ecological role model. Most importantly, all versions shared a historical, or more precisely, *a*historical quality, and were completely at home in the natural environment. Actual Native residents were treated rather differently.

The presence of an Aboriginal people confirmed the primeval state of the archipelago. "The Indians and the half-Indian guides, with their primitive outlook, add to the sense of difference," one visitor declared in 1936.[21] First Nations were seen as a people of the past, a living artifact from an earlier age. Writers borrowed passages from historical romances such as Henry Longfellow's *The Song of Hiawatha* or James Fenimore Cooper's *The Last of the Mohicans;* or created fictional Amerindian heroines – beautiful, innocent, and noble – to symbolize the virgin landscape.[22] Natives also were integral to landscape art as models of a "traditional" wilderness lifestyle. Paul Kane, Cornelius Krieghoff, William Armstrong, Lucius O'Brien, and Thomas Mower Martin all featured Amerindians in their paintings of Georgian Bay, poised gracefully in birch-bark canoes, fishing or trapping, in small shoreline encampments, or engaged in commercial or ceremonial exchanges. In 1879 the Muskoka and Parry Sound atlas suggested "the possession of pagan Indians, real *bona fide* and tolerably picturesque pagans" might pique interest in the district, and soon tourist attractions at Parry Sound included excursions to the reserve on Parry Island.[23] Purchasing islands "from the Indians" (via the Department of Indian Affairs [DIA]) enabled whites to feel almost as though they were re-enacting first contact; the DIA issued a grandly worded certificate of an Indian Land Sale that managed to sound very much like an eighteenth-century surrender.[24]

Once they had purchased the islands, whites adopted a different attitude. Cottagers might respect Natives' historic association with the area but not any claim in the here and now. The Madawaska Club at Go Home Bay appreciated the "picturesque and vivid reminder of a vanished era" but not "the annoyance of their camp being too close to a member's cottage, and the fact that they often landed to pick blueberries close to other cottages."[25] To the DIA, the value of the islands for traditional subsistence activities such as fishing paled in comparison with their new value as real estate.[26] There was also some lingering distrust of the uncivilized, amoral Indian. Historians were particularly harsh in their characterization of the malevolent savage. Basing their accounts on sources such as the Jesuit *Relations*, writers imagined in colourful and gruesome detail the abuse suffered by missionaries "at the hands of their ill-humoured conductors."[27] In novels and magazine serials, Native characters were often aggressive or sulkily taciturn, untrustworthy at best and traitorous at worst. Public testimony expressed similar suspicions. Fishermen, such as Gilbert Peter McIntosh, a local "fish dealer" who testified before the commission, complained that Amerindians exploited their right to a self-sufficient fishery by poaching dwindling stocks: "They can catch fish for their own use, and for some reason or other they use a great many ... The Indians are under the impression they can catch fish anywhere they like."[28] Staff at Camp Hurontario remember having little contact with local Natives, apart from boat drivers "who were almost always either hung over or drunk at the time" or residents of the nearby reserve who were suspected of draining the camp's fuel tanks over the winter.[29]

More often than not, however, Natives were largely invisible. Once whites found the archipelago valuable, whether as an industrial or a recreational resource, they found it remarkably easy to ignore any competing claims to the area. Natives were railroaded – literally – into surrendering land under a variety of pretexts. J.R. Booth built a railway port and company town at Depot Harbour on land carved from the Parry Island Reserve. Mining claims disregarded reserve boundaries from La Cloche to Parry Island; eight years after the Robinson-Huron Treaty had supposedly created a reserve at LaCloche/Spanish River, the Special Commission on Indian Affairs reported 13,000 acres marked as mining locations on reserve territory. In 1908 the Dokis First Nation lost a thirty-year battle to defend its territory on the French River from loggers.[30] In fact, non-Natives were as dependent on the local Native, mixed-blood, and French-speaking populations as they had always been. Local residents guided campers and sportsmen, and performed all manner of construction and maintenance work for cottagers: laying fireplaces and docks, cutting wood and ice, and running marinas. Because of their practical knowledge of the area, tourists and cottagers continued to cast locals as a kind of romantic "folk" close to nature.[31] Natives are depicted as simple, superstitious, capable outdoorsmen who are inseparable from the landscape: "Old Pete was as much a part of the Painted Waters as the age old rocks, and the sun and sweet winds. He was just that clean and simple, too."[32] In Frederick Varley's *A Wind-swept Shore, Georgian Bay* (1922), a line of Natives emerge from a cut in the rock almost as an extension of the rock itself, their robes blending with the sworls of granite. Newspaper columnist Greg Clark satirized Toronto society by having his woodsy guides question the inanities of modern urban life. Guides tagged with a single name – Pete, Dave, or Michel, for example – dispense laconic observations about the landscape in broken English. The local French-speaking population was seen in a similar light, as a historical fragment with a hint of the voyageur about them. (The francophone community at Penetanguishene descends in large part from families that accompanied the British garrison when it relocated from Drummond Island to Penetang in 1828. Families up the shore still bear names such as Trudeau, Robitaille, Le Page, Roi/King, and Dion.) As late as 2002, John Bentley Mays wrote that these descendants of "children of the wilderness" remained vaguely foreign, "mysterious and different" from southern Ontario.[33] But more and more, seasonal residents were simply indifferent to the Native presence – or more precisely, to Natives *as* Natives – especially when they appeared less romantic than the "old-time Indians" who had peddled crafts in fringed buckskins. "We didn't really have anything to do with them," says a former cottager from Cognashene.[34] Two solitudes had developed between the reserves and the cottagers, for whom Georgian Bay meant the physical landscape. As far as the cottagers were concerned, Natives had faded into the background and into history.

That is not to say that whites were uninterested in Aboriginal culture – far from it. But their enthusiasm was reserved for a specific, romanticized, and largely anachronistic idea of Native heritage. The national story depended on First Nations as inhabitants of the wilderness. As accomplices in exploration and the fur trade, they helped begin a dramatic story of continental expansion and nation building. As a New World people replete with artifacts and legends, they provided a young country of immigrants and internal disunity with a single, ancient, and place-based cultural heritage. Stella Asling-Riis referred to the creation myths in her 1931 book *The Great Fresh Sea* as "the Homeric tales of Canada."[35] And as children of nature, they were role models for the back to nature set. Ronald

Perry exhorted young canoeists "to preserve the tradition of our ancestors who traveled Canada in the canoe before us."[36] The *Globe*'s Mr. Kedgery
announced to his fishing companions in 1889, "We want the sensation of
having unlimited space to move around in ... we want to be free to go
round in any clothes or none. We want to live as uncontrolled as savages."[37] In all these roles First Nations remain locked in history, because
they had to harmonize with the principal image of nationalist iconography, a boreal wilderness. So only a certain type of Native belonged in
Georgian Bay: the sort who might have guided explorers two or three
centuries before. This idealized Indian cast an ever-larger shadow over
the Bay and the Natives who lived there.

The idealization of the Native resulted from a search for a sense of
place as well as a sense of nationhood. North Americans embraced the
back to nature movement as an alternative to the rapid pace of change
and alienation from the natural world associated with urban life. The "so-
called savage, at least, loved God's opens and worshipped him in the wind
and thunder," reflected poet W.W. Campbell.[38] Whites sought to adopt a
Native ancestry in hopes of acquiring a similar relationship with the land.
Raymond Spears wrote in 1913 that "the Canadian people need the Indian faculties and instincts in their effort to conquer that wide desert of
stone and starved wilderness that lies along the North Shores of the Western Lakes ... those of Indian instincts are rejoicing that in that great wilderness 'civilization' cannot make wheat fields and town sites."[39] This
became a pronounced theme in twentieth-century Canadian literature,
epitomized by John Newlove's 1968 poem "The Pride":

... they become our true forbears, moulded
by the same wind or rain,
and in this land we
are their people ... we are no longer lonely
but have roots.[40]

This desire has generated an entire pseudo-Native subculture pieced
together from what whites knew of or believed to be Aboriginal wilderness philosophies and practices. Outdoor recreation in North America
relies heavily on this borrowed culture to promise a return to a state of
ease in nature. In Georgian Bay, cottages were decorated with sweet grass
and quill handicrafts. The story of Kitchikewana – who furiously gouged
handfuls of earth out of Muskoka and the Penetang Peninsula and flung

them into the lake, where they became the Thirty Thousand Islands, before lying down at Giant's Tomb – was circulated as a quaintly entertaining explanation of the geography of the archipelago.[41] Stone Inukshuks are erected out on open rock, a practice which, although done for amusement, acknowledges the Native presence, permits a signposting in a relatively unobtrusive way, and claims a link with the North. Hurontario director Birnie Hodgetts told campers to respect Gitchi Manitou, a sort of omnipresent spirit of nature – that is, to approach an unpredictable landscape soberly and sanely.

Playing at being Native can appear embarrassingly hokey or as blatant cultural appropriation. But it goes deeper than that. As Daniel Francis has argued, the enthusiasm for "going Native" stems from the belief that the First Nations provide a model for living in the North American environment. The environmental movement of the late 1960s celebrated Aboriginal spirituality as a model for ecological thought. The "wish to be Native," Margaret Atwood has said, is based on the longing for a feeling of unity with the land and of cultural authenticity. "In so far as there is such a thing as a Canadian cultural heritage," she concluded, "the long-standing white-into-Indian project is part of it."[42] But the popularized version of Native culture has always been based on a selective vision, seeking to emulate only the "good" Native, the woodsman of an unviolated wilderness. (This may explain the feelings of resentment when Natives do not appear to be outstanding ecocitizens.) Canadian culture has a deeply entrenched habit of seeing Amerindians and landscape as reflections of one another.

∾

Whites used the physical landscape to tell their history as well. Except in Georgian Bay, history (like the First Nations) was interpreted according to what seemed appropriate to a pre-modern wilderness. Selected images of the past buttressed the enduring identity of the Bay *as* wilderness. The northeast shore was often overshadowed by centres of missionary, trading, or military activity at Huronia, Manitoulin, and Mackinac. But the rugged, unsettled landscape of the archipelago was perfectly suited to the historical imagination, as Hugh Cowan demonstrated with *La Cloche: The Story of Hector MacLeod and His Misadventures in Georgian Bay and the La Cloche Districts,* a melodramatic tale of piracy and blood feuds among fur traders on a lawless frontier, published by the Algonquin Historical Soci-

John Elliott Woolford, *Rapid of La Dalle, French River, Ontario.*
Watercolour over pencil with scraping out, 1821.

Toronto Public Library, T-18168.

Paul Kane, *Indian Encampment on Lake Huron*. Oil on canvas,
48.3 x 73.7 cm, c. 1845.

Art Gallery of Ontario, Toronto. Purchased 1932.

J.E.H. MacDonald, *Inhabitants of Go-Home Bay, Times Past* (left) and *Inhabitants of Go-Home Bay, The Present* (right). Oil on beaverboard, c. 1915-16. Part of a series of murals painted for the walls of Dr. James MacCallum's cottage, West Wind, at Go Home Bay.

National Gallery of Canada. Gift of Mr. and Mrs. Jackman, 1967.

A.Y. Jackson, *Terre Sauvage*. Oil on canvas, 1913.
National Gallery of Canada, Ottawa. Acquired 1936.

A.Y. Jackson, *Night, Pine Island*. Oil on canvas, 1924.

National Gallery of Canada, Ottawa. Bequest of Dorothy Lampman McCurry, 1974, in memory of her husband Harry O. McCurry, Director of the National Gallery of Canada, 1939-55.

Arthur Lismer, *A Westerly Gale, Georgian Bay*. Oil on canvas, 1916.
National Gallery of Canada, Ottawa. Purchased 1916.

Arthur Lismer, *The Guide's Home.* Oil on canvas, 1914.
National Gallery of Canada, Ottawa. Purchased 1915.

J.E.H. MacDonald, *The Elements*. Oil on board, 71.1 x 91.8 cm, 1916

Art Gallery of Ontario, Toronto. Gift of Dr. Lorne Pierce, Toronto, 1958, in memory of Edith Chown Pierce (1890-1954).

Elizabeth Wyn Wood, *Passing Rain.* Marble, 1928.
National Gallery of Canada, Ottawa. Purchased 1930.

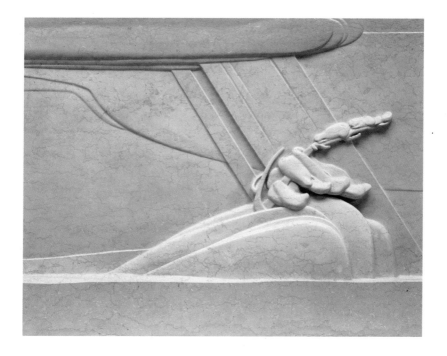

F.H. Varley, *Stormy Weather, Georgian Bay.* Oil on canvas, 1921.

National Gallery of Canada, Ottawa. Purchased 1921.

Frank Carmichael, *A Grey Day*. Brush, pen, and black ink on wove paper, 1924.

National Gallery of Canada, Ottawa. Purchased 1925.

Yvonne McKague Housser, *Little Clearing, Georgian Bay*. Oil on masonite, 1952.

National Gallery of Canada, Ottawa. Royal Canadian Academy of Arts diploma work, deposited by the artist, Markham, Ontario, 1953.

12

Edward Bartram, *Precambrian Point # 2.* Etching and photo engraving, 1986.

Courtesy of Edward Bartram.

Aerial view of the gardens of Little Jane Island.

Photograph by Freeman Patterson, in Nicole Eaton and Hilary Weston,
In a Canadian Garden (Markham, ON: Viking Studio Books, 1995).

Pointe au Baril lighthouse.

Supply ship *Trimac* at Camp Hurontario.

ety in 1928. Historical imagery, though, was not confined to literature. Artists, academic and local historians, the tourist industry, and cottagers all mined local history to create a portrait that inextricably linked human actors with the natural setting. In the twentieth century, Georgian Bay was constructed to be a conduit to the past; a place in which to re-create the spirited adventures of a voyageur, and to absorb solemn lessons about national history.

In this, it was shaped by a mode of nationalism prominent on both sides of the Atlantic in the nineteenth century. Infused by Romantic ideas about primeval wilderness, this nationalism sought to locate the origins of distinct nation-states in distinct landscapes. Nature, in other words, was seen as the formative influence on a nation, or its "geographic crucible." The best-known example is the American frontier, which, although it proved remarkably fluid, was ultimately located and fixed in the West, thanks to a cultural industry predicated on such heroic figures as the cowboy. In Canada, on the other hand, the definitive landscape traditionally has been the North, the defining historic era the fur trade, and the popular hero the voyageur.[43] In both countries, however, there was a conscious attempt by artists, writers, and politicians to identify those essential landscapes and to make these accessible *to* the nation. This way, wilderness would be a site of physical and spiritual renewal for the individual, but also a site of secular pilgrimage for the citizen. Nowhere is this more evident than in the creation of national parks, which selected either "wilderness" spaces to approximate the effect of contact with the frontier, or grand physical monuments to provide impressive cultural icons. (Thus, there has always been a great deal of ambivalence about evidence of human history *in* natural parks, as they represent two clashing expectations.)[44] The fundamental principle, however, was the same: claiming a historical relationship with a certain landscape affirmed the nation's existence. Nature was the key to writing a usable past.

In the Bay, history provided both an epic story of nation building and a more personal model of wilderness life. Historians such as Francis Parkman canonized the "fearless, sturdy, firey-hearted explorers" and missionaries, who represented, as E.J. Pratt later wrote, "the very incarnation of the age – / Champlain the symbol of exploring France ... Brébeuf / The token of a nobler chivalry."[45] Drawing on works such as the Jesuit *Relations*, they accented the contrast between pious self-sacrifice and savagery, portraying the missionaries as lights of Christian civilization against a dark wilderness. In the case of the Huronia martyrs, the bleak background suitably foreshadowed their doom. Explorers, gentlemen traders such as Alexander Henry, and British naval officers from the War of 1812 provided material for larger-than-life adventure and drama on a grand stage of imperial conflict and continental expansion. Like the missionaries, these men gave the Canadian story a heroic ancestry of heroic action and accomplishment in an arduous environment. The pathfinders were given a human face by C.W. Jefferys, who played a key role in the development of Canadian nationalist art in the early twentieth century. *Champlain Taking an Observation with the Astrolabe, on the Ottawa, 1613,* shows the explorer wielding the instrument, to the astonishment of his guide, who stands in open-mouthed awe at European technology; *Champlain on Georgian Bay* has Champlain gazing west over the islands, with his guide gesturing as if welcoming the explorer to appropriate the entire continent. That Champlain only briefly mentions *la Mer Douce* in his journal was conveniently overlooked – what mattered was that the first generation of nation builders had passed this way. In 1920, the Historic Sites and Monuments Board designated two of Ontario's first national historic sites at Victoria Harbour, site of the Jesuits' Saint-Louis Mission, and on Christian Island, site of "Fort Sainte Marie II," where Jesuits and Huron had fled in 1649. The Georgian Bay Association later erected a series of stone

Figure 6　Charles W. Jefferys, *Champlain on Georgian Bay, 1615.* From Charles W. Jefferys and Tom W. McLean, *Picture Gallery of Canadian History,* vol. 1, *Discovery to 1763* (Toronto: Ryerson, 1942-50), 93.

crosses as monuments to Champlain along his presumed route down the archipelago.

The tourism industry thrived on such colourful stories. As one visitor wrote in 1922, "the attractions of such localities is not complete until the local legends and traditions have been collected and made accessible to the sightseer."[46] However, the desirable "local legends" belonged to a specific era, and encouraged visitors to see the landscape as an eighteenth-century frontier. Cottagers and campers imagined the experience of the explorers as their wilderness ideal, tripping by canoe through uncharted waters in near solitude. Tourism material proclaimed the entire region "Champlain country" to suggest a return to an age of New World innocence and the opportunity to rediscover the same wilderness grandeur. The excavation and reconstruction of Sainte Marie among the Hurons provided a major tourist attraction for Midland when it opened in 1967, and revived "Huronia" as a term for the south shore and Simcoe County. Penetanguishene followed suit with its "Discovery Harbour," a re-creation of the early nineteenth-century naval outpost. Archaeological sites on Beausoleil were the original justification for a national park when first proposed in 1914 by representatives from the Historic Sites and Monument Board and the Royal Ontario Museum. The park – courting the tourist trade and hosting several children's camps – eventually developed an extensive if fanciful presentation of the island's seventeenth-century history. Soon Beausoleil was criss-crossed with hiking paths to destinations such as a Treasure Trail, Pirate's Cove, and Champlain's landing place.[47]

Beyond Beausoleil and the didactic writings of historians, a sense of history shaped popular recreation in Georgian Bay. Middle-class urbanites on their two weeks' vacation found a rose-coloured appeal in the mobility and seasonal rhythms of the lives of fur traders and loggers. Obligatory working expeditions and prolonged isolation in *un pays étranger et dangereux* now seemed more like a life of freedom and adventure. The hero of popular mythology was the *coureur de bois,* "celui qui préfère la mouvance à la fixité, celui que l'inconnu et l'aventure attirent," admired for his woodsman skills and envied for his personal autonomy.[48] Campers imagined themselves reliving this unconfined, supposedly untroubled existence, "paddling up and down the Bay, jumping in and out of the water whenever you wanted to, being a free spirit to go out and build a fire, travel as a group rather nomadically along."[49] Even the rough and raucous resource frontier was viewed fondly as a boisterous world of charismatic steamer captains such as "Black Pete" Campbell, defying the elements with intuitive skill and daredevil confidence, and rowdy lumberjacks, with their "hell-raising tomfoolery" at "Parry Hoot." (Parry Hoot grew up just outside the limits of Parry Sound to cater to those who did not share the town's policy of prohibition.) As Ian MacLaren has said of Jasper National Park, tourists celebrate past industrial activity romantically and uncritically, yet dislike any sign of its present-day counterpart.[50] Unlike the great-explorers narrative, which implies the arrival of civilization in the New World, this version of history did not suggest a permanent human presence in or transformation of the wilderness. Its heroes are woodsmen in an unspoiled *pays d'en haut*. Fishermen and lightkeepers were sometimes described as having a weathered quality, to indicate they were at home in nature. Herbert Harley and his sailing companions spent one storm-stay day with an old lightkeeper on Gereaux Island, delighted to have discovered this "most picturesque old child of the sea" who added a touch of local colour.[51]

Claiming ancestors in these men was also a heartening commentary on the virility of the Canadian race. Coureurs de bois, loggers, traders, and explorers all conveyed a powerfully appealing image of masculinity. An excursion into the wilderness meant instant membership in a heroic and reassuringly masculine version of history. A former Hurontario camper remembered his excitement before one canoe trip in 1979: "This was awesome; this was like the *coureurs de bois* being sent off from Hochelaga to conquer the northwest territories for France. We were going into God-forsaken territory ...! This is going to make men of us!"[52] Both the history and the landscape of Georgian Bay were considered inherently masculine, a rugged and powerful place that demanded strength and endurance. This made it an ideal site for the "strenuous life," prescribed to toughen and revitalize urban middle-class men and counter the degenerative effects of civilization – not unlike the argument made concerning First Nations decades before. Harry Symons declared that "there is a lusty, two-fisted, mannish atmosphere about it that smacks of ruggedness and desirable male independence."[53]

In point of fact, women had been patronizing the tourist trade since the first steamer excursions up Lake Huron. "I have had a good many of both sexes to visit me recently," Henry Schoolcraft wrote in his diary at Mackinac on 16 August 1837. His visitors included one Mrs. Anna Jameson. In 1843 Sarah Ossoli shot the rapids at St. Joseph's with gusto: "I was somewhat disappointed in this being no more of an exploit than I found for it ... I should like to have come down twenty times."[54] Even the most conservative of social authorities could sanction elements of the outdoor life, particularly if Canadians claimed a special rapport with the wilderness. "The Canadian Girl" was touted as a child of Nature, "an authority on outdoor sports," a "well-rounded, well-muscled, firmly nerved, and mentally alert and vigorous creature ... ruddy as the rose, robust, hardy, and athletic."[55] Girls' camps, such as the Canadian Girls in Training camp on Beausoleil, demonstrated that the benefits of the outdoor life – physical and moral health, character development, group cooperation – were not restricted to boys. Photographs show women at the fishing clubs in irreproachably proper attire wielding canoe paddles, fishing rods, and outdoor stoves. Records of the Department of Indian Affairs contain a barrage of correspondence from women seeking to buy islands as early as the 1880s. One cottager, shaking his head with admiration, related how his wife's grandmother and great-aunt boated to Nares Inlet at the turn of the nineteenth century: "There was nobody there at all. They sent up a tent on the island, ate blueberries and fish. Amazing women, when you think about doing that, in those days. They got the deed of ownership for the island for ten dollars."[56] Women also predominated in cottage communities; so-called summer bachelors were stranded in the city and commuted to the cottage on weekends. Nevertheless, the mythology of wilderness, and the heroic pantheon of Canadian history and Georgian Bay, has remained strongly masculine in tone.

Cottagers embraced the role of neo-pioneers. They spoke of "exploring" the islands and "staking" a claim, often tenting as "squatters" before buying property. Informality and "woodsiness" were elevated to the highest virtues. "The ability to manage a boat safely, consideration for one's neighbours and the environment – these are the attributes most appreciated," insists the history of Pointe au Baril. "Material possessions are relatively unimportant."[57] Conventional wisdom already prescribed a rustic appearance for outdoor camps, as in the American northeast. In Georgian Bay, though, simple cottages built by local handymen of "seconds" from lumber mills and propped up by boulders commemorated the practical constraints of building in the archipelago as well as an admiration of the pioneer virtues. In 1964 Tom Lawson built his own cabin at Nares Inlet, with pine planks laid tongue-and-groove for floors, hand-hewn spruce beams, and slightly crooked windows. "You can do anything if you put your mind to it," he said philosophically, "and if you get stuck, just ask for a bit of advice."[58] Self-reliance, self-denial, and hard work were considered physically and morally healthy, a kind of rejuvenating therapy for weary urbanites. Disdaining modern conveniences produced impassioned debates over the introduction of hydro and telephone lines, and

Figure 7 Go Home Bay, 1908. Library and Archives Canada, PA-29360

a nostalgic attachment to kerosene lamps, propane, and CB radios.[59] Cottagers still insist that they seek a "stripped down life" in the Bay, and even those cottages profiled in tony lifestyle magazines are praised for their weathered facades and rapport with "an uncompromising land-scape[,] not overwhelming it with imported technology ... learning to adapt to it, to belong in it."[60] This helps explain the ambivalence, even simmering resentment, toward the display of wealth in newer and grander homes in a region where properties themselves can now cost several hun-dred thousand dollars. It is a question of "image congruity": place attach-ment and place identity are heightened when residential design accords with prevailing ideas of place.[61] Ostentation challenges a core element in local place identity because it is inappropriate to a frontier environment, which is how cottagers traditionally have seen Georgian Bay.

Its wilderness identity was deeply entwined with nationalist mythology. It gave Canadians a heritage that was both scholarly and accessible: an official history of famous names and grand, national design, and a cul-ture of the outdoor life as lived by anonymous voyageurs. These two tra-ditions are manifested explicitly in the West Wind murals. In the winter of 1915-16, Tom Thomson, Arthur Lismer, and J.E.H. MacDonald painted a series of murals for the walls of Dr. James MacCallum's cottage, West Wind, at Go Home Bay. MacCallum was a supporter of the future Group of Seven and often hosted them at his cottage. The murals are a group portrait of outdoorsmen from different eras sharing a single backdrop of the islands around Go Home. MacDonald's *Inhabitants of Go-Home Bay* arrays Champlain, a Récollet priest, and a Huron brave on one panel, and a trapper, a fisherman, and a logger on another (Plate 3). Canoeists and cottagers inherit the landscape in MacDonald's *The Supply Boat* and Lismer's *The Fishermen, The Picnic, The Campers,* and *The Skinny Dip.* The murals suggest a longing to join this heroic wilderness fraternity and to experience nature as they had, to recapture the chance to "[move] at last at ease in a world he never dreamed, / this new world, ours."[62]

Clearly there was something in the landscape that evoked thoughts of the past. For Douglas LePan, this "Rough Sweet Land" exists in a "time without tense," as canoeists recreate the sight of "paddles flashing / as a brigade of *canots du nord* drives upstream ... Two hundred years are nothing, nothing."[63] Even at the height of the industrial era, the archipelago was described as "only inhabited by the wild denizens of the forest, a few Indians and white men who may be engaged with lumbering operations, and practically all the eastern shore of Georgian Bay is a great wilderness."[64] Thanks to the Shield and the open water, the human imprint remained relatively shallow. This made Georgian Bay not just a latent frontier but a durable one. Certainly it conforms beautifully to the frontier as defined in Canadian cultural theory: a labyrinth which swallows the explorer; a hinterland set apart from the metropolis; a world of potential danger not "set over against you but from everything surrounding you"; a place where artists might explore new styles as well as subject matter.[65] Because it appeared impervious to the passage of time, it offered an immediate return to the age of the voyageur. "At a single bound a man is as his ancient forbears were and knows their mind," said canoeist Marlow Shaw in 1926.[66]

As powerful as the historical images are, they have been rivalled by descriptions of Georgian Bay as *outside* of history, outside of time. This tradition emphasizes its *pre*-historic, non-human qualities, the "Huge and ancient boulders lying / Where dissolving glaciers left them."[67] This is the Georgian Bay that dominates visual art, most famously in works by the Group of Seven but persisting through the twentieth century. Contemporary artist Clive Powsey blithely admits, "I take out the junk – the cruisers, the buoys – I'm a pure landscapist."[68] Since the late eighteenth century, Americans have used dramatic natural scenery to emphasize their difference (and independence) from Europe; the natural heritage of the New World versus the cultural heritage of the Old.[69] Canadians, lacking a revolutionary agenda, adopted this view somewhat later, but it is evident in the nationalism that emerged after the First World War and urged a turning away from the battlefields of Europe. Georgian Bay – with its ancient granite and waterways – was an obvious choice for a symbolic New World. As Cory tells her English hosts in *Angel Walk,*

> If you went, you'd say there was nothing there. No history, not like here. But there was ... Grander signs of civilization, like your manor house here, were never attempted, nor would they have lasted. Oh no, it was a trackless expanse of scalped rock, the skull of the earth, the oldest land and the youngest state.[70]

The wars of the twentieth century crystallized this opposition between nature and culture, between the ahistorical wilderness of the New World and the civilization of the Old. Of course, Georgian Bay was never entirely cut off from the outside world during either war. Art historians have argued that the western front, with its shell-blasted trees, prepared artists to see the bleak Shield as paintable material after the First World War.[71] The Second World War rejuvenated shipbuilding in the southern ports, the production of explosives at Nobel, and shipping at Depot Harbour. Sea Cadet camps were opened on Beausoleil in 1942 and Minnicognashene in 1943, and after the war a couple of Fairmile-class ships were sighted in the Bay, still in camouflage paint.[72] But this activity did not undermine its image as a place unaffected by the chaos and destruction of modernity. Arthur Lismer's poem "To Georgian Bay" expresses his generation's disillusionment with Europe and hope of finding refuge and restoration in the wilderness interior: the "park, and chateau, and velvet lawn ... none could your great, untamed / Grandeur excel ... I have dreamed of you when the smoke of Hell / Has blackened the sky."[73]

This sense of escape owed much to the physical isolation of the islands. There are many anecdotes of being surprised by news of the outside world: learning that Queen Victoria had ascended the throne, that the atom bomb had been dropped, or that William Lyon Mackenzie King was dead.[74] Writers have used it as a site from which characters contemplate events from a distance, whether during the Second World War (*Angel Walk* and *The Age of Longing*) or Vietnam (*A Prayer for Owen Meany*). During the Cold War, it offered a sense of security. "I'm going to get an Island myself someplace," one man told the Department of Indian Affairs, "even if only to duck the Atom Bombs." Wendy Willoughby was working at Camp Hurontario during the summer of the Cuban Missile Crisis and remembers thinking "'Well, if anything happens, they're never finding us here!'"[75]

Not only were the islands themselves still difficult to reach, but the age of the Shield seemed remote from that of human time or the rise and fall of nations. "In wartime," literary scholar Kathleen Coburn reflected, "as one looked at or lay on its smooth glacier-carved contours, one could not fail to think piteously of the frailty of the world's cities being leveled to dust and of the ephemeral existence of man."[76] Its permanence made Georgian Bay a perfect foil to the Old World, as Michael Ondaatje demonstrates in *The English Patient.* The damaged Italian villa, its grounds laden with land mines, represents the wreckage of Europe: it is "a temporary thing, there is no permanence to it." Hana, the Canadian nurse, longs to flee the ruined landscape:

> I am sick of Europe, Clara. I want to come home. To your small cabin and pink rock in Georgian Bay. I will take a bus up to Parry Sound. And from the mainland send a message over the shortwave radio out towards the Pancakes. And wait for you, wait to see the silhouette of you in a canoe coming to rescue me from this place.[77]

Juxtaposing Georgian Bay with modernity was a predictable response to the turbulent and traumatic experience of war. But it was hardly original – the same opposition of wilderness and civilization had pervaded nineteenth-century attitudes toward the region and its Native peoples.

Since the 1960s, when the proliferation of cottages and boats began taking its ecological toll, history has assumed a very different and a very political dimension. The accelerated rate of change within memory is contrasted with a more distant and supposedly timeless wilderness past. The history of Georgian Bay becomes the story of the decline and fall of a wilderness:

> It was a remarkable place naturally, and I underline was ... It's not a wild place, anymore ... I would have loved to have been [there] in the forties ... It has a sad, inevitable quality to me ... there's this kind of nostalgic, it's a looking back, it's very much a black and white, sepia image of the past that I have ... I don't see a big future for Georgian Bay.[78]

One time camper/canoeist, now president of World Wildlife Fund Canada, Monte Hummel's comment betrays the short and selective nature of popular memory. The Bay he remembers of the 1940s was relatively undeveloped but no primeval wilderness (and Hummel might have been more disturbed by the industrial activity in the Bay circa 1900). By now, however, its centuries-old *image* as wilderness was being eroded because it was increasingly difficult to imagine oneself back in time. It undermines a regional mythology built on historical references that had sustained the anti-modernism and popular nationalism of twentieth-century wilderness myth. The postwar generation is especially conscious that its coming of age coincides with widespread development in the Bay, amplifying feelings of loss and guilt. For the baby boomers, historical time has been compressed into a single lifetime. "It hadn't changed at all for decade[s]," David Macfarlane has said, "and suddenly at the point that *I* come along, with all the rest of my fellow baby boomers, the place changes, in many ways in a very sad way." In the novel *Summer Gone*, he elaborates on the change:

> "The Reaches remained daunting," Elizabeth had said. "They remained gorgeous. But as the summer population increased steadily, as the boats multiplied, as the electric lights came, as the loons and the mink and the rattlesnakes disappeared, as the wildness of the place diminished, the Reaches became – for those watchful enough to see the change – not quite so extraordinary as they once had been. And that was truly heartbreaking."[79]

This new definition of history, as a process of change over time, would energize political debate about Georgian Bay and environmental quality at the end of the twentieth century. Yet, the distant past remains a kind of imaginative reference point for the wilderness ideal, and Canadian history remains predicated on our wilderness stories. The mythology of this *terre sauvage* is still infused with historical references – "always the backward view."

FOUR

Rocks and Reefs:
The Culture of an Inland

It is occasionally necessary for me
to tell Torontonians of the presence of the Atlantic and Pacific oceans;
they tend to think of the Great Lakes as the waters of the world.

– John Irving, *A Prayer for Owen Meany* (1989)

For centuries Georgian Bay was seen as a remnant of the original, impenetrable wilderness of the New World, one which could not (according to the nineteenth-century surveyor) or should not (the twentieth-century cottager) be remade into a settled landscape. This reaction was more than frustration or nostalgia – it was a reasonable response to the physical realities of the archipelago. A.Y. Jackson was not entirely wrong when he saw "a country in outward appearance pretty much as it had been when Champlain had passed through its thousands of rocky islands three hundred years before."[1] Billion-year-old granite and open water do appear immune to change, especially to city-dwellers who spend most of their time in extensively manipulated and largely artificial environments.[2] To understand the history and identity of the Bay, then, we need to examine the role of its two most enduring and fundamental elements: rock and water.

We encounter them again and again in the historical record: in art, the design of boats and cottages, folklore, and patterns of activity. Isolating these two elements provides a coherence and unity to the history of the Bay. Environmental history sometimes comes across as a fragmented series of unrelated episodes, changing agendas, and opposing perspectives. (According to this line of thinking, a fisherman – or a lumberman or a cottager – automatically sees nature a certain way by virtue of who they are, or when they are, as opposed to where they are.) But in Georgian y, people from different backgrounds were continually forced to respond to the same basic elements and adapt to the unusual conditions of an inland sea. By framing all human activity on the Bay, rock and water form physical and historical bedrock for regional identity and tradition.

Sometimes it doesn't seem right to call the archipelago a "landscape" at all. As Katherine Govier writes, "There is more 'scape than land, less land than a surfeit of energy beating against shores."[3] Seen from the air, it is an extraordinary sight: the unobstructed horizon follows the gentle curvature of the earth; the forested mainland fractures first around inland lakes, then into small islands ringed with rock, and finally into shoals gradually disappearing as darker shadows deeper and deeper beneath the water. Water has always been the most important environmental factor for those on the Bay, and the mark of difference between the archipelago and the interior. Lighthouses, Coast Guard patrols, and boat channels all suggest a maritime community. Water provided the means of transportation, determined the patterns of settlement and types of activity, and affected all perceptions of nature, for those who lived "up the shore" could never remain unaware or unconscious of their surroundings. So water is ubiquitous in Georgian Bay culture.

The natural starting point is the Open, as the main body of Georgian Bay is commonly known. The Open is immediate and unavoidable, a vast expanse of sea matched by an equally vast expanse of sky, the space emphasized by the flattened topography of the outer islands. It is the unobstructed view of the Open that distinguishes the archipelago from the rest of the near north, which feels closed in by comparison. On the outer islands, one is constantly exposed to the fast-changing weather and the supra-human scale of a Great Lake:

> Almost endless, going on into the horizon. So that humans, and human life, in comparison seemed dwarfed ... If I felt dwarfed, it was okay, this is how I *should* feel, this is a good proportion for me to feel like a little tiny human being in this enormous seemingly endless rocky landscape. It was just – right. That was my place, and my position in relationship to this landscape.[4]

It is vaguely disorientating to be disconnected from the usual *land*marks we use to locate ourselves. The familiar, safe world of dry land here consists of mere patches of solidity floating between sky and sea. This is heightened by an optical illusion, caused by the warm air of summer, in which the outer islands appear to shimmer and hover above the water. The idea that the earth floats on water is integral to the creation myths of Eastern

Woodland First Nations, which tell how the first animals dove for mud at the bottom of a primordial lake to fashion dry land. Ojibwa cosmology sees the universe as two worlds of sky and water, while "suspended between these layers lies an island, the earth upon which humans spend their days."[5]

Artists have always been fascinated by the Open and the way it inverts the usual pattern of land broken by lakes and rivers into patches of rock floating on water. Paintings persistently face outward, exchanging the security of land for a feeling of distance and contemplation of the unknown. Sometimes a thin silhouette of island or shoreline forms a narrow horizon dividing sky and water, as in Lawren Harris's *Georgian Bay* (1911), Tom Thomson's *Giant's Tomb* (1914), and Margaret Rossiter's *Island and Weather* (1996). The land exists tenuously, compressed, between the sky and water, emphasizing their boundlessness by its definite lines. An elevated vantage point, as in Goodridge Roberts' *Rocks, Pines and Water* (1968) and Franklin Carmichael's pen and ink *A Grey Day* (1924; Plate 11), sinks the islands farther into the expanse of water until it threatens to engulf them. John Hartman tried to capture the phenomenon of the mirage by leaving out the line of the horizon entirely so that the shapes of islands float in open space.[6] Even when there is an object or enclosed space in the immediate foreground, the Open is almost always visible or implied somewhere in the background. The *Stormy Weather* (1921) of Fred Varley's painting blows in from the Open to buffet a single pine, while Charles Comfort gazes out on *The Edge of the World* (1972) from behind a screen of pine. The channel in Bruce Steinhoff's *Islands and Channel, Georgian Bay* (1998) draws the viewer's eye around to the back of the painting, where the trees end abruptly, promising an exit from the maze of islands. In the Bay we are always aware that the Open is out there – if not seen, then heard, as the waves break against the outer shoals.

To Europeans, the Open called to mind the expanse and force of an ocean. As Frances Simpson, wife of Hudson's Bay Company governor George Simpson, wrote in her diary in 1830, "This Lake had all the appearance of an open Sea: no land visible to Sea-ward, thousand of Gulls flying about, the Waves rolling long & heavily like those of the Atlantic ... with the noise of distant Thunder." "It must have made / Champlain think that he was coming to a sea," agreed Douglas LePan.[7] In the mid-nineteenth century, the Great Lakes inspired a school of marine painting with sweeping waterscapes and wave-tossed vessels. William Armstrong is probably the best-known painter of this genre on the Canadian side of the Lakes, using his engineer's eye for detail in portraits of individual

ships set against the dramatic action of a churning seascape.[8] The Open
seemed the perfect realization of the Romantic sublime, an illimitable
expanse that defied human comprehension and the imposition of human order:

> ... this realm
> Of sky, and wide
> Bleak sweep of tide,
> Gray, tossed, scarce plowed by keel or helm.[9]

From the vantage point of a Georgian Bay island on the edge of this expanse, cut off from the physical and psychological security of the mainland, the coastal temperament of this "bay" is inescapable and immediate.

We may think of Superior as the greatest of the Lakes, but historically
many travellers considered Huron the fiercest. "The greatest fury of the
wide Atlantic is mere mockery to Huron's maddest moods and roughest
shapes," said Calvin Colton, aboard a steamer on Lake Huron in 1830.
"The most experienced mariner of the former has been filled with wonder, and stood aghast at the terrors of the latter."[10] Weather systems pool-
ing in the Great Lakes Basin are trapped in a relatively confined space of
Georgian Bay. In 1926 Lewis Freeman was completely unnerved by
"weather that was bad by every standard of comparison of my experience
of the Seven Seas."[11] The Red Rock lighthouse near Parry Sound was described by the Department of Marine and Fisheries as "one of the most
exposed stations on our inland water, on a bare, rounded granite rock,
exposed to the full force of all westerly storms, and to the full sweep of
Georgian Bay. In bad weather the sea breaks completely over the whole
building."[12]

Bad weather *was* a constant concern, since all travel was by water, and
often through fairly exposed areas. Communities along the northeast
shore resembled coastal outports in their isolation and their dependence
on boat traffic. Activity evolved in seasonal rhythms to match the Bay's
changing conditions, following schools of whitefish in the fall, driving
logs on swollen rivers in the spring, or camping in the mild summer nights.
Even then, movement was largely confined to periods of decent weather
from May to November. Miller Worsley, running supplies to Mackinac
after his busy summer of 1814, wrote on 6 October that he would soon
"lay my vessel up for the winter it being out of my power to navigate this
Lake after the 1st of November it being covered with ice and so very cold

that you can scarce show your nose out."[13] Late ice can delay shipping until June; fall storms, winter ice, and fluctuating water levels play havoc with docks and wharves. Andy Hamelin recalls why he persuaded Birnie Hodgetts to replace Camp Hurontario's fixed crib docks with floating docks:

> the whole top was just floated off in high water, the storms, you know, would come in and just lift that top right off. And it completely disappeared ... It would drift off and the whole thing would break up in the water pounding. I don't know how many times we would put new crib docks in that same area. They cost a lot of money.[14]

In the 1930s local residents invented the scoot, a powerful if dangerous combination of engine, airplane propeller, and flat-bottomed hull. It provided a speedier alternative to horse and cutter over a frozen Bay and, perhaps more importantly, permitted travel across broken or thin ice and open water during freeze-up and break-up. Use peaked after the Second World War, when surplus airplane parts were available but snowmobiles were not yet common. The scoot and airplane helped end the isolation of the winter months. National Park staff on Beausoleil reported in 1965 that their scoot was used constantly for hauling supplies and transporting children to school at Honey Harbour.[15] Accidents, sometimes fatal, were accepted as the price of convenience and speed.

Other types of transportation were adopted for the paradoxical conditions of a maritime environment situated in a continental interior. This was the territory of the *canot du maître*, a large workhorse of a canoe, thirty to forty feet long, paddled by up to sixteen men, built with sea-going sturdiness for the northern shores of Huron and Superior, yet light and deft enough to run rapids on larger rivers such as the French. (The smaller *canot du nord* was used west of Superior for carrying trade goods on the lesser rivers of the interior.)[16] Logging companies constructed splash dams and chutes to manoeuvre logs on narrow rivers during the drive, and then used the surface of the lake to transport huge log booms. Tugs and motorboats enabled fishermen to reach their seasonal stations in the remote northeast corner; government overseers followed to crack down on fishy behaviour. Sportsfishers attached motors to canoes in order to cover long distances of exposed water and still venture into the shallows nearer the shore. Owning a decent outboard, or more commonly an inboard/outboard, is an absolute prerequisite to owning property in the archipelago – a condition unique to Georgian Bay in cottage country. As one camper pointed out, "it was only the advent of the dependable outboard motor that populated Georgian Bay."[17]

Conventional motorboats, though, were of little use. The glamorous pleasure launches of Muskoka boat-building companies such as Minett-Shields, Ditchburn, and Greavette were ill-suited to Georgian Bay. Most of these boats had a small, open cockpit in the middle or rear of a long, low, narrow-beamed hull. With its polished brass and mahogany finish, the package looked pleasing at boat shows but was meant only for recreational touring in sheltered, shoal-free waters. By the 1940s, with cottages mushrooming up and down the shore, permanent residents and local handymen saw a need for a new type of supply boat. The result was the Georgian Bay launch, a utilitarian inboard designed and built by local residents. This sturdy workhorse had a deep and broad V-hull for better stability in rough waters, and was steered from the bow, better to watch for shoals. The large, partially covered cockpit could carry passengers and bulky loads of cargo of building materials. These "tough boats" were used as all-purpose water taxis, to transport supplies, tow fishing boats, or ferry cottagers to and from the mainland. "They'll call and say, 'It's too rough for my boat, come and get me,'" explained a taxi operator at Honey Harbour.[18] Cottager Michael Mitchell gleefully resurrected one:

> If anything other than an aboriginal canoe belonged to the landscape of the archipelago, it was a wooden boat like mine. It somehow just looked right against the humpback rocks of the Canadian Shield and the little wind-warped pines ... made of trees that had grown on those same granite rocks. I made it clear that I wasn't defending all wooden boats, no Ditchburns or Greavettes made for puddles in Muskoka. I was talking about serious boats built for serious waters, boats built right on the bay.[19]

A commercial line of similar V-hulled boats known as Limestones were later built specifically for cottagers braving the open waters of the Bay. The Georgian Bay launch will not appear in any history of boat design, but like the other functional, multipurpose vessels that served the east shore, it was a pragmatic response to the challenges of life in the archipelago.

While the open water demanded propulsion and durability, the confined waters between the islands imposed different limitations. Father Sagard asked the Huron "why they used such small craft, but they gave me to understand that they had a number of channels so troublesome to follow and narrows between rocks so difficult to pass."[20] Double-ended, flat-bottomed bateaux were used as all-purpose transport boats in the eighteenth and nineteenth centuries, as they could be rowed or sailed. Bayfield relied almost entirely on such boats "amongst Islands so numerous where ... we could make no use of our sails." He concluded that from "the immense number of the Rocks and Reefs, this part of the coast is almost unnavigable for Vessels of any size, but ... large Boats might row for many miles in the Channels within the Islands in any wind or in any weather."[21] An early proponent of a French River canal noted that rapids prevented schooners from travelling much more than eight miles upriver from the Bay, but "traders at present frequently bring up to Lake Nipissing bateaux of a light build."[22] There was always the danger of running up on a shoal or grounding in shallow waters. Storm winds (a "wind setup"), changes in barometric pressure, and natural cycles (from high to low water takes seven years, according to local folklore) can push water up or down over two metres, changing the entire topography of the archipelago in the process. One solution was the shallow-draught skiff; another was the disappearing propeller (or "DP") motorboat, patented in 1915, whose propeller could be pulled or bumped up into the centre of the boat (presumably pulled if you saw the shoal beforehand, bumped if you didn't).[23] Unlike canoes and kayaks, yachts and cruisers have been confined to designated channels and mooring bays.

Despite the treacherous nature of boat travel, water remained the only viable means of access. The interior was too rocky, uneven, and hilly for road construction. The colonization road to Parry Sound was delayed by the "extreme difficulty" of hiring workers in so remote a locality.[24] (Highway 141 today follows the old road, between Bracebridge and the Sound.) The expansion of Ontario's road system in the 1920s did not extend easily onto the Shield. For decades the only road access to Georgian Bay was along the old Highway 69 (now 169) from Gravenhurst at the southern end of Lake Muskoka. It was not until the late 1950s that there was a concerted effort to provide a more direct route north nearer the archipelago. A new road finally circumvented Gravenhurst by joining Port Severn to Foote's Bay in 1958. Highway 69, the only highway along the east shore, pushed through to Sudbury in 1955.[25] Short roads branched off 69 to a handful of communities on the shore to provide cottagers with access to marinas: Honey Harbour in 1941, Twelve Mile Bay in 1963, and the lengthy run (Highway 637) into Killarney in 1962. These twisting and undulating roads, though paved, call to mind an older account of the road to Parry Sound in 1930: "a pair of tracks winding, roller-coaster fashion, through the forest."[26] The inconvenience of the archipelago was alien to southern Ontario, and to the North American automobile culture of the twentieth century. However, this turned out to have some unexpected benefits. In at least three cases, natural areas were protected only when the Ministry of Natural Resources was unable to construct the road-access facilities considered part of normal park "development."[27]

Travelling by water profoundly affected how people saw Georgian Bay. Anna Jameson left the steamer at the Sault and travelled through the archipelago in a birch-bark canoe; sitting *in* the Bay she was fascinated by "the transparency of the water ... I could almost always see the rocky bottom, with glittering pebbles, and the fish gliding beneath us."[28] One of the great advantages of boat travel is that it literally immerses us in the environment, and forces us to be more aware of our surroundings than we would be if encased in a car on a highway or city street:

> [The Bay] required a mindset that was connected to the surroundings, and that connection could only be made after a lot of learning about the environment and learning to appreciate [it], and understanding how best to fit into it ... So that was the motivation for learning how to paddle a canoe, because if you didn't know how to paddle a canoe properly, you would find Georgian Bay a very hostile and unfriendly environment because you basically couldn't go anywhere.[29]

Of course, there are other types of non-mechanized travel – like sea-kayaking or cross-country skiing – that share the canoe's "direct, uncomplicated, economical, and silent passage"; but in a canoe we can pattern ourselves after Champlain "committing himself to the tender / skin of a birch-bark canoe, and a new continent, / and a new world, where anything might happen."[30] There are also practical reasons for staying low to the water. Canoes and other light craft skim over shoals lying silently at arms' length, as well as over the debris of larger and less fortunate ships; there are few experiences as eerie as gliding over the upright hull of the *Waubuno* lying in a sheltered cove of Wreck Island. Local steamers, top-heavy with upper decks above a narrow hull, could be tipped off balance when broadsided by the wind and waves. This was the explanation given for the sinking of both the *Asia* and the *Waubuno* (whose upper works were ripped off and never found). In J.E.H. MacDonald's painting of *The Supply Boat* (1915-16), the steamer manoeuvres awkwardly into the dock, its bow wedged into a rocky narrows.

Local architecture developed a maritime orientation as well. Besides their practical considerations — sheltered docking, boat access, a boat house – cottages offer the opportunity to re-create the intimate perspective of being on the water. David Milne described his cabin on Six Mile Lake as "about the best painting place" he ever had, "like the bridge of a ship, right out into the light – and the weather." Architect George Robb's cottage, featured in a 1989 *City and Country Home,* has the simple rectangular shape of a scow, with decks at both ends and a front room walled on three sides with sheet glass; the view, with water in all directions, gives the impression of piloting a ship. The Parry Sound Harbour Marine Building was designed to resemble a ship's prow out of a "concern for regional authenticity," complete with a deck and lookout covered by a sail hoisted on a wooden spar. As in other maritime communities, the local tradition of boat building has influenced house design.[31] Since 1999 Georgian Bay has been the centre of controversy over floating cottages, a stationary variation on the houseboat: two-storey houses on pontoons towed and moored in unoccupied corners of the archipelago. But houseboats make a certain type of sense in this setting – as much for the inconvenience of building on land as the benefits of living on water – and were used here over a century ago. Surveyor J.G. Sing reported houseboats tied to islands

Figure 8 Parry Sound Marine Harbour Building. From *Viewpoints: One Hundred Years of Architecture in Ontario, 1889-1989* (Ontario Association of Architects/ Agnes Etherington Art Centre, 1989).

Figure 9 Cottage on Deer Island.

opposite Gibson and Baxter townships in 1898; James MacCallum, the patron of the future Group of Seven, lived with his family in a houseboat at Go Home for six years before their cottage was built.[32] Even cottages anchored firmly on land are designed to feel open to the water and the elements. Those on Deer Island near O'Donnell Point, and on Carolyn Island at Pointe au Baril have entire walls of windows supported only by granite columns or wood posts. There is a feeling of continuity between inside and out, so that the living space is not insulated from nature.

Underneath the cottages, and underneath the Bay itself, lies a weighty, immovable foundation. With the water, this floor of granite defines Georgian Bay and pervades all understandings of the landscape. The Canadian Shield is exposed to an unusual degree here, in part because this shore was submerged under a series of glacial lakes for longer than most of the Great Lakes Basin, in part because advancing glaciers scraped the shore and left a rocky outwash of boulders, and in part because wind and water continually wash clean the exposed bedrock. The linear inlets and rivers that flow into the Bay, like the Moon and the Seguin, follow basic fault lines in the gneiss. Land here consists more of bare rock than soil, which is why explorers and surveyors universally concluded that the Bay was desolate and barren. Alexander Murray's opinion was cool, blunt, and entirely typical: "All these channels and outlets flow through a barren and desolate waste. The greater part of it is either perfectly bare rock, or a surface made little better than such by a scanty covering at intervals of small stunted trees and bushes." Tourists were often taken aback by the hundred of miles of "rocky and desolate" coast.[33] (Unfortunately, air pollution has made "barren", a term more accurate than ever, for acid rain falling on the naturally acidic gneiss has proven to be an efficient means of killing marine and bird life in the Shield's lakes.) Yet, even a rocky and "barren" coast could be intriguing to the studious traveller, for geology was one of the most prominent of the natural sciences in a scientifically-minded age. At the same time, the popular Romantic doctrine of the sublime nurtured a fascination with ancient, massive rock formations, where "the viewer could hope to glimpse a spiritual infinity through a geological one."[34] Anna Jameson noted the stages of natural succession as plants colonized the rock face, while other tourists marvelled at the mass of shoals, which they likened to the submerged peaks of a giant underwater mountain range.[35]

In the twentieth century, artists began to emphasize the aesthetic features of the rock. The Group of Seven used bright colour, heavy outline, and simplified forms (techniques from Post-Impressionism and art nouveau) to draw out the shapes and colours of the granite from the confusion of the archipelago. Arthur Lismer and A.Y. Jackson juxtaposed islands of coral against skies of cobalt in an electrifying contrast in *Happy Isles, Georgian Bay* (1924) and *Night, Pine Island* (1924; Plate 5), respectively. Franklin Carmichael washed the exposed crests of the La Cloche hills in the luminous mauves, blues, and greys of a watercolour palette. Today, Ed Bartram's prints and oils replicate the streaks and ribbons of different minerals covered with the splotches of green and black lichen, and the discolouration of underwater shoals seen through green-blue water (see Plate 13). In Katherine Govier's *Angel Walk*, Cory, a photographer, explains that she finds all her inspiration in the granite: "I have all this – miles of ancient rock twisted in lava layers, laid out in swirls, rolled into boulders, opened in crevices ... Every shape known to humans was first imagined here, in some creative tantrum of the earth's molten state."[36] Writers too have been attracted to the worn, curved silhouettes of the shoals. Several have created characters that embody the qualities of the islands in human form. Cory in *Angel Walk*, Jake Jacobsen in *Crossing the Distance*, and John Wheelwright in John Irving's *A Prayer for Owen Meany* are all depressing, solitary figures, isolated and emotionally crippled (even "barren"). But Cory, at least, demands respect for her fierce will to defy convention and live independently; it is as if she draws on the granite for its strength, and is an extension of her island in her solitude.[37]

The incredible age of the Precambrian rock also became a source of inspiration. Georgian Bay rests against two provinces of the Canadian Shield, the older Huron or Southern province along the North Channel and the younger Grenville Province down the east shore. The Thirty Thousand Islands and the quartzite hills at La Cloche are the eroded roots of two ancient mountain ranges formed around 2.5 billion years ago. (The higher peaks of the Algonquin uplands farther east belong to a younger range about 1.8 billion years old.)[38] The bedrock dwarfs the human conception of time the way the Open does in space, the temporal dimension of what a nineteenth-century commentator would call the sublime. As a cottager from Pointe au Baril explains, "you're aware that this rock was for all those millennia under a thousand feet of ice and that it's the oldest rock in the world ... you feel so small in it ... you can actually feel centuries, or millennia ... I get it in the Bay only." Such resilience and such permanence can be oddly reassuring:

> And still they lie, indifferent to my love ...
> I touch them reverently,
> knowing that when I die, being so frail,
> though waves erode, though colours bloom
> and change these rocks will still endure.[39]

I was struck by the similarities with Susan Toth's description of the Scottish Highlands, its "glaciated hills with barren crests of gray, lichen-covered rocks rising like islands from a broken sea of purple heather" sounding very much like the Thirty Thousand Islands. "Although it had no human scale or feel to it," she writes, "I somehow was comforted – in a world becoming so thoroughly tamed and colonized – by its sheer resistance to assimilation."[40]

It would be particularly heartening for a young country (and one perpetually insecure about its identity) to be able to claim the oldest rock on earth as its foundation. Historians and poets alike dwelt on the Shield's "beauty / of strength [and] beauty of dissonance," its continental reach, and its role in national expansion, finding a basis for national unity and identity in "the barren Shield, immortal scrubland and our own ... / barbaric land, initial, our / own."[41] Shield country was undeniably distinct from the long-cultivated landscapes of Europe, whether the Romantic English garden or the formal French promenade. It also bore little resemblance to the ideal of American mythology, the so-called Jeffersonian landscape of small, independent farms newly wrested from the wilderness in the character-building process of settling the frontier.[42] Since the Shield could never be pastoral, and its frontier would never be closed, it provided a tangible symbol of Canada's difference, both from the Old World and in the New.

The unyielding surface demanded a more pragmatic response when it came to building. Ambitious canal engineers might propose to blast through the Shield, but what actually *was* built was adapted to the rock. The granite is hardly ideal building material, but it is available, so boulders have been used to fill dock cribs and foundations, and to prop up floors built over uneven surfaces. Cottages are surrounded by patches of granite and built around the broken and jumbled rocks. Contemporary architecture also respects the qualities of the granite in more formal ways. The unique home on McBrien Island, at the north end of Sans Souci, mirrors its setting in angular shapes. Cement pillars blend with the grey gneiss, while black timber supports echo the vertical line of the surrounding pines. Of course, not everyone in the Bay shares such a rapport with the landscape. The neat floral garden on nearby Little Jane Island, featured in the glossy coffee-table book *In a Canadian Garden,* looks painfully *out* of place (see Plate 14). The owner – who freely states she prefers "symmetry, classical statuary, hedges and vistas" – has soil barged in and deposited on the exposed rock. From this she segregates vegetable, herb, and flower gardens, each neatly delineated by borders of stones. Unlike the island or the archipelago, everything here – the gardens, the plants – is designed to a small, human scale. It creates a manageable, comprehensible landscape as an alternative to the unnerving vastness of the open Bay that surrounds the island. Within these plots she determines what grows, when, and where; she can introduce familiar plants to heighten her enjoyment of her space. Such control and power of selection is reassuring, but the attempt to transplant an urban aesthetic is neither successful nor convincing. The square plots jar with the asymmetry of the pines and the island shoreline; the uneven granite thrusts upward to throw the border stones out of line; and the plots of barged-in soil look like illfitting toupees. The Bay has its own way of "managing" such efforts: the soil is washed away over the winter and must be replaced each spring. Doggedly denying the landscape in favour of the familiar and the ideal, Little Jane is oddly appropriate as a metaphor for the Canadian settlement experience and the colonial imagination.[43]

But the Bay can inflict far more damage than simply eroding the top-soil of a cottager's ill-placed flower garden. The combination of rock and water makes Georgian Bay an extremely dangerous place, and the prospect of danger is as integral to the Bay as its two fundamental elements. A shoal ripping the bottom out of a boat, fluctuating water levels concealing granite shore, thin and broken ice, furious storms that blow up on the Open – all put the "blue lake and rocky shore" of idyllic campfire lore in a rather different light. Canadians are no strangers to dangerous environments, and as scholars from Northrop Frye to Margaret Atwood have reminded us, surviving a hostile nature is a central motif in Canadian literature and culture. A community that evolves in such close proximity to a dangerous environment naturally develops a "cult of catastrophe" in its collective memory.[44] In Canada we might think first of fishing villages or mining communities on Cape Breton, with their mournful Gaelic-tinged lament in music and storytelling. But this extends to the maritime communities of the inland seas as well. Its inherent potential for violence has informed all understandings and representations of Georgian Bay.

Risk and peril pervade records of the Bay. The first European ship on the upper lakes, the *Griffon,* mysteriously disappeared in 1679 after leaving Michilimackinac. Fr. Louis Hennepin watched it depart from "the Lake of the Illinois" (Lake Michigan). "It was never known what Course they steer'd, nor how they perish'd," he reported, "but the Ship was hardly a League from the Coast, when it was toss'd up by a violent Storm in such a manner, that our Men were never heard of since." If caught in a storm out from Lake Michigan, in northern Lake Huron, it could very well have blown into Manitoulin Island or the Bruce Peninsula along the west shore of the Bay.[45] Voyageurs and loggers related the hazards of working in the wilderness in a tone of sorrowful resignation rather than hearty masculine assurance. Their songs warned of "rapides très dangereux," or places where "les Sauvages [qui] te feront la guerre"; the journey west to Grand Portage on Lake Superior promised "Tou's les peines et tous les tourments ... Il s'en ira sur la mer périlleuse." Le Bûcheron fared little better, for "L'arbre le menace en tombant / Il faut donc penser à la mort." Ballads tell of fatal accidents in the woods and on the logging drive, of men such as Sandy Gray, who broke a jam on the Muskosh River and fell to his death.[46]

Nineteenth-century shipwreck lore paints an even grimmer picture. The Bay shared ships and storms with the rest of the Great Lakes – "the cruellest, most fatal seas without exception on the face of the globe," writes poet W.W. Campbell. The Lakes' fall storms are legendary, especially the storm of 9 November 1913, when 235 people were drowned and ten ships (eight on Lake Huron and two on Superior) lost, with no survivors.[47] Many fishermen appearing before the Ontario Fisheries Commission in 1893 had no problem with closing the fishery in November because they had lost their nets in autumn storms anyway. The Bay has what are called "three-day blows," storms that rage for three full days against an unforgiving coastline. "Georgian Bay coast is probably the most dangerous of any on the great lakes," warned the report on Georgian Bay Ship Canal.[48] Travellers in open boats risked exposure and disorientation in the maze of islands. "It was a marvel how any escaped death," said one woman of the "three weeks of terrible suffering" she endured making the trip in 1838.[49] In this, at least, the different shores of the Bay had something in common: the "Thirty Thousand Islands thy wrath attest ... To wreck, and wreck, and slay," but it was the southern port communities that mourned the loss of "steer, and gear, and brave seafaring men" every year.[50]

The dangerous environment heightened the exploits of these "brave seafaring men." The historical figures associated with Georgian Bay – voyageurs, loggers, fishermen – coupled with the frontier imagery has constructed a thoroughly masculine landscape. As one camper explained, "The Georgian Bay was always, to me, very wild, slightly dangerous, quite isolated ... a little bit, well, macho, basically, because it had all these uncharted rocks and everything."[51] Scholars have suggested that wilderness exploration was constructed as a manly combat against a hostile terrain, while nature was gendered female to provide a more appropriate conquest. In the Canadian context, though, the story tends to be darker, nature more vicious, and the goal one of mere survival rather than conquest.[52] The attitude is a pragmatic resignation, the heroism an heroism of endurance: months at isolated light stations or logging camps, years scratching a homestead when "five acres of soil in a forty of rocks [is] a good farm," to borrow Herbert Harley's 1899 phrase. The Shield defeats settlers, wears them down, forces them out. Men came to the Bay intent not on settling there but simply on getting out in one piece. One would not survive a trip paddling supplies down the French River or piloting a fishing tug across to the Minks by approaching the Bay with overbearing confidence and brute force.

Despite their knowledge and experience, year-round residents remain especially vulnerable because they do not have the luxury of dealing with the Bay only in the placid days of summer. In his 1999 novel *Crossing the Distance*, Evan Solomon describes a Bay only a few ever see, and for good reason:

In the winter it reverts to more primordial rhythms. It's not a place that speaks of holidays or summer boat rides, but of endurance and the dumb intentions of survival. The land is silent, as if sound might awake the ire of the brutal wind blowing in off the frozen horizon and wreak destruction. Winter is always longest on the shores, which are protected neither by the contours of land nor the warmth of cottagers, and Terry Hogan likes to say that there's no summer season at all, just a few months of détente before winter once more attacks.[53]

In winter, and during the unstable periods of freeze-up and break-up, the water becomes carnivorous. "The Bay is savage with locals," writes Katherine Govier. "Water swallows them. Ice breaks and they fall through with their snowmobiles ... [The] community is not shrinking because it's impractical to live on islands; it is shrinking because the water is swallowing people."[54] A scoot's open propeller could amputate a limb, or its hull catch on the ice at high speeds and flip. One victim was Phil Trudeau of Manitou in 1949:

> Phil went out to check the ice to see if an airplane could land bringing his oldest boy, Eddie ... And he never come back ... so they went out looking for him and they never could find him. They found the hole where he had gone through, but he must have gone through the ice and into a patch of water and the thing must have tipped, or maybe he gunned it or something and it flipped over, whatever. But he drowned, anyway, and they found his body near the camp island the next spring.[55]

Since the landscape has not been tamed by time or technology, the "traditional" beliefs of the Ojibwa have never become anachronistic. According to the Anishinabe, storms indicate that the Thunderbirds (the sky manitou) are hunting the Mishebeshu (the underwater manitou). So "one simply doesn't venture onto a lake in a storm. Anyone can see that the Thunderers are hunting and only an idiot would place herself between them and their prey." Humans who do venture onto the water must do so cautiously, asking Mishebeshu's forbearance, for this is his realm, not theirs.[56] This does not inoculate the Ojibwa against accidents on the water; a prudent and respectful attitude born of long experience remains the most sensible approach to such an environment.

Art and literature reflect the immediate and ever-present threat of danger and the brutal force of the elements. MacDonald's *Lonely North* (1913) and *The Elements* (1916; see Plate 8) are filled with bruise-coloured storm clouds that are "threatening, engulfing, overwhelming."[57] Gaile McGregor singles out *Elements* and Fred Varley's *Stormy Weather* (1921; Plate 10) to argue that Canadian painting offers a harsh, disturbing, even nightmarish, portrait of nature, made hostile with "impenetrable barriers."[58] In literature, images of the landscape are consistently violent: trees wracked and twisted, as ghostly shadows of the Bay's victims, "like masts of wrecked ships drifting," or "bent westward like suspects thrust against a wall, strip-searched by the rough hands of the wind"; granite that "screams from the tortured banding of the minerals"; shoals that materialize silently from beneath the waves, menacing and unforeseen, like a flotilla of submarines.[59] While ice, water, and rock all play a prominent role in Canadian literature, the storm is a particularly convenient literary device. A "big blow" starts off the 1928 novel *La Cloche: The Story of Hector MacLeod and His Misadventures in Georgian Bay and the La Cloche Districts,* scattering the members of a camping party in their first major "misadventure." Sometimes a storm is the act of an animate Nature, a demonstration of her power; sometimes it is "just a howling, lashing impenetrable inferno."[60]

Why would cottagers and campers – middle-class urbanites on vacation – be drawn to a landscape with such a history of death and destruction? Some visitors were frankly taken aback at this face of their holiday spot. Lewis Freeman was appalled to discover that what he "had mentally pictured as veritable Islands of Enchantment, fairy bowers of sylvan loveliness, had metamorphosed to humps of half barren rocks, with flattened trees waving despairing signals of distress above breaking surges of sullen white."[61] But the Romantic influence had planted a taste for the sublime in North American tourism. Whereas the rational philosophies of the Age of Reason were manifested in manicured gardens of geometric precision, Romanticism rejected reason for the extremes of emotion to be found by experiencing wild nature "on so monumental a scale as to exceed our powers of framing and control, and to produce in their place a sense of overwhelming magnitude and awe."[62] Georgian Bay's wild, rocky, stormy shore was the perfect site to experience heightened feelings of fear and exhilaration, as Douglas LePan discovered sitting in the Red Rock lighthouse in a November gale, "cowering and exultant."[63]

This atmosphere of peril set the Bay apart from the more sheltered existence of urban life, the pastoral landscapes of southern Ontario, and even the adjacent cottage regions of the near north. As artist John Hartman points out, "no matter how hard it blows, you're probably not going to get into trouble on Lake Muskoka."[64] While it might seem irresponsible to deliberately seek out dangerous situations, by *experiencing* the Bay at its fiercest, people learned to respect it in a way much like the Anishinabe do. As former resident Ulla Elliott explained,

> My favourite thing to do would be to go out ... in the Open ... with the waves and the wind and the rain, and just experience the storm. Dangerous as all hell ... just crazy all around us, and you just sit there and enjoy it. God's power is just everywhere. It's the most fabulous feeling: exhilarating and awesome and powerful, and you feel little by all this magnificent wildness around you ... When it's really wild it makes you so aware of how little and insignificant *you* are as a person, and yet to be there when it's tranquil and more pastoral you still remember the strength of it and then you can appreciate it ... You know you've got to respect it ... but to be in Georgian Bay in a storm is the most exhilarating thing I've ever done.[65]

An uncontrollable landscape erases the sense of distance or protection from nature afforded us by modern urban life. LePan sitting in Red Rock Light or Elliott in her inboard can imagine how Black Pete Campbell or Alexander Henry might have felt toward the inland sea a hundred or two hundred years before. "It can be very frightening out there," admits Go Home Bay cottager John Lord. "But I don't view that as negative, necessarily. That's nature. We're protected from that in the nice, sort of septic city environment."[66] In such a landscape we discover the limits of our control. We may be able to ignore nature in the "septic city environment," but we simply can't in Georgian Bay. A lifelong cottager and boater says it best:

> It has to do with fast-changing weather, with size, with dramatic weather, dramatic water. Right away, it removes the grip of organized society in the form of cities and towns and the normal way you live life every day, and puts you right in touch with [it], so quickly ... Because it's changing so dramatically, it's talking to you, getting your attention immediately. You are obliged to pay attention, and talk and listen, because it's powerful and you are small in it ... You can't, no matter where you are, ignore the Bay when you're there. You have to be aware of the weather and the water and where you're putting your feet and what you're touching, of whether it's alive or not and whether it's changing. Even though it isn't life and death most of the time for a human up there anymore, there is a danger to it ... you know you're in a place where you need to pay attention, because if you don't there will be some consequences.[67]

The Bay has always demanded this respect, whether expressed in Ojibwa propitiation rituals or mournful shipwreck ballads. Yet, rarely is the concept mentioned in terms of wilderness and the Canadian imagination. The First Nations are often romanticized as living at one with nature, but non-Natives are usually placed in one of three camps: the stark terror of pioneers in the bush, the pragmatic and calculated economic interest of loggers or fishermen, or tourists seeking picturesque "thrills."

By studying one landscape in detail, we can appreciate its complexities and contradictions, and the range of responses it evokes, for a subtler, more sophisticated, picture of environmental attitudes. For example, the "wrathful" Thirty Thousand Islands have long been known for the shelter they can offer *from* the dangerous elements. Local knowledge of waterways has been integral to travel through this inland sea. The island is a natural symbol of asylum, and even the inhospitable islands of Georgian Bay offered some protection from both human and natural assailants. After the Iroquois invasion of Huronia in 1649, Fr. Paul Ragueneau reported that some Huron "took refuge upon some frightful rocks that lay in the midst of a great Lake"; some Jesuits were sent back to Quebec by "small bark canoe, [used] for voyaging along the coasts, and visiting the more distant islands"; and others, including Ragueneau, built a log raft "that should float on that faithless element [, and] voyaged all night upon our great Lake," landing a few days later at the Island of St. Joseph. (The refugees built a fort on what is now Christian Island, but mass starvation and Iroquois attacks forced them to abandon "Ste. Marie II" in 1650.) The rocky shore acted as a defensive border or buffer between the Ojibwa and Iroquois, who regarded it as the edge of their respective territories.[68] Natives introduced Europeans to the sheltered passages through the archipelago, taking advantage of the outer shoals acting as breakwaters against the full force of the waves. Steamers attempting to cross the Open were likely to get into trouble, but the so-called inside passage "dissipated the terrors of Georgian Bay."[69] Ironically, the shoals that provided a natural fishery and spawning ground also proved a natural haven for poachers, much to the frustration of overseers. In *Crossing the Distance*, Jake flees to the Bay for a different type of refuge, seeking peace from his chaotic life in a landscape whose very hostility promises him solitude, like a bird in a thorn bush. His brother, too, is running from a murder and thinks, "It's the perfect place to hide."[70]

Just as shelter can be found amid the shoals if one knows where to look, there is a landscape of detail and delicacy hidden in the crevices of the granite. A wild orchid "veined with tenderness, / that reaches down to glacial rock" symbolizes a landscape "where savagery and sweetness melt as one."[71] These contradictions help explain the longstanding ambivalence toward Georgian Bay. People have rarely been sure if the northeast shore of the Bay should be described as "beautiful." J.J. Bigsby waffled back and forth, alternately attracted and repelled: "The scenery is truly beautiful. Fir-clad hills all around – rocky islets and open basins ... so uncompromising and desolate ... a rock formation of great beauty and clarity ... There is little use in minutely describing these monotonous wastes."[72] There are the fragile, ephemeral wildflowers — waterlilies and meadowsweet and cardinal flowers. And there is the brilliant colouring in ribbons of black basalt, white quartz, and pink and grey granite, dotted with sage-green lichen and darker moss; deep blue water with wind-whipped white caps, covering pale green and salmon-coloured shoals; or dark evergreens against the autumn colours of sumac, poplar, and maple. These details are beautiful in the conventional sense. Colours intensified by the clear light have long been considered a distinctive quality of the Canadian north, different from the soft and misted English light or the warm and golden Mediterranean sun. But the raw northern light also reveals why beauty is elusive here: cool and honest, the light "makes nothing beautiful that is not already so."[73]

So the word "beauty" has never fit easily here. It suggests a benign, pastoral loveliness, a rose rather than a twisted pine or a stunted juniper, qualities not immediately apparent in a "landscape of exposure" framed by rock and water, mass and power. (This probably explains why the Bay has been less popular than more sheltered inland locations, such as the Muskokas.)[74] The word must be qualified by an explanation of what specifically is beautiful but why the landscape as a whole is not:

> Sure, there's beauty: my eyes water to think of it. But it's not really beauty, because of the stark and even threatening nature. For me the feeling is more awe, feeling very small and vulnerable in a majestic but somewhat overpowering surroundings. I remember paddling into the open, looking a half mile to the right and seeing a vein of quartz disappear into the water, then a half-mile to the left and seeing it wave itself up out of the water again, a reminder of millions of years of geological age ... looking down and seeing ghostly pale shadows of a shoal, then nothingness: a touch of scary tinging a whole lot of wonder.[75]

This is different from the disappointment of people schooled to expect a Wordsworthian Nature confronted with the Canadian reality, and different again from the misfit of a Romantic language of the picturesque projected onto a landscape for which it was never designed. Paul Simpson-Housley and Glen Norcliffe suggest it is common in Canadian culture to "juxtapose a motif of struggle against elemental forces with a leitmotif of hidden beauty among the hostile elements – for those who persevere in their search for it."[76] I wonder how many other wilderness landscapes have been viewed this way: not in the black-and-white oppositional categories of predatory frontier or picturesque fairyland but rather as something of a savage garden. The struggle is partly one of adaptation: Canadians have had to learn to see their environments as beautiful for their "beauty of dissonance" and "beauty of strength" despite, or separate from, any conventional wisdom or learned expectations (a process of acclimatization discussed further in Chapter 5). People have had to learn where to see beauty in the archipelago, just as they have learned to find safe passages between the shoals. Govier captures this perfectly:

"Lawrence, you told me this was the most beautiful place in the world," wailed the little woman with the dog.

"It is," he tossed over his shoulder at her, "but it takes getting used to."[77]

In other words, finding beauty in Georgian Bay reflects an intimacy of locality rather than a generic aesthetic. Those who learn to love this difficult landscape often convey a sense of belonging to a select group, based on a kind of insider knowledge about the place. Even summer residents learn that they can appreciate wildflowers and marvel at sunsets, but only if they never forget, as O'Brian says, to pay attention to "the weather and the water and where you're putting your feet." Place attachment and community identity are based on a feeling of difference from other types of landscapes.

If the light is honest, the rock and water have nourished an illusion, because they have allowed us to think of Georgian Bay as a kind of permanent, inviolable wilderness. "There will still be elbow room when a thousand resorts are located along that wonderful shore, and a million people take recreation on its channels and isles," Herbert Harley promised in 1898. A century later, the Ministry of Natural Resources (MNR) concluded that "the area can be described as being unspoiled," and the

Georgian Bay Association told the minister of the Environment that Georgian Bay was a "largely pristine resource."[78] Clearly there is some myth-making at work here, with a healthy dose of escapism. People heading back to nature want to find nature in a primeval state. Part of the appeal of canoeing, for example, is that we hope to return to a technologically and ecologically innocent age in which the human presence is transient and benign. A landscape with an apparently infinite capacity for abuse forgives our ecological trespasses, giving us the luxury of not worrying about environmental consequence.

Such thinking is, of course, delusional and dangerous, for nowhere is nature immune to human actions. But to be fair, this image of invulnerability is grounded as much in fact as in myth, and as much in the specific qualities of Georgian Bay as in romantic ideas about wilderness. Granite that has survived for hundreds of millions of years might well be strong enough to withstand a century of human occupation. The autumn storms wash pollutants out into the Open so when summer residents return it looks as though the Bay has cleaned itself up. Settlement has not altered the general configuration of the landscape. "The very roots and rocks that we stumbled over and walked around in our first days there have not changed," a camper remembers of his childhood on Hurontario's island.[79] And the storms, the rattlesnakes, the shoals, and the rock shears continually remind us that even if Georgian Bay is no longer virgin wilderness, it can still be wild. So, people have modified the concept of "wilderness" to suit, just as they have qualified the term "beauty." The Bay is described as "comparative wilderness," "almost wilderness," or "threshold wilderness."[80] Neither wilderness itself nor our ideas about this quintessentially Canadian subject then have remained static or monolithic. And yet, Champlain visiting the archipelago today might well recognize the rose-coloured gneiss, blue water, and rocky islands of his *Mer Douce*.

FIVE

Our Dear North Country:
Developing a Sense of Place

When A.Y. Jackson first visited Georgian Bay, he was intrigued but unsure as to how to respond as an artist. "I have done very little sketching, this country does not lend itself to it," he wrote in 1910. "It is a great country to have a holiday in ... but it's nothing but little islands covered with scrub and pine trees, and not quite paintable." Yet, three years later, he would paint it as *Terre Sauvage* (1913; Plate 4), an oil of dark spruce and red sumac on a curve of open rock beneath an ultramarine sky. *Terre Sauvage* became one of his most famous paintings, marked the start of a lifetime of painting these "little islands," and signalled a new era in Canadian art.[1] The history of Georgian Bay is the story of what happens when imported perspectives confront local conditions, or what happens when people arrive in a new place with certain agendas and ideas about landscape in mind. The meeting of expectation and place, of ideas and geography, produces a series of intellectual and practical adjustments – in ways of thinking about nature and ways of living in a difficult environment. This process of adaptation is integral to the construction of a regional identity. First, coming to terms with a new and strange environment forms the genesis of place attachment. Second, recognizing how this new environment is distinct – why it demands adaptation – leads people to delineate its borders and gives the community a sense of difference. The relationship between place adaptation and regional identity touches on some of the most fundamental questions in Canadian history. How do cultural values get attached to certain landscapes? What type of exchange occurs between metropolis and hinterland, between an outside authority and the immediate setting? How do regional differences develop within a wider political framework? What role do regional landscapes play in a national culture?

In this chapter, I examine the process of regional definition from three angles. First, I look at the evolution within what I call imperial paradigms, or some of the most common ways of seeing nature in British North America in the nineteenth century. In the Romantic rhetoric that infused tourism literature and the visual arts, people struggled to apply what they knew to what they found here. As Margaret Atwood has observed of Canadian literature, there is a "tension between expectation and actuality ... between what you were supposed to feel and what you actually encountered when you got here," as well as a "desire to *name* struggling against a terminology which is foreign and completely inadequate to describe what is actually being seen."[2] Next, I trace the emergence of regional identity in vernacular culture, or how those living "up the shore" developed a sense of place and defined among themselves boundaries for Georgian Bay as a distinct region. Finally, I look at how a specific landscape can be invested with national meaning and marketed to a larger audience. Once defined as a specific place, paradoxically, Georgian Bay was distinctive enough to become a recognizable symbol of central Canadian values. On all these levels, Georgian Bay offers insights into the chronology of Canadian culture as a whole.

The confident, utilitarian outlook of Enlightenment science framed most reactions to the North American wilderness from the age of exploration. However, a rival point of view existed since the late eighteenth century, when Romanticism emerged as a philosophical critique of the Age of Reason. Like science, though, Romanticism provided a ready-made vocabulary and a set of expectations as to what landscape should look like, and it was extremely influential in tourism in the nineteenth century.[3] Thus, an imperial mode of thinking filtered or interfered with encounters with a new place. Romanticism has been widely criticized for framing landscape as picturesque scenery, and for framing the proper response. Anna Jameson is rarely at a loss for words – W.J. Keith, in his literary history of Ontario, calls her the "most intelligent as well as acid-tongued of the travellers in Upper Canada" – yet in the Thirty Thousand Islands, she resorts to an overblown, saccharine outburst to demonstrate that she has had the sort of overwhelming emotional experience prescribed by the Romantic poets:

> I wish I could give you the least idea of the beauty of this evening; but while
> I try to put in words what was before me, the sense of its ineffable loveliness
> overpowers me now, even as it did then ... overcome by such an intense feeling of the beautiful ... I must have suffocated if – [4]

That same year Henry Schoolcraft complained from Mackinac that English visitors "look on America very much as one does when he peeps through a magnifying glass on pictures of foreign scenes ... to regard our vast woods, and wilds, and lakes, as a magnificent panorama, a painting in oil."[5]

It is often difficult to sense any genuine or original response to the specific features of Georgian Bay in nineteenth-century tourism literature. Instead, we have endless accounts of a place that fits very neatly with the Romantic ideal. Everywhere, it seems, is a standard vocabulary of effusive and floridly sentimental language, describing enchanted isles and virgin groves, sunsets of rose and waters of crystal, "splendours of nature in their sublime solitudes ... the rapture of the moment is profound ... we move as in a dream through straits of blessedness."[6] The endless superlatives make for reading that is wearying instead of inspiring, and hearing that the archipelago is among the most beautiful island scenery in the world or the finest scenery east of the Rockies actually tells us very little about Georgian Bay. Comparisons to the Old World seem a transparent attempt to match the cachet of the Grand Tour. The "highlands" of Parry Sound are compared to the hills of Killarney, the inland lakes to the lochs of Scotland, and the islands to the Hesperides, while Pointe au Baril is christened the Venice of the North.[7]

This mismatch of language and place has long interested literary scholars, who have wondered if these visitors really "saw" Canada at all. W.H. New argues that writers were so conscious of the verbal eloquence expected of them that they reached for "the elevated word and the distancing phrase." Perhaps this is why they often quote passages from literature, preferring a ready-made language of the ideal to the reality. The *Globe*'s Mr. Kedgery borrowed a lyric from English poet A.G. Swinburne's "On the Sea," then brusquely told his bemused fishing companions, "never mind who it's from. Just now it's Kedgery on Georgian Bay."[8] Greg Gatenby, on the other hand, suggests that tourists saw nature in a benign way because they had such limited exposure to the wilderness. Someone travelling by train through the Ottawa-Huron Tract might well see simply "lakelets that reflect nothing but the fragrant acres of the white water-lilies that cover them from shore to shore," since they were not trying to farm it.[9] Visitors to Georgian Bay brought with them a great deal of cultural baggage, and sometimes what they claimed to see in the Bay was what they had learned to expect of such a wilderness. And the archipelago conformed beautifully to the picturesque's requirements of "natural" asymmetry, partial concealment, and the unexpected. The endless variety of shoals and islands, the "primeval" unsettled landscape, the "thrills" of boating through narrow passages: "we see nothing but an island before us, outlet there is none to be seen, surely we are not going to run on shore! involuntarily you hold your breath" – all were qualities which, as Calvin Colton wrote in 1830, "made fancy more vivid, romance more romantic, and the very wildness of nature more wild."[10] So it was easy to toss about platitudes of the picturesque because here they seemed fitting.

To be fair, the hyperbole was often a product of the tourism industry rather than Romanticism. Romanticism has played a principal role in modern ecological thought by providing an essential alternative to the utilitarian mentality of the industrial age. Even scientists were divided between those who subscribed to the mechanistic view of nature as a series of systems and those who saw nature as a singular, organic whole characterized by interdependence between species. Romantic literature popularized the idea of an intimate relationship with one's local landscape, based on an emotional attachment and a knowledgeable study of its ecosystem. Writers such as Henry David Thoreau and Ralph Waldo Emerson made this genre of nature writing and the solitary reflection it celebrated a central part of the nineteenth-century American literary imagination.[11] Nostalgia for the primeval became popular by the mid-nineteenth century in eastern North America, where wild spaces were rapidly being eroded. So visitors to Georgian Bay *were* schooled in a way of thinking and an ideal experience of wilderness, but the Romantic perspective was sympathetic toward nature and encouraged a spontaneous emotional reaction to immersion in its "sublime solitudes." As Patricia Jasen concedes in her study of nineteenth-century outdoor tourism, "we need not take too deterministic a view of their experiences." Even steamer passengers felt pulled into the archipelago: not passive and detached, but exhilarated participants. Passing through a narrow channel, "one feels tempted to jump on the rock and take a run across one of the rocks."[12]

That Romantic conventions persisted through the twentieth century further suggests there is something more to it than just clichés. There are admissions, much like Anna Jameson's, that "one finds it difficult to put into words anything at all that does not seem trivial in the perspective of the total experience": thrilling tales of peril in the manner of the sublime, and images of a mysteriously animate Nature where "Rock and stones and trees are living / Stir like shifting restless spirits."[13] The entire culture of cottaging and camping relies on a Romantic view of nature as a place of beauty and solitude, of restoration and spiritual communion. Contrary to conventional wisdom, the Great War – in Canada, at least – does not seem to have marked the death of nineteenth-century thought with a crisis of modernist disillusionment. Rather, Canadian modernism emerged as a unique hybrid of stylistic innovation and Romantic sensibility, preserving the older emotional attitude toward wilderness through the 1920s and even later. The Group of Seven became the public's perennial darling in large part because they perfected this formula. Elizabeth Wyn Wood, a contemporary of the Group, was praised by critics for marrying traditional landscape themes with modern form in stunningly beautiful sculptures of Georgian Bay islands in tin and marble (see Plate 9).[14] Romanticism permanently shaped how Canadians see wilderness, but this may be more than the durability of a colonial heritage, of learned or conservative taste. In *Summer Gone*, David Macfarlane alludes to the Group's "romantic" style, a style that he credits to the Bay itself:

> For many years the Waubuno Reaches had a fabled quality. There were people who thought it a romantic invention: a poetic fiction of the north. Its blustery water, wind-rowed skies, glacier-smoothed granite shores, and starkly galed pines were well known as a result of a few adventurous artists who traveled and painted there. But there were many people who assumed – since these tumultuous scenes looked quite unlike northern lakes anywhere else – that the artists were making the landscapes up.

"The fable proves no cul-de-sac," agrees Douglas LePan in his famous poem "Canoe-Trip."[15] We are not seeing illustrations purely of an ideal or an agenda, no matter how well or how poorly the Bay conformed to such expectations. Ultimately, these were attempts to describe a real place, not a fictional one.

This was the colonial dilemma, and an unavoidable step in the evolution of the Canadian imagination. People had to learn how to communicate a new and foreign experience, and how to describe the unfamiliar in familiar terms. For over two centuries, science and Romanticism provided different ways of understanding wilderness. They acted as frameworks for interacting with nature intellectually and emotionally, and they provided different standards of measurement, one functional, the other aesthetic. When people described Georgian Bay in these terms, they were expressing both what they thought they should see and what they actually saw; actively (in scientific policy) or imaginatively (in romantic imagery) manipulating nature to resemble the ideal of the day. But the paradigms never quite fit and so had to be adapted to the unique contours of the

Bay: retaining what seemed appropriate, adjusting what did not. It was in the arts, though, that this process would become most apparent, perhaps because here was enough creative freedom to carve a regional language out of an imported tradition and "to *name* ... what is actually being seen."

Art and literature so clearly represent this process that they have become a principal symbol of the colony-to-nation arc that pervades Canadian history. Coming to terms with a strange environment, creating a homeland in a foreign space, is the narrative told again and again in histories of Canadian culture. Toward the end of the nineteenth century, literature began to reflect an interest in local identities, local places, regional vocabularies, and a more positive view of wilderness.[16] Writer-naturalists such as Charles G.D. Roberts and Ernest Thompson Seton instructed readers about wildlife and regional habitats in a form of nature writing that was recognized as distinctly Canadian. W.W. Campbell tended to succumb to rose-and-azure rhapsodies about Lake Huron, but he has been called Ontario's first truly local poet, and some of his lyrics do sound as though, as Mr. Kedgery declares, they were "written of a morning near Moose-Deer Point":

> Behind, the wild tangle of island,
> Swept and drenched by the gales of the night;
> In front, lone stretches of water
> Flame-bathed by the incoming light ...[17]

The pattern of inherited Romantic motifs set in distinctly Canadian landscapes created another characteristically Canadian literary genre, known as romantic realism or the regional romance. Fair Indian maidens or villainous pirates may be par for the course in a conventional romance-adventure story, but it is the setting – Beausoleil Island or Killarney – that is distinctive and actually directs much of the plot in *The Great Fresh Sea* (1931) and *La Cloche: The Story of Hector MacLeod and his Misadventures in Georgian Bay and the La Cloche Districts* (1928). This extended to nonfiction as well. *Mer Douce*, a historical magazine published in the 1920s, for all its talk of verdure and nostalgia for the primeval, is bursting with enthusiastic interest in local history and detailed accounts of communities in northern Ontario. As late as 1960, Percy Robinson was using lines such as "Dawning day and flaming sunset, / Stainless and immortal beauty" and references to isles Elysian and Dian's nymphs in his lengthy poem "Georgian Bay"; but the unremarkable purple prose is overshadowed by stanzas embedded in the archipelago: "the winding channels / Linking ancient pine-clad islands / Each to each – wave-worn, windblown ... where the gneiss and granite glitter – huge and ancient boulders lying / where dissolving glaciers left them."[18]

Artists first sought out places that seemed uniquely Canadian, such as rural Quebec and the Rockies, to provide the new country with impressive visual symbols. This also introduced the phenomenon of elevating regional landscapes to a national art. By the 1890s prominent painters such as Lucius O'Brien, C.W. Jefferys, George Reid, and J.W. Beatty were exploring rural Ontario and the near north, painting Muskoka, Burk's Falls, Temagami, and Algonquin. Painting regional landscapes was a way to establish a cultural independence from Europe.[19] In fact, the new interest in "native" landscapes owed much to contemporary developments in European art. French Impressionism, for instance, advanced the concept of painting *en plein air* to study and depict local sites. J.E.H. MacDonald and Lawren Harris, attending an exhibition of Scandinavian art in 1913, were struck by how the painters were "not trying to *express* themselves so much as trying to express something that took hold of *themselves*," as MacDonald later explained. "The painters began with nature rather than with *art*."[20] In Canada, ironically, this resulted in a departure from conventional Impressionism, a departure most often associated with the Group of Seven and most evident in their work in Georgian Bay. The sun-dappled glade and gentle pastel colouring of Arthur Lismer's *The Guide's Home* (1914; Plate 7) is a world away from the tumult of driven rain and bent pines in *A Westerly Gale, Georgian Bay* (1916; Plate 6); the calm blue horizon sunlit and sparkling in J.E.H. MacDonald's *August Haze, Georgian Bay* (1912) quickly gives way to the dark, threatening storm clouds piled high in his violent *Lonely North* (1913). There was clearly something about the *place* that promptly forced these artists to question the artistic language they had been taught; to realize it had been created for an entirely different sort of place and so was inappropriate here; and to experiment with different styles in order to find a way of depicting the stark elements of Shield country. Still, this was a process of adjustment, not an epiphany about the northland, as the myth about the Group of Seven suggests. After all, the member who became perhaps the best-known painter of Georgian Bay had found it at first glance "not quite paintable." ·

By the time of the Group's official debut in 1920, all its members had visited the Bay. Jackson's initial reservations were quickly proven wrong, for they found it an excellent site for sketching and would make it one of most painted places in Canada. What is remarkable about these early paintings – for instance, Jackson's *Terre Sauvage* (1913; Plate 4), Tom

Thomson's *On Georgian Bay* (1914), MacDonald's *The Elements* (1916; Plate 8), and the West Wind cottage murals (1915-16; Plate 3) – is that they are the first clear, consistent expression of what we recognize as the Group's trademark style of portraying landscape. Art historian J. Russell Harper singled out *The Elements* as an example of "the essential style of the Group of Seven during the finest years."[21] So much has been written on the Group that I hesitate to add anything more; but more *does* need to be said about their relationship with Georgian Bay. Algonquin Park traditionally has been credited as the site of their earliest and formative experience of the northland. I would suggest that the pivotal years came in Georgian Bay, from the first visits of Jackson, Harris, and MacDonald in the early 1910s (predating all but Thomson's trips to Algonquin) for the next decade or so. Here the Group cemented their preoccupation with the Shield country (which they identified as the northland) and its anti-pastoral interpretation of this raw land. The Shield's characteristic features – its clear light, heavy mass of the rock, stark and angular trees, and turbulent skies and waters – were all more exposed and more dramatic in Georgian Bay than in the interior; and this encouraged simpler lines, bold colours, and energetic motion in paint. The paintings done before 1920 suggest that Georgian Bay was the formative site of the Group and, given the Group's influential role in Canadian art, one of the key sites in modern Canadian art as a whole.

Much of the mythology surrounding the Group stems from the claim that they were the first to express a "native" perspective in Canadian art – to offer Canadians a view of the true spirit of their own land rather than scrutinizing it from a distance and through a screen of foreign expectations. In fact, the Group was not overly original in either their art or philosophy. Their European influences have been well documented, for they studied and painted abroad, and borrowed techniques from Impressionism and Post-Impressionism. They also followed a distinctly Canadian tradition, the explorer/surveyor sketching from a canoe to document an unknown landscape. They still presented the Bay as an ideal of wilderness, for an urban audience. They were not the first in either Europe or North America to present the land as a national symbol and the experience of nature as a patriotic exercise.[22] But there are two important differences between these paintings and earlier Canadian art. Theirs was not (not yet, at least) an aesthetic ideal; no one, and certainly not the audience of the day, would consider these images beautiful in the conventional sense. And it was more an interpretative representation of the

Bay than any other art had been, interested in the vibrant and unforgiving temperament of the Bay and the emotional reaction this rugged landscape could evoke. This gave their work a new degree of authenticity. The Group represented "the nineteenth century's final, crucial intellectual arrival in Ontario."[23]

This may be the clue to their perennial and unmatched popularity. A great deal has been written about the Group's impregnable place in the Canadian canon, but lately historians have given more credit to political machination than to artistic method. Some point to a calculated and relentless marketing campaign led by the National Gallery of Canada, which inundated homes and schools with copies of *The West Wind* and *The Solemn Land*, accompanied by a mythic narrative defining this rugged style as the expression of a national spirit.[24] Undoubtedly these efforts affected popular awareness of art, but other factors shaped public taste. The popular nationalism generated by the First World War spilled into an expectation that the arts should represent a distinctly Canadian experience. The Group also capitalized on the back to nature movement of the early twentieth century and the growth of outdoor recreation among urban Canadians. Here were artists who lived the wilderness ideal: canoeing and camping in search of material, sketching pines and birch, northern rivers and lakes. Douglas Cole attributes the success of the Group to a "commonality of experience" between the artists, their patrons, and the public, for they shared similar ideas about the north country.[25] It is perhaps more accurate to say that there was a commonality of belief, a public will to believe in such images, for the number of Canadians summering in Ontario's near north was relatively small in 1920 and has remained disproportionate to the Group's public influence. Nevertheless, there is a consensus that accepted these images as the Canada we *want* to have. It was not a picture of a foreign and unattainable ideal, like the strained examples of the picturesque, the softened pastoral images that preceded it; nor was it divorced from the known world, like the abstract art that would follow. As Margaret Atwood has observed, this is why the principal images of Canada in art and literature are physical ones, based on historical reality.[26] If the style suited the subject, then the subject suited the audience. The art popularized by the Group of Seven is a rare example of high and popular culture – too often treated as unrelated phenomena – overtly and deliberately converging.

❧

While MacDonald, Lismer, and Jackson produced a visual identity for Georgian Bay, another type of regional identity was emerging within the archipelago in a less spectacular but no less interesting fashion. As in the arts, it was a process of adjustment, of negotiating with local conditions, as the practices of everyday life gradually conformed to this part of the Bay. With this came a feeling of difference from other places, thinking of it as a particular and separate entity, distinct from the rest of the Great Lakes Basin and adjacent areas in Ontario. The archipelago was becoming an "imagined community," but one based on the very real environmental concerns that inform everyday life.[27] Much of what has been recorded about Georgian Bay is not indigenous or local in the usual sense, since it was produced by travellers, seasonal workers, or seasonal residents who were really urban refugees. But scholars from different disciplines have suggested that regional culture is a kind of littoral, a meeting of indigenous content and imported form in literature or "polite" and "folk" styles in architecture.[28] Cottages built from local materials by resident handymen for summering urbanites are thus a quintessential vernacular. More generally, anything that occurs *in that place,* in the hinterland, contributes to a body of local culture and the process of regional definition.

For Europeans in Georgian Bay, that process began in the mid-seventeenth century. Monique Taylor examined references to landscape in the Jesuit Relations and concluded that the missionaries' views of this unfamiliar environment "reflected new reactions to a new landscape." An eight-hundred-mile canoe trip to Huronia, for example, was well outside the European frame of reference. She even found glimmerings of a feeling of place attachment in writers such as Father Ragueneau, who in 1650 lamented the loss of the country "which had possessed our hearts and engaged our hopes." E.J. Pratt pictured Father Brébeuf as a convert to the New World, carrying his new way of seeing back to France:

> At Rouen he gauged the height of the Cathedral's central tower in terms
> Of pines and oaks around the Indian lodges.
> He went to Paris ... but his mind ...
> Rested on glassless walls of cedar bark.[29]

As long as the Bay was treated primarily as a means of passage, however, it would be conceptualized in linear terms. Until the mid-nineteenth century, the urge to move through it or around the Bay discouraged people from thinking of it as a separate place.

The industrial era brought a new cohesion and a new prominence to the Great Lakes but kept Georgian Bay subsumed within the wider boundaries of the Lakes region. As the "north shore of Lake Huron," it was one of the inland seas. Like other frontier communities, its lumbering songs and ghost stories commemorated heroic or tragic encounters with nature. The ghost of log-driver Sandy Gray haunts the falls on the Muskosh River, where he flouted the Sabbath by declaring on a Sunday morning, "We'll break the jam, boys, or breakfast in hell." (He, at least, did both.) When the *Asia* sank in 1882, a morbid epitaph in a local newspaper lamented, "in the deep they're fast asleep ... on the beach, their bones will bleach / Along the Georgian shore."[30] The old pull outward was still there, for the traffic in natural resources relied on the physical unity of the Great Lakes Basin and bound Georgian Bay to the other Lakes. Consolidation of industrial production required a consolidation of geography. Tugs towed log booms from rivers mouths on the Bay to American mills, fishermen from both countries chased fish to the northeast shore, and American tourists colonized the islands. Common environmental problems, from species depletion to water pollution and acid rain, would migrate just as easily across the undefended border.

The archipelago also shared a history with central Ontario. Greg Halseth's *Cottage Country in Transition* (1998) finds a parallel scenario in the Rideau valley: quickly abandoned by farmers and loggers, colonized early by cottagers, and by the 1970s seeking a mechanism of coordinated management along the waterway. As part of the Ottawa-Huron Tract, Georgian Bay was seen simply as the poorest section of a difficult frontier. And as a frontier not yet settled by vacationing southerners, it had more in common with northern Ontario than with the south. As the House of Commons debated the Georgian Bay Ship Canal contract, MPs read endorsements from communities located along the proposed route and argued that a canal would benefit "the people of northern Ontario, the red-blooded pioneers who have helped to develop that northern country."[31] This section of the province, a swath from the upper lakes to the Ottawa Valley, has since coalesced as cottage country, with benign images of the outdoors such as white pine and loon calls. As with the Great Lakes, this wider regional sensibility re-emerged in public concern with environmental quality. In the 1960s municipal and MNR planners, along with conservation authorities, began to speak in terms of watersheds; the 1969 municipal government review, for example, recommended against severing the Bay shoreline from the District of Muskoka because most of the district drained into the Bay.[32] Occasionally, then, the shore has identified its interests with the interior.

But with permanent settlement came a wider awareness of the unique conditions of the islands and a narrower regional sensibility. Of course, the Bay's northeast shores had never been confused with the shores of Lake Huron proper, the gentler southern shoreline of Nottawasaga Bay's sandy beaches, or the limestone ridge of the Bruce Peninsula. Reaching Huronia in 1615, Champlain remarked with relief, "Here we found a great change in this country, this part being very fine ... [and] very pleasant in contrast to such a bad country as that through which we had just come."[33] Township surveying by an expansionist Canada West quickly carved up the Huron Tract after 1815 but was stopped dead by the Shield at the Severn River for decades afterwards. As late as 1926, Lewis Freeman emerged from the archipelago and was jarred by the sight of towns, roads, and train smoke in Severn Sound. The "transition from the wilds of 'The Thirty Thousand Islands' to the civilization of the head of the bay," he writes, "was almost startling in its suddenness."[34] Paintings of Lake Huron and the Bruce by contemporary artists Jack Chambers, Roly Fenwick, and Bruce Steinhoff show a rural tranquillity of fields of wildflowers and unbroken horizons of serene blue skies. Although pulled into the social and economic orbit of southern Ontario, geological fault lines keep the archipelago on the fringe of the north.

More surprising is the feeling of distance between the Bay and the rest of Ontario's near north. After all, this is a single and relatively small area of the province, with common geological and biological features, dominated by the same industries (once forestry, now tourism and cottage services) and inhabited by seasonal and permanent populations of comparable socioeconomic status. But speak to anyone from the Bay and in a very short time you will be confronted with certain basic tenets of local culture, all predicated on its "wilderness" landscape of rocky topography and exposed shoreline. There is an unabashed pride in their "woodsiness" – a comfortable knowledge of shoals and rattlesnakes, a preference for informality and solitude. There is an intense feeling of place attachment – unusual because it is based not on deep historical roots (with a hundred years' residence at most, cottagers hardly constitute an "indigenous people") but on the ways cottagers imagine and use the Bay, as a landscape associated with the enjoyment of nature through recreation. Finally, there is a caustic contempt for other parts of cottage country. Folklorists who study regional identity call this the esoteric-exoteric factor: labelling insider and outsider, pitting those who feel they belong to a place against those who do not.[35]

These boundaries in the near north are not recent inventions. Georgian Bay never attracted sightseers or campers on the scale of Niagara Falls, the Thousand Islands, or Algonquin Park.[36] The broken and rugged topography, the navigational hazards, and the inconvenience of water access all discouraged mass tourism and cottage settlement. Those who favoured Georgian Bay quickly attached moral value to their choice and its voluntary hardships. It was easy to disdain the opulent summer homes of the St. Lawrence or the "populous watering-places" of Muskoka as a "vortex of fashionable dissipation." Far better for the serious nature lover was "an excursion to the wilderness beyond, to that *ultima thule* known as Georgian Bay."[37] Tenting and fishing groups clustered around clapboard clubhouses, such as those of the Iron City Fishing Club. Four or five generations later, the descendants of those who deliberately sought out this "wilderness beyond" believe that their preference for Georgian Bay reflects an expertise and appreciation of nature either lost or redundant in other parts of the near north. A Bay person is one who knows that "you can't, no matter where you are, ignore the Bay when you're there."

So they insist passionately that Georgian Bay is *not* like Muskoka. To an outsider this may seem puzzling: not only are the two communities side by side, but Georgian Bay today suggests money and old society as much as Muskoka. But to those in the Bay, the difference is a simple fact, as obvious as the difference between the exposed granite and the open water of the islands and the grassy lawns and flower gardens of the interior. The physical differences symbolize differences in values and in attitude, differences that date to the fishermen of the 1870s who preferred the seclusion of the islands to the genteel hotels of the mainland. The hotels offering lawn tennis and croquet gave way to mahogany launches and expensive summer homes, but the assumption remained the same: those who favoured Muskoka preferred comfort, diversions, and pretty scenery over the challenge of wilderness. To Georgian Bay cottagers, Muskoka (and to a lesser extent Haliburton and the Kawarthas as well) signifies a busy social schedule, expensive resorts, green lawns, and crowded cottage neighbourhoods – an essentially urban way of life and a status-conscious public display that has crowded out the natural environment. (Bay people will tell you that in Muskoka you need a cocktail outfit, but in the Bay cut-offs and bare feet will suffice.) This is more than good-natured sniping. Bay residents make a distinction between what they see as a physical landscape and a social one. This is partly an echo of the old image of fashionable watering places, but the social identities are maintained by the daily and immediate reminder of the underlying physical differences. Water-based communities in the archipelago must operate within its physical perimeters: getting supplies, getting to the cottage from the marina, marking shoals in high water and through passages in low water, sheltering from winds. "It's the ruggedness, it's the openness, it's the vastness of space," explained camper and cottager Tim Stinson. "I think you're truly closer to nature in Georgian Bay, versus Muskoka, which is kind of a transplanted downtown Toronto. I don't feel that you're in the outdoors when you're in Muskoka." (Even the District of Muskoka now tries to exploit this distinction by claiming its section of Georgian Bay shore, from Severn Sound to Twelve Mile Bay, as its "most rugged and wildly beautiful ... a striking wilderness environment.")[38] Residents use the unique qualities of Georgian Bay to distinguish themselves from similar communities. Collective memory idealizes a common history in, and affection for, a difficult environment.

Older cottages help sustain this memory, for they are read as an expression of both community values and the Bay's physical constraints. The Madawaska Club prizes the "plain, inartistic and inexpensive" austerity of its oldest cottages as a reminder of the club's century-old history and the members' traditional self-image as voluntary pioneers.[39] Vernacular architecture has been described as "a concretization of its environment."[40] It is dependent on the ability of local builders and is built to on-site constraints, using available materials and designed to draw attention to the setting rather than the dwelling. For several decades, cottages in Georgian Bay were precisely this. Even cottagers who brought in building kits and supplies (mail-order designs were widely available by the late nineteenth century) needed the help of local builders to assemble them, and had to adjust the layout to the uncooperative terrain. Each community on the shore relied on a few permanent residents who acted as general handymen, such as Wellington Welsh, Johnny Clorac, Joe Naught, and generations of Rois, Trudeaus, and Robitailles. Although North American architecture was often sympathetic to regional styles and natural settings, architectural fashions were far less important than local ability. Typical was the cottager whose testimonial appeared in the 1912 brochure issued by a Penetang contractor: "I had no architect, but simply got a practical builder to confer with yourself." Andy Hamelin, a Midland carpenter who built most of Camp Hurontario, recalls his first meeting with the camp director in 1947:

Figure 10 Cottage at Go Home Bay, built c. 1911.

The architect had drawn up the design for a cabin which was 18 by 50 ... But [there] are only the three that are designed that way, from the blueprint. From then on, we sat down and we talked over what type of building Birnie wanted. Then we sort of designed it between the gang of us ... and that's what we went by.[41]

A cottage depended largely on what these men were capable of doing, what they chose to do, what they could boat in, and what they could fit on the site. Building materials were found around the shore. Lumber companies stacked cheap seconds and shingles, boulders on the shore filled dock cribs, fishing stations and abandoned buildings were dismantled and recycled or renovated. Local materials and simple plans produced precisely the rustic style favoured by cottagers who wanted to get back to nature: low sight lines, natural colouring, unpolished assembly.

These were practical considerations, but they sent a clear message about the community's sense of itself. Resorts such as Muskoka Sands or the

summer homes of Lake Muskoka's Millionaires' Row mould the landscape to suit with infusions of money and cosmopolitan tastes. Rustic and
modest structures amid granite crevices suggest a community of
outdoorsmen adapting to a rugged landscape. As May Leigh Otto, a summer visitor to Cognashene, maintains:

> The people who liked it up there so much did so because they appreciated
> and responded to outdoor life and physical activity. They weren't show-off
> rich and they weren't filthy rich. And they didn't give a damn – even if they
> had been, they wouldn't have had the biggest showiest cottage. Not in those
> days, anyway. They do that in Muskoka. They've always done that in Muskoka.

(One could argue that the preference for rusticity is itself an urban taste,
but at least it seems less arrogant an imposition on the tangled Shield
country.) Perhaps this is why there is such resentment toward new arrivals who, by building larger or more ostentatious homes, violate traditions
that kept the Bay aloof from other areas. Grand homes blur sacred dis-
tinctions by treating Georgian Bay like generic cottage country. Community memory remains strong, for there are generations of families here;
and the topography has not changed. A fourth-generation cottager in
Cognashene lambasted new neighbours for building a "monstrosity ...
plunked straight out of Muskoka" and sodding the rock. "There are these
morons out mowing their grass lawn," he said. "You don't mow lawns in
Georgian Bay!" The negative reaction also reflects a kind of Loyalist mentality among old cottage families that awards a certain status to those who
can claim a longer connection to the Bay, and gives a greater legitimacy
to their sense of how the Bay should look. Yet, some of these newer buildings do harmonize with the place in more aesthetic or formal ways. That
on Carolyn Island at Pointe au Baril is expensive and eye-catching, but
there is an unmistakable congruence between its pink granite columns
and the island rock, and between its transparent glass walls and the
Open. Hence, Gillian Michel's ambivalence toward it, typical of "old" Bay
families:

> If you're up in a place that has that kind of beauty, then you sure ought to
> build it so that you see the beauty. So even though I think it's pretentious in
> the way of spending money to have a front window that goes into the granite
> [and] the granite is your floor and there's nothing between you and that
> view, [it] is in a way appreciating the environment.

Figure 11 Summer home on Carolyn Island, 1992.

Environmental constraints maintain regional differences. As Michael Hough writes, "irrespective of the civilizing influences of cultural tradition, it is the native landscape that is a primary determinant of regional identity."[42] Emphasizing the physical differences between the archipelago and adjacent regions was essential to defining its borders, detaching it from the rest of Ontario, and thereby creating a distinctive identity for the place in the public imagination.

∾

But Canadians rarely think of Georgian Bay as a region, even – perhaps especially – those who live there. The Prairies and the Maritimes and the North are *regions*; Georgian Bay, with its leaning pine and waterways and Canadian Shield, crossed by Champlain and the voyageurs, abstracts the *national* experience. Or so Ontario likes to think, for Central Canada has always had a tendency to assume that its experience speaks for the country as a whole. The Group of Seven would produce the nation's art, while the Laurentian school explained its history, framed by the Canadian Shield, the Great Lakes, and the St. Lawrence. This meant a larger picture was drawn from local details; place-specific images were exported to a wider audience. Georgian Bay art has achieved one of the highest profiles of any body of art in Canada – but never as *Georgian Bay* art. In the art of the Canadian Shield, the lone pine bending in the wind on a rocky shore becomes an archetype of the northland. A cottager sees in the pines' "opportunistic growth" on rock scraped bare by glaciers a metaphor for Canada as a whole.[43] Poems such as "Canoe-Trip" and "Country without a Mythology" become fixtures of literary anthologies because they address the theme of national identity in the sanctioned "Canadian" setting, the rivers of Shield country.

Yet, if this poetry, like Group art, is read not as official Canadiana but as a response to a specific place, it resonates far more truthfully. Here it is a single island:

> A few acres of pines and cedars where he knew
> almost every tree. Abrupt granite rising from the clearest
> water in all the world. Crowned with a tangled diadem
> of blue green foliage ...
> And always beneath birdsong the sound of water.
> Tonight he is imagining the calm after a three-day blow,
> the roar of the open from the swell on the outer reefs,
> while here on the inner islands the waves are gently lapping ...[44]

These celebrated figures of the national canon actually show us that Canadian culture is not a single portrait of the Canadian Shield but a collection of individual responses to different landscapes, a patchwork culture that owes more to physical diversity than to nationalist aspirations. Over time the Laurentian school became one of many, as artists and writers found meaning outside the centre. As Northrop Frye said, the emergence of regional expression marked the maturation of a nation's culture.[45] If any single trend characterizes the history of the arts and letters in modern Canada, it is regional fragmentation and the growth of regional arts groups across the country. This pattern accelerates *after* the Group's appearance. (Even the Seven quickly found themselves pulled in different directions, gravitating toward landscapes that inspired them: Varley to British Columbia, Harris to the Arctic, Jackson to Quebec, Carmichael to La Cloche.) As Harris later explained, the Group sought to depict the "form and character and spirit" of distinct regional landscapes.[46] Though Ontarioans might point to the Group or the Laurentian writers as the culmination of a colony-to-nation process, their very skill in depicting Ontario's Shield country reveals the regionalism, or provincialism, of this central Canadian nationalism. Georgian Bay is significant not as a national ideal but as an example of regional definition: the successful (other parts of Canada might even say *too* successful) expression of a regional environment and its *genus loci,* or spirit of place. The rock-and-pine iconography has, and rightly so, become "national" in another way. It provides an image of northern Ontario in the mosaic of images – lighthouses, seigneurial strip farms, wheat fields, mountains, tundra – that makes up any visual representation of this country.

Nonetheless, Georgian Bay has always inspired nationalist rhetoric because it exemplifies the type of wilderness landscape so central to Anglo-Canadian nationalism. In large part this is because it lies on the edge of the huge northward sweep of the Canadian Shield. The Shield is Central Canada's example of making specific landforms into national icons or national monuments, like the Grand Canyon in the United States, Denmark's Himmelbjerget, or England's White Cliffs of Dover. Promoters of Canada West, looking to expand northward, proclaimed that finding "new elements of natural strength in the occupation of our interior country" was a subject "of national importance." Nation building required timber and ore, but they expected the frontier to expose "elements of natural strength" in the Canadian character as well. Many in English Canada believed that a pure, rugged, unsettled northland would produce a superior Canadian race. W.W. Campbell declared of Lake Huron that there was "no cradle for the rearing of a great people ... more fitting than the shores of this vast lake."[47] This variety of nationalism was eventually discredited for its racist overtones, suggesting as it did the superiority of a northern people. But a milder form of environmental determinism persisted. The idea that the land defines the Canadian experience, that we are in some way a product of our environment, proved remarkably durable. Ninety years after Campbell, historian and Australian expatriate Jill Ker Conway was greatly impressed by Shield landscapes such as "the sparkling waters and black rocks of Georgian Bay." These, she felt, gave Canadians their sense of place: a respect for the power of "untrammeled nature" and a toughness born of living "close to natural forces beyond human control [so] they had learned never to give in." Wilderness remains the most powerful of Canadian myths, our "secret ideology."[48] As poet F.R. Scott notes with a touch of irony, we believe that the land "will explain ourselves to ourselves ... And prove that something called Canada / Really exists."[49]

Georgian Bay tapped into nearly every aspect of the enormously powerful myth of wilderness. On the edge of the Shield, it evoked Canada's longstanding romance with the north as the country's defining landscape, a vast and unknown wilderness, a place of exploration and adventure.[50] The past was unusually accessible in this "unchanging" landscape: catch a glimpse of fur-trade country only a few hours from Toronto! Any era from the nation's history could be imagined here – the life of indigenous peoples, European exploration and the fur trade, the bustle of logging and fishing. Even cottaging and "the outdoor life," which now resembles urban life more than that of the voyageur, is treated by Canadians (and especially Ontarioans) as their patented creation. To a nation perennially insecure about its identity, this landscape, this history, and these types of environmental experience represented a distinctive heritage. "It became my definition of what I love about Canada," admitted a former minister of Tourism. "I went all over Canada and the world and I never saw a place more beautiful than O'Donnell's [Point]."[51] It also emphasized the differences between Canada and other countries. One March morning in 1936, artist Doris McCarthy was in London when she found herself homesick for "our dear north country." Lying on the grass in Hyde Park in the watery sunlight of an English spring, she "felt what exile could be, how I would cry with longing for the feel of a paddle in my hand, the warmth of flat Georgian Bay rocks, the cool ripple of blue water."[52] Writing about the Canadian landscape suggests that by this point attitudes toward "our interior country" had altered somewhat: still a pride of possession, but now also a feeling of belonging, of coming *from* the place. The spirit of LePan's "A Rough Sweet Land" comes not from a race of Indian braves "before the white man came," nor civilizations of "the far Aegean [and] fertile Tuscany,"

> his is a new treaty with a new environment,
> his eyeballs catch the glint of ice in August
> as ours do. And he is new now, new and ours, and always.[53]

For as long as Canada has struggled with its colonial complex, in the shadow of Britain or the United States, geography has been viewed as an agent of cultural patriation.

Little wonder there is such resentment when people discovered our Canadian landscape already occupied ... by Americans! Critics accused American interests of invading Canadian territory and carrying off resources properly belonging to "the people." Georgian Bay offered a particularly dramatic illustration of foreign investment; sufficiently remote from Canadian cities and exposed along its entire flank to American boat traffic, the Bay practically invited American fishermen and loggers into Ontario. In 1886, for example, US companies (many from Michigan) controlled 1.75 billion feet of standing timber in the Bay. But supplying American factories and American markets with raw materials did not suit the political agenda of Empire Ontario for industrial development at home. The fight to direct Canadian wood to Canadian mills came to a head with the United States' controversial 1898 Dingley Tariff, which taxed imported sawlogs at a lower rate than lumber already processed in Canada, and which drew intense criticism from Georgian Bay loggers.[54] Advocates of the Georgian Bay Ship Canal argued it would provide a shipping route entirely within Canadian waters and spur the development of northern Ontario's resources. Royal Commissions into the fishery denounced the dominance of American "syndicates" and "unscrupulous combines," and recommended managing fish populations "for the benefit of Canadian fishermen and Canadian subjects." They also blamed American tourists (or "aliens from the United States") for overfishing sport species. (This did not, however, discourage the province from later courting those same American tourists with advertisements promising unpeopled northern lakes with matchless fishing and hunting.)[55] A century later, public opinion on environmental issues such as acid rain and bulk water export is coloured by a proprietorial attitude toward natural resources, an attitude that shades into anti-American sentiment. If a stay and bulwark of Canadian identity is that we are not American, this reaction is not surprising. Canadians have invested tremendous significance in this type of northern environment as a mark of difference from our neighbours. In John Irving's *A Prayer for Owen Meany*, Georgian Bay represents all that a "MORALLY EXHAUSTED" America has lost, with the innocent pleasures such as "lots of leaky canoes, and the smell of pine needles" at Pointe au Baril in sharp contrast to the violence of Vietnam and the debacles of the Reagan administration.[56] Canada and the United States do share the continent, and the forty-ninth parallel is not always a convincing divide; but a distinctive landscape laden with added symbolic weight – a northern orientation, historical associations, artistic images – provides a physical foundation for arguing national difference. Landscapes such as the Bay are used to affirm Canada's right to exist, its "natural" autonomy when politically this autonomy is not always evident. We are different because it is different – the same argument by Georgian Bay residents about their relationship with the rest of Ontario.

Yet, we cannot claim to have ever shared a single idea of the land or experience with it. Place identities can be highly localized, even to a twenty-five-kilometre-wide corridor of islands and inland lakes, as long as the physical environment is noticeably different. Nevertheless, these constantly draw from and contribute to our "secret ideology" of wilderness myth. So Canadians face a very federalist sort of nationhood: a collective ideology of outdoor life and the centrifugal counter-pull of history in a thousand different landscapes. The "national" identity and visual iconography of Georgian Bay were constructed in the arts, aided by the cultural authority of Central Canada, even as the Bay's identity as a region was more narrowly defined in vernacular culture. The development of cohesive and coexisting identities was gradual, however, and opinions have varied widely: a means of prosperity and industry ... the most beautiful and extensive island scenery in the world ... nothing but little islands covered in scrub ... my definition of what I love about Canada. Little wonder scholars tend to assume that people "were making these landscapes up." In fact, each reaction was a combination of preconceptions, expectations of nature, and experience; all of the agendas – economic, scientific, artistic – brought to Georgian Bay were forced to accommodate "what you actually encountered when you got here." A regional identity is not really discernible until the arrival of a settler population that attributes its perspective and a sense of difference to its history with the Bay. As Kent Ryden concludes in his study of folklore, literature, and place identity,

> a sense of place results gradually and unconsciously from inhabiting a landscape over time, becoming familiar with its physical properties, accruing a history within its confines ... Any setting can become a symbol of and element of personal and group identity through sufficient familiarity and propinquity. Through extensive interaction with a place, people may begin to define themselves in terms of their relationship with and residence in that place, to the extent that they cannot really express who they are without inevitably taking into account the setting which surrounds them as well.[57]

There is a subtle shift as people consider themselves citizens of the place rather than commentators on it, speaking from rather than about it. As we will see in Chapter 6, the community's confidence in its local ecological knowledge would be a key factor in its political activism as it lobbied for better regional planning. Environmental planners and policy makers would face a similar struggle to adapt familiar practices to unusual conditions as they encountered the unexpected in the archipelago.

SIX

Some Proper Rule:
Managing and Protecting Georgian Bay

We like to think of Canada as a country of wilderness and wild places, a great expanse of fast-flowing rivers and pine forest stretching north for thousands of kilometres, uninhabited and untouched by humanity. This is, in large part, a myth. People have sought to preserve, conserve, manage, and otherwise interfere with Canada's environment for close to two hundred years. Generally, their actions have been motivated less by concern for specific ecosystems than by political norms and competing economic agendas. Our sense of the environment as a political issue tends to be reactionary and crisis-oriented (save the ozone layer! clean the oil spill!), so the cultural foundations of environmentalism often go overlooked.[1] To understand when, where, and why governments had the political will to enact environmental policy we need to have some sense of contemporary ideas about the role of the state, economic growth, and the meaning of nature. This chapter also examines the political implications of the transition from resource to recreational use, characteristic of many natural areas, in the Bay. All these factors affected how people approached Georgian Bay or understood it as a place to be managed. The concern with industrial sustainability, as in fish reserves and provincial forests, was supplanted by park creation, municipal reforms, and cottage politics. As a major centre for fishery, forestry, and recreation, and the subject of regulation by all levels of government, the Bay has mirrored trends in environmental management in the Great Lakes region and across North America.

As a case study in environmental history, however, it poses a challenge. Conventional records such as legislative debates, planning studies, or departmental memos give us only a partial picture. The subject is far from being safely dead and archived, and issues of environmental quality, land use, and political boundaries dominate discussions of the Lakes today. The other problem is that such records paint environmental history as the story of policy in a vacuum; they are filled with instructions, plans, and proposals but not always with records of what actually happened next.

Geography does not always cooperate with political agendas. Environmental management is a constant negotiation between political intent and physical reality. At times the archipelago chafed against standard practices. When the landscape is such a significant factor in shaping historical events, as it is in Georgian Bay, we need to assess what role – if any – it played in the formulation and implementation of policy. The interaction between policy and place is also integral to explaining the evolution of a regional sensibility. There is a learning curve as people (for instance, park managers) learn to recognize what programs will be feasible in the archipelago. At a community level, cottagers become active in "defending" their beloved Bay from other users. In short, influence runs both ways. When standard conservation policies are imposed on a unique ecosystem and its resident communities, it forces people to reconsider the efficacy or appropriateness of the usual types of management. How did instructions drawn up in Toronto or Ottawa envision Georgian Bay? How did they play out in the archipelago? In other words, how was policy interpreted at ground level?

Originally, the instructions were imported from much farther afield than Queen's Park, the seat of the Ontario provincial government. Canada's history has often been influenced by its position in the North Atlantic triangle, and Canadian ideas about environmental management were no different.[2] From eighteenth-century Britain came a tradition of natural history in the sciences, a Romantic aesthetic appreciation of landscape, and, most importantly, demands of Mother England for the natural resources of her overseas colonies. As a result, colonial policy kept ownership of lands and resources with the Crown. Gradually the emphasis shifted from reserving these resources for the state's own use to selling rights to the resources in the form of timber limits or mineral licences in order to raise revenue. This system of management would pose two problems to conservation efforts. As long as the Crown benefited directly from the sale of resources on its lands, the value of conservation would not be apparent until the supply of these resources was seriously diminished. Historians have concluded that Crown ownership merely created a cozy relationship between industry and state that allowed private interests to fully exploit public lands.[3] Numerous conflicts of interest were also inherent in "the Crown" itself, because the responsibility for different types of land and resources was divided among different branches and levels of government. For example, the Crown was simultaneously responsible for Indian reserve lands (at the federal level) and resource development (at the provincial).

Canada may have inherited its policy framework from Britain, but ideas about nature, especially wilderness, and environmental management were · more easily imported from the United States. "Environmental" issues differed across the pond: coal and other mineral resources were critical to an industrialized Britain, whereas Canada was more inclined to share the American concern with forests and watersheds. Indeed, some of the earliest examples of modern ecological research came out of the Great Lakes Basin, notably George Marsh's work in watersheds and deforestation in the 1860s.[4] The advanced erosion of wilderness in the eastern United States gave rise to a literary tradition celebrating wild nature, led by writers such as Ralph Waldo Emerson, Henry David Thoreau, and Walt Whitman. By the latter part of the nineteenth century, the opening of the American West brought to public attention large-scale resource issues such as water and grazing rights. The American government was forced to devise conservation policies somewhat earlier than in Canada. Environmental thought in the United States thus emerged as a strange hybrid that wanted to idealize wilderness and ensure its perpetual use at the same time. As Richard White has observed, Americans thought of themselves both as children of nature and children of the machine (or Yankee technological know-how).[5] It is not a coincidence that American "environmentalism" was typically divided into two streams, one advocating a utilitarian conservation of resources and the other a more stringent preservation of natural spaces. Canadians may have complained about rapacious Americans invading our resources, but our attitudes toward nature borrowed heavily from intellectual developments and public policy south of the border, from forestry programs and park legislation to grassroots support for wilderness protection. One Georgian Bay outfitter credited tourists "imbued with the vigorous conservation programs of the United States" with bringing a new sense of environmental responsibility to Canadians.[6]

But what put the environmental health of Georgian Bay on Canada's political radar was a much more immediate and identifiable concern: the exhaustion of fish and forest stocks. The country could not afford to lose two of its most lucrative industries. Although quick to condemn the "theft" of their "natural heritage," nineteenth-century Canadians assumed that resources could be used indefinitely as long as they were properly managed. In 1892, for example, Ontario's Game and Fish Commission professed revulsion at "the same sickening tale of merciless, ruthless, and remorseless slaughter," but concluded optimistically that if "pains [were] taken to preserve and propagate the supply, the community would benefit materially thereby."[7] Conservation entered the political mainstream around the turn of the nineteenth century as part of the eclectic reform movement known as Progressivism. Cottagers transferred early the contemporary interest in urban zoning, sanitation, and public health to the water quality of Georgian Bay. Part of the Georgian Bay Association's original mandate in 1916 was to "encourage boats and cottages to have proper sanitary arrangements and appliances so that the water in the inner channels may not become polluted." Central to Progressive ideas of conservation was the idea that resources constituted part of the nation's natural heritage and should be protected and managed by government for the benefit of society as a whole. But Progressivism was itself very much a product of industrial North America. Its idea of conservation emphasized productivity, scientific study, and centralized management; in short, it saw nature as a resource, to be used for the betterment of society. In practice, natural resource management in the late nineteenth and twentieth centuries in North America was governed by what we would think of as economic principles: a "doctrine of usefulness." Conservation meant the efficient use of existing resources and the cultivation of renewable ones in order to maintain economic growth.[8] The emphasis on sustainability – sustaining resources in order to sustain the industries dependent on them – thus suited those with the greatest interest in Georgian Bay at the time. Fishing and logging companies would concede state regulation if it guaranteed profitable use in perpetuity, not if it declared a working landscape off limits.

A classic example is the response to the failing Georgian Bay fishery, where the key commercial species of whitefish and lake trout were threatened by overfishing as early as the 1870s. Surplus catches were left to rot in abandoned nets, further polluting the water. "It is possible, under the present plan of fishing Georgian Bay and contiguous waters, to exterminate those fish in a season," warned the 1893 Royal Commission into Ontario's inland fisheries. Fifteen years later, a second commission struck specifically to investigate the Georgian Bay fishery reported that the process was well under way: "owing to the immense slaughter carried on by man, the balance of nature was destroyed, the supply has declined,

and the extinction of the whitefish will inevitably follow."[9] Everyone seemed
to assume it was the state's responsibility to repair and restore the fishery.
The commissioners admonished the fishermen testifying before them:
"There is no doubt the Department [has] to care for [the fish] in Georgian Bay ... The Government [has] to consider the people of Canada and
the country ... it would be as well for them to understand that the time
has come that some proper rule should be made to prevent the extermi-
nation of the fish."[10] Ontario had a long history of "proper rule" when it
came to regulating the fishery, for regulations pertaining to fish catches
and habitats (notably Lake Ontario salmon) were among the earliest pieces
of conservation legislation in Canada. Since 1857 the federal government
had imposed fines for polluting water or fishing in closed seasons, and
maintained a system of licences and overseers. The usual, and most conservative, approach was limiting the size of catches by means of closed
seasons, prohibited technologies, and licences. For example, pound nets
were banned in the waters east of Cape Hurd in 1885 because this type of
net caught immature whitefish fingerlings too indiscriminately. Setting
limits on harvesting overshadowed attempts to protect habitat from pollution. Prohibitions on damming rivers or dumping wastes at sawmills
were less effective, and met with opposition from the powerful timber
lobby.[11]

"Some proper rule" in Georgian Bay consisted of two types of regulation: comprehensive or geographical reserves, and restrictions on certain types of fishing practices. It was hoped that setting spatial and technical
limits would be enough to permit natural regeneration. None of the proposals was entirely new or original; similar measures had been attempted
in Great Britain as early as the 1820s.[12] But what is interesting about the
diagnoses of the Georgian Bay fishery is that they recognized the unique
conditions of the archipelago, and the relationship between habitat, fish,
and fishermen. The northeast shore was known to be "the haunt and the
home" of valuable game and commercial species because the rocky shoals
made it "peculiarly adapted" as a spawning ground.[13] Its remote, exposed
location offered the fish additional protection, as the fishermen who suffered annual losses of nets and boats well knew. Both commissions recommended closing the entire east shore, from Matchedash Bay to
Thessalon, first under a complete ban that excluded only licensed rod
fishing, and then to commercial fishing as a game fish reserve. Overseers
and fishermen agreed that in "these numerous groups of islands every
facility is afforded for illicit fishing on the most extended scale," and it

was easier to declare the entire area off limits than to chase all boats and try to distinguish between poachers and licensed fishermen.[14]

No one agreed on much else. The politics of regulation and enforcement in the fishery laid bare the political weakness of Progressive-era conservation, which assumed that a central authority could and should impose control over local stakeholders. Fishermen were suspicious of measures that might interfere with their livelihood; they welcomed restocking from hatcheries but were divided over restrictions on the size of mesh in netting, catch quotas, and closed seasons. Loggers were loath to reduce mill waste or the number of booms crossing the lake. Overseers struggled to patrol huge territories of open water and groups of islands where poaching flourished; reports of the North Channel suggested that the Noble brothers of Killarney were able to run the local fishery as a personal fiefdom. Homesteaders demanded exemptions from fishing and hunting bans in order to provide for their families. First Nations resented the system of leases and licences that encroached on their traditional fishing grounds, just as the government was pressuring the bands on Manitoulin to surrender the island for white settlement. In 1863 tensions exploded into armed confrontations at Wikwemikong and the mysterious murder of a government fisheries inspector. The government was also beginning to align its conservation policies with the increasingly influential sporting lobby. In Toronto the second commission heard from "prominent citizens of the province, who are interested from the sportsmen's point of view in the protection and repletion of the game fish of Georgian Bay." Well-to-do and well-connected sportsmen would become more and more successful at staking their rights to the Bay against both industrial and Native fishers (though residents had harsh words for tourists who left fish "upon the rocks to spoil, sheer, wanton waste").[15] Outfitters, too, worried about the impact of commercial fishing on bass, pickerel, and pike in inshore waters.

Two fishery reserves did exist briefly, but for the most part only on paper. In 1899 the province closed the area between the French River and Sturgeon Point to commercial fishers in order to protect game species. Ten years later a second reserve was created between the French River and Grondine Point, but only for spawning season and only "for the purpose of securing an adequate supply of fish eggs for artificial Fish Breeding purposes." Faced with opposition from local fishermen, the Fisheries department neither collected the eggs nor enforced the ban, and the order was rescinded three years later.[16] The anaemic infrastructure of nineteenth-century regulation was thus further weakened by physical obstacles and political interests. It is hardly surprising that measures that were in place the longest were simply those that proved the most unobjectionable: hatcheries or research stations, such as that operated at Go Home Bay by the Department of Marine and Fisheries between 1902 and 1913.[17] To add insult to injury, there were already ominous signs of the ecological threats that would pervade the fishery within half a century. Exotic species were beginning to migrate from the St. Lawrence system through the Erie and Welland canals into the upper lakes. Fishermen reported evidence of "the ravages of the lamprey eel" on the bodies of fish caught in Georgian Bay as early as 1893 (which is particularly interesting given that the lamprey itself was not discovered on Lake Huron until 1937).[18] Later the Bay was invaded by zebra mussels, which cluster on ships' hulls and water pipes; purple loosestrife, which suffocates wetlands; and cormorants, which nibble away at young fish populations. These non-human trespassers are proving just as difficult to police as the poachers of the nineteenth century. There are also concerns about the growing industry of aquaculture that thrives in the "natural fishery" of the North Channel. Environmentalists worry that these fish farms deoxygenate and pollute the water until their operators obtain new leases and move on – a familiar pattern given the migratory harvesting of nineteenth-century industry.

Ironically, given the problem of uninvited exotics, the practice of introducing new species or attempting to reintroduce native ones into an ecosystem has been one of the most common in environmental management. One study counted seventeen successful and seventeen unsuccessful introductions of fish species in the Great Lakes dating back to 1870. Such introductions have been motivated primarily by commercial demand, and to a lesser degree, recreational demand. The fishery is thus an excellent illustration of the profit motive of the managerial approach, which seeks to preserve or enhance "the biotic capital while maximizing the income," as Donald Worster has said. This remains ingrained in environmental policy; the Ministry of Natural Resources, for example, justifies its current restocking program, including lake trout and muskie in Georgian Bay, as both rehabilitating native species and aiding the recreational fishery. Since the 1960s the rhetoric of "stewardship" has been used by environmentalists and environmental managers to characterize their approach to environmental management. The term is meant to be flattering, for it suggests a responsible, considered, almost benevolent kind of management – humanity caring for the earth like a gardener would care for his or her garden. But stewardship also connotes governance, and

historically governance of the environment has entailed deliberate intervention with the intent to direct and shape that environment according
to an agenda. Stewardship has been criticized for implying that we have
the ability and the right to manage creation – for preserving the idea of
Adam's cultivation of Eden – when our management may do more harm
than good.[19]

On land, the concept of forest conservation had entered the political
mainstream by the 1870s. Fires from pine slash, wasteful harvesting, and
massive timber auctions all energized popular support for better forest
management, as demonstrated by the formation in 1882 of the American
Forestry Congress. Cleared watersheds suffered flooding, drought, and
erosion, providing graphic evidence of the detrimental effects of extensive logging. While surveying Baxter in 1878, James McLean found that a
sawmill had dammed Six Mile Lake, flooding the area, altering the flow
of rivers and streams, and making his job "exceedingly troublesome."[20]
The consequences of large-scale deforestation were too dramatic to ignore, and potentially damaging to agricultural areas as well. The first
major park reserves in the United States, Yellowstone (1872) and the
Adirondacks (1885), were also justified as a way of protecting the watersheds that sourced the headwaters of important rivers. The same argument was used when Ontario created Algonquin National Park in 1893.
Conservation was far more likely to receive support from legislators and
industrialists when it was presented in such pragmatic terms. (After all,
the government received timber dues and therefore had a vested interest
in timber sales.) In Ontario, though, forest fires were the paramount concern, because they consumed such vast areas of mature stands and because the damage was especially severe in the thin-soiled Shield country,
where fire could sterilize the bare rock and regeneration took much
longer. In 1878 the province declared a wide swath of the Lake Superior,
Lake Huron, and Georgian Bay shores a Fire Protection District, wherein
settlers, surveyors, logging parties, and railway companies could be fined
for failing to "exercise and observe every reasonable care and precau-
tion" with fire.[21] Unlike restrictive legislation such as diameter minimums
(which permitted logging only of trees that had reached a certain age
and size, a system which proved largely unenforceable), firefighting was
politically benign because it did not attempt to rein in logging crews.
These measures were directed at mainland stands; the offshore islands
were already protected to some degree from spreading fires, and according to surveyors, grew pines that rarely reached the legal minimum of six-
or eight-inch diameters anyway.

In the twentieth century, the province used its authority as a landlord to assume a more interventionist role in the industry.[22] Ontario established the first provincial forest reserves in 1898, essentially to reserve timber stands on Crown land from homesteaders cutting down the forest cover. (Unfortunately, political intervention occasionally worked at cross-purposes to forest regeneration. That same year, at the urging of the timber lobby, the province enacted legislation that required all pine cut on Crown land be processed in Ontario – which promptly and dramatically revived production in the mills on Georgian Bay.) Reserves were an entirely utilitarian concept; in the words of Ontario's 1929 Provincial Forests Act, their purpose was "to preserve the said forests according to the best forestry practice, and to gradually bring them under a sustained yield basis." Georgian Bay Provincial Forest, a 677-acre reserve stretching from Shawanaga to the French River, was one of three created in 1929, bringing the total number of forest reserves to eight. The reserves may have protected some old-growth stands but more likely allowed second growth to reach a decent size. The Department of Lands and Forests (DLF) also opened pine plantations at Lost Channel and Little Blackstone Lake, but the topography did not encourage rapid or large-scale reforestation, so most of the Department's energy went into keeping forest fires under control. The state had assumed responsibility for managing the forest, an arrangement that obviously benefited the private sector.[23] Like their associates in the fishery, loggers expected the government to guarantee their industry's sustainability by replenishing the supply of resources, not by interfering with their consumption of that supply. Georgian Bay Provincial Forest, however, was fairly short-lived because recreation had already become the dominant industry in Georgian Bay. Recreation was legally permitted in forest reserves anyway, and as early as 1936 the DLF granted a licence of occupation for a tourist lodge in the forest. In 1958 the reserve was opened to sales of "summer resort lands," and the system of provincial forests was abolished entirely six years later.[24] Still, foresters – along with their counterparts in the wildlife and mining branches of the department – continued to voice opposition to park or recreational zoning that excluded industrial activity.

By the 1960s the DLF knew that commercial foresters and their argument for "intelligent utilization" of timber stands faced an uphill battle against public opinion.[25] For over half a century, urban North America had been developing another relationship with wilderness spaces, prizing them for aesthetic and recreational use. Landscapes such as Georgian Bay were recast from a resource hinterland to a play land, a place for recreation rather than work, for public consumption rather than industrial production.[26] Parks usurped resource conservation as the government's primary concern in (or for) the Bay, and popular environmental thought likewise evolved in new directions. By the latter part of the nineteenth century, sportsmen and naturalists were fronting what historians term the preservationist wing of North American environmentalism, arguing for more stringent protection of habitats and critical of the utilitarian and businesslike approach to nature implicit in conservation's wise-use philosophy. Preservationists maintained wilderness landscapes should be protected for their natural beauty, as wildlife habitats, and for their spiritual and recreational value. Despite their differences, conservation and preservation shared an essentially positive view of nature (as opposed to the overt hostility toward wilderness we ascribe to the frontier generation), a realization that nature's resources were not unlimited, and the conviction that nature could serve the needs of modern society, whether as a limited storehouse of resources or a spiritual restorative.[27] Preserving wilderness was as much a product of an industrial society – a projection of its desires, a construction for its needs – as managing its forest stands or fish stocks for harvest. The story of national and provincial parks in Georgian Bay reveals the contradictions of managing nature for human use and the multitude of demands placed on any single landscape.

Today we tend to think of parks as sanctuaries of nature, where fragments of an original landscape are preserved for future generations. Books on Ontario's provincial parks have titles like *Islands of Hope* and *Protected Places*. But parks have always been shaped by a tension between protection and development, and between diverse demands on natural landscapes. Scholars have devoted more and more attention to the cultural construction of wilderness, and the definition and delineation of nature in parks: which landscapes were idealized as "wilderness" or considered worthy of preservation at certain times and places in history, and what purpose these landscapes were meant to serve. William Cronon, for example, argues that the idealization of an unsullied, uninhabited wilderness is an urban fantasy, created by people who have never had to work the land and "whose relation to the land was already alienated."[28] A park is a paradox: a natural space – and in North America, the ideal is a *wilderness* space – which assumes, indeed requires, a human presence, even in its very creation. Well into the twentieth century – at least until the popular environmental movement of the 1970s, and arguably up to the present day – parks have been imagined not as a way of *preserving* nature but of *reserving* it for various uses. Wild nature was set aside, in a phrase typical of nineteenth-century park legislation, for "the benefit, advantage, and enjoyment of the people." National parks such as Yellowstone or Banff were designated because the terrain appeared worthless for industrial purposes, or because it contained striking physical monuments that provided scenic national icons.[29] Parks promised revenue and voter support by catering to the burgeoning industry of outdoor tourism. But this user-oriented approach would encounter opposition from two sources in Georgian Bay. By the 1960s concern for wilderness protection made the public far more critical of traditional park management. As well, conventional development was repeatedly thwarted by the physical obstacles of the archipelago. This is an aspect of park history that has been largely unexplored: how external plans or standard agendas fared in specific landscapes.

Both public officials and cottagers expressed interest in an archipelago park by the time of the First World War. In 1916 six cottagers' associations formed the Georgian Bay Association, in part to address issues of water pollution and protection for fish and game. The association was soon lobbying for parks on Beausoleil and Franklin islands, arguing they would make good wildlife sanctuaries as well as ideal sites for boating, picnicking, and scenic appreciation.[30] The prospect of a Georgian Bay Islands National Park certainly satisfied the pragmatic criteria of park planners. With the national park system concentrated in large western parks such as Banff, Jasper, and Wood Buffalo, James Harkin (the first Dominion Parks commissioner) and the Parks Branch hoped to acquire properties in eastern Canada more accessible to the majority of Canadians. Support for parkland was bolstered by the popular idea of wilderness as Canada's natural heritage and a resource for public benefit; a park in the archipelago would ensure "our own Canadian public of having these islands available for public use forever."[31]

But there is an echo of the old bafflement of explorers and surveyors among government officials, who wondered just what else they could do with such a landscape. Robert Stead, director of publicity in the Department of Immigration and Colonization, suggested to Harkin that "there are some beautiful summering places out in Georgian Bay and as it would seem to me that many of them are of no great value for any other purpose

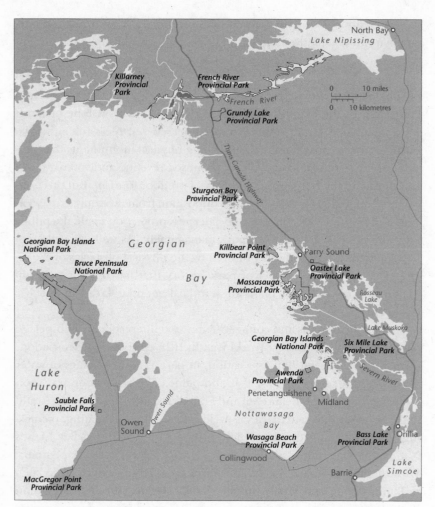

Map 10 National and provincial parks

it might be found practicable to establish a national park in that district."[32] Archaeological sites on Beausoleil dating to the seventeenth-century Jesuit missions at Huronia had already attracted the attention of the chair of the Historic Sites and Monuments Board and the director of the Provincial Museum, who first broached the idea of a park before the First World War.[33] Parks at this time often had an archaeological or historical dimension. The United States' 1906 Antiquities Act allowed for the designation of archaeological sites as federally protected national monuments; in eastern Canada, local historical societies and national learned organizations such as the Royal Society of Canada were lobbying for federal commemoration of forts and battle sites by the turn of the nineteenth century.[34] As a site of "national" historic interest, Beausoleil was worthy of preservation and a promising tourist attraction. Fifty years later, the same idea of a landscape as a "heritage resource" fuelled the campaign to create a park on the French River, a voyageur route whose archaeological sites included pictographs, portages, and underwater canoe spill sites. By 1986 the French was both a provincial waterway park and one of the first Canadian Heritage Rivers.

Beausoleil was reserved for park creation in 1923 and formally purchased by the Dominion Parks Branch from the Department of Indian Affairs in 1929, along with twenty-eight other islands. (Today it is Canada's fourth smallest National Park, with satellite islands up to Manitou.) The Parks Branch then set about developing Georgian Bay Islands National Park for "the enjoyment of the people." Indian Affairs had granted a lease on Beausoleil to the Simcoe YMCA as early as 1920, and camps brought the majority of visitors to the park through to the 1960s. Organized camps were actively encouraged, including YMCAs and YWCAs, Boy Scouts, Lions Clubs, and church groups. Wardens acted as caretakers for camp buildings, and every year constructed additional "improvements" such as pavilions, stoves, shelters, and wharves. By 1960 Beausoleil (an island only seven kilometres long and less than two kilometres wide) boasted thirty-one trails, twenty-two campgrounds, tennis courts, and a playground. Landscape management ranged from the fairly innocuous (laying sod at park headquarters) to overt manipulation: dredging channels, thinning brush, spraying poison ivy, fogging public areas with the pesticide DDT, and planting DDT "bombs" in stagnant ponds.[35] Operating a national park meant making a dangerous landscape safer and more appealing for the general public.

In particular the park staff tried to nurture popular game species. Generally, "scientific management" in Georgian Bay involved the manipulation of individual species rather than care of an ecosystem. Staff planted species of fish popular with fishermen, particularly black bass, in Beausoleil's small lakes. The deer population caused especial concern in the late 1950s and early 1960s when a drop in the numbers of this valuable sport species coincided with the collapse of the commercial fishery. At Killarney, deer yards were maintained to assist herds through the winters, while park staff undertook a massive slaughter on Beausoleil Island in 1951 in an attempt to halt the spread of a parasite and stave off widespread starvation in the herd. (The slaughter appears to have been the result of some crossed wires with Ottawa; the Dominion Wildlife Service planned a live trapping in January, but a month later the chief of the

Parks Branch coolly told the park superintendent to dispose of the herd "by such arrangements ... as you may consider most suitable.")[36]

On the other hand, there were some efforts to maintain the natural state of the park. Superintendents, at least in the early years, resisted calls to develop and promote the park from local communities hungry for tourist dollars and work projects. They instructed park staff to leave rattlesnake populations and shoreline undisturbed; and they urged the Parks Branch to acquire neighbouring islands in order to protect migrating deer, and to hire additional staff to chase away poachers. In 1964 the park began employing a naturalist. More often than not, however, development was moderated less by design than by geography. Since the park consisted entirely of islands, it was more difficult to access and consequently attracted fewer visitors. The earliest attendance count in 1931 reported 2,800 visitors; by 1962 that number had risen to only 19,195 (though annual counts were always approximate because staff had no way of tracking visitors to other islands or even the north end of Beausoleil). According to park officials, such low attendance did not warrant substantial development. By way of comparison, Killbear Provincial Park on the mainland received 43,000 visitors its first year of operation (1960).[37] Those who made the effort to reach the islands tended to be campers who wanted a natural setting anyway. When a local MP proposed a golf course on Beausoleil, the Parks Branch pointed out there was neither space nor demand for it.[38] But even this was a response to the park's users rather than a policy of protection.

At the provincial level, parks are merely one of several uses for Crown land, so the provincial government traditionally has favoured a policy of multiple use. With the competing agendas of its member agencies, responsible for forests, wildlife, mines, and parks, the Department of Lands and Forests – later the Ministry of Natural Resources – has often resembled a forced marriage. Park policy has reflected this internal rivalry. Ontario's first Provincial Parks Act in 1913 stated that Crown land "not suitable for settlement or agriculture" might be set aside "as a public park and forest reserve, fish and game preserve, health resort and pleasure ground, for the benefit, advantage and enjoyment of people, and for the protection of fish, birds, game and fur-bearing animals therein." The act was concerned largely with game laws within park boundaries; it still permitted leases for timber and mining, as well as leases for facilities "for the accommodation of visitors." The first park reserve in Georgian Bay exemplified this idea of a park as "pleasure ground" and wildlife preserve. Franklin Island, along with forty-two nearby islands, was reserved in 1923 to provide a wildlife sanctuary, a site for "excursion parties," and an attractive facade fronting the boat channel. But like Beausoleil, these islands were too inaccessible for a "public park," and the park suffered from benign neglect until the reserve order expired in 1944.[39] Elsewhere, though, things would get worse before they got better, for parks would be nearly exhausted from intense recreational use before the province began moving cautiously in the direction of ecological protection.

Wider prosperity, an automobile culture, and the baby boom propelled a massive increase in outdoor recreation after the Second World War. State, provincial, and federal governments all sought to cater to this new demand for nature as accessible recreational space. After 1954 highway parks mushroomed across Ontario: located convenient to urban centres, designed for car camping, and often with no particular ecological value. Georgian Bay was soon ringed by such parks. Sturgeon Bay Provincial Park – originally Sturgeon Bay Provincial Camp and Picnic Grounds – was one of four parks opened along Highway 69 in the space of two years (1958-60), with Killbear, Six Mile Lake, and Grundy Lake, joined by Oastler Lake in 1964.[40] These were matched by the Chutes, La Cloche, and Fairbank along Highway 17 west of Sudbury. Meanwhile, the DLF, swamped by sales of recreational properties, began to think about reserving land from sale in order to guarantee public access to wilderness. In 1962 the province reserved 25 percent of the lands fronting any body of water "for recreational and access purposes," and the next year all remaining Crown land fronting the Great Lakes was withdrawn from sale.[41]

That year the province also targeted Georgian Bay for a planning experiment on an unprecedented scale when it amalgamated the old Georgian Bay Provincial Forest and the newly expanded Killarney Provincial Park into the North Georgian Bay Recreational Reserve. It seemed an ideal solution to the booming demand for recreational wilderness: a large amount of Crown land already in public hands; near Toronto, London, Hamilton, Ottawa, and Sudbury, as well as cities in the Midwest states; a picturesque landscape of water and pine, exactly the sort that would appeal to canoeists and campers; and an environment thought to have scant capabilities for other uses. The reserve's official plan brought a huge area of shoreline (a million hectares) under zoning and insisted that "environmental quality was the overriding factor in all decisions and recommendations concerning the development of the Reserve." Still, the reserve *was* meant to be developed, to make recreation the dominant land use.[42] In fact, the development proposed by the reserve's plan was never really carried out, discouraged by the same uncooperative landscape that had confounded settlers a century earlier. The decision to leave this portion of the shore as Crown land was probably the best thing the province could have done to minimize human use of the area.

Parks, on the other hand, *did* develop recreation as an industry, and introduced an era of unprecedented pressures on Georgian Bay. By the mid-1960s, parks were under siege, victims of their own popularity. Wasaga Beach Provincial Park on Nottawasaga Bay became the consummate example of the overcrowded, car-camping playlot, and exemplified the weaknesses of postwar park planning. Road-access parks with extensive visitor facilities encouraged this trend toward intense and prolonged use. If you build a park, people will come; it makes nature an attraction, and funnels people into the environment. People had also been culturally primed to think of Georgian Bay specifically as a desirable destination. A landscape idealized (in art, for example) as a resilient fragment of primeval wilderness was enormously attractive. Killbear Provincial Park, as mentioned above, attracted 43,000 visitors its first year, and 212,456 five years later.[43] A park also entails a certain level of development for the "accommodation of visitors." By this point even the Thirty Thousand Islands could no longer expect the laissez-faire treatment that Beausoleil and Franklin had enjoyed in the 1920s and 1930s. In 1977 one resident pleaded that access to the proposed Blackstone Harbour Park "be made reasonably difficult so that only those seriously interested in experiencing the benefits of the natural environment will use the facilities."[44] Such is the fundamental irony of twentieth-century park policy: designation as a park tended to be hazardous to a place's ecological health. Given the fragility of the Georgian Bay ecosystem, with its shallow soils, exposed vegetation, and interconnected bodies of water, this pack-'em-in approach was especially dangerous. Residents – particularly summer cottagers – became increasingly concerned about the heightened activity and volume of boat traffic that parks brought into the archipelago. Parks, they feared, were eroding the Bay's wild character.

Enthusiasm for parkland and concern about environmental wear and tear both stemmed, ironically, from a new interest in the environment. The culture of youth camping and outdoor recreation had long promoted the idea that nature study and "woodcraft" engendered a sense of environmental responsibility toward a particular place. Hurontario campers, for example, participated in a range of wildlife studies, banding birds, and tagging threatened species such as monarch butterflies, spotted turtles, and massasauga rattlesnakes, in conjunction with universities, the MNR, and Parks Canada.[45] But the 1960s saw new public and political attention devoted to what has been called the second generation of environmental issues: air and water pollution, nuclear and toxic waste, energy resources, and biodiversity. These were problems on a much larger scale, potentially far more destructive; and they coincided with a new era of social activism. Whereas the language of conservation borrowed industrial objectives and terminology (production, efficiency, sustainability), the new environmentalism characterized a healthy environment as a quality of life issue. Not surprisingly, it won a broader and more deeply rooted base of support. Canada and the United States responded with unprecedented legislative frameworks, including the formation of national environmental agencies (the Department of the Environment and the Environmental Protection Agency) and comprehensive anti-pollution legislation (such as the Canada Water Act and the US Clean Air Act). Park planning was also becoming more bureaucratic and subject to a much wider range of government agencies and environmental NGOs (nongovernmental organizations) such as the Canadian Parks and Wilderness Society and the Algonquin Wildlands League. Energized by the political climate of the 1960s, environmentalists adopted a more aggressive political persona and used the shift in public sentiment to pressure governments into devising more explicitly protectionist designations.[46]

Ontario was moving tentatively in this direction in small areas of Georgian Bay. As early as 1945 the province withdrew from sale all islands under a third of an acre in size, "to preserve natural scenic beauty and to eliminate possible water contamination."[47] Amendments to the Public Lands Act in 1960 enabled the minister to designate restricted areas, closed to disposition or development. In 1973, at the urging of cottagers' associations, the province used the act to impose an Area Control Order on the archipelago. Two years later the shore was designated the Eastern Georgian Bay Interim Development Control Area, freezing development on the islands and mainland to 305 metres inland. Another legislative tool was the Wilderness Areas Act of 1959, which allowed for "the preservation of the area as nearly as may be in its natural state." Within six years small wilderness areas were established at Crater Lake (Killarney), Eighteen Mile Island, and McCrae Lake (1960), Whitefish Lake and Blair Township (1964), and Gibson River (1965).[48]

Nature reserves were a common choice in Georgian Bay because a landscape that resisted extensive development left the DLF with relatively few options. Not only did the archipelago appear conveniently unsettled and undisturbed (and therefore legitimate objects of preservation), it was thought to have "practically no capacity for other uses," as a district forester said of the French River. McDonald Bay, McCrae Lake (the western outlets of Six Mile Lake), and Go Home Lake were proposed as wilderness areas as early as 1958 because they were considered too rugged and too remote to support the standard amenities of a park. In fact, the term "nature reserve" could mean either a site under study for selection as a wilderness area or a category of park involving no or minimal development. Reserves were declared up and down the shore, at McDonald Bay, O'Donnell Point, Round Lake, South Bay, Bigwind Lake, Shawanaga Basin, the Limestones, and Moon Island. Some reserves, such as the remote Limestone Islands, were selected for explicitly ecological and aesthetic reasons. But old habits died hard. The proposal to reserve Moon Island in 1968 was matched with one for a road and bridge to make the island accessible by car. Luckily, the DLF concluded the island had too little "development potential" to warrant a road, but reserved Moon anyway.[49]

As the Moon Island case suggests, park management in the 1960s and 1970s was less a green revolution than a dogged struggle between two schools of thought. The US National Park Service, for example, was wedded to its ten-year Mission 66 development program, which sought to improve park services for the rising number of visitors during the postwar period, and had to be prodded toward more ecological practices by outside critics such as wildlife biologists.[50] In Georgian Bay, the challenge to the idea of parks as service centres was first made in Killarney Provincial Park. The dramatic scenery of the quartzite La Cloche hills had long invited an aesthetic appreciation of nature; early supporters of a park included the Federation of Ontario Naturalists and the Ontario Society of Artists, notably members A.Y. Jackson and Franklin Carmichael. Yet, in the early 1960s, the DLF entertained proposals for, among other things, a ski hill and golf course in the newly enlarged park. As Roderick Nash has argued, this "access-resort philosophy" indicates "a confusion of purpose" by injecting inappropriate types of recreation into wilderness areas.[51] In 1967 the government attempted to resolve this confusion by revising the process of formulating park policy and introducing steps for public consultation, intra-park zoning, and park classification. With Algonquin and Quetico, Killarney became a focal point of the battle over the meaning of wilderness status in Ontario's park system. Its reclassification in 1971 as the province's second "primitive" class park has been counted as a victory for the new confrontational techniques of the media-savvy preservation lobby.[52] But the pro-development faction had already encountered an effective opponent in the rugged La Cloche terrain. Feasibility studies as

early as 1963 concluded that such projects were unworkable and un-
advisable.[53] The "achievement" of wilderness status was also undermined
by the fact that the wilderness of park policy remained a well-managed
one. One delegate at a 1968 conference on national parks defined wilderness purely in terms of planning categories: "A wilderness is a designated area where for specified reasons development control permits only
those activities that are compatible to the specific reasons for designation."[54] As a wilderness park, Killarney was still subject to human use, simply a particular type. Announcing the park's reclassification, Ontario
premier William Davis declared that "we now have a 140-square-mile area
in a wilderness setting easily accessible to the majority of the people of
Ontario," and urged MPPs to visit it.[55] Zoning and classification, while
allowing for more site-specific management, were actually predicated on
the old concept of multiple use.

But now, at least, ecological integrity was beginning to be acknowledged
in park planning and management. The story of Massasauga Provincial
Park encapsulates the evolution of park planning over the past century,
and exposes the mechanics of negotiation involved in the creation of
parks and in environmental politics generally. In 1973 the MNR announced that a new park would be established in the Thirty Thousand
Islands about ten kilometres south of Parry Sound. The park, to be called
Blackstone Harbour, would encompass 6,070 hectares south to and including the Moon Island Reserve. The original rationale for the park was
a classic example of the parks-are-for-users philosophy: officials wanted
to deflect campers from other parks in the near north already operating
"at capacity," from as far east as Algonquin. From the outset, the ministry
laboured under the delusion that thanks to the magic of zoning, objectives as diverse as shoreline preservation, wilderness harbours, and car
camping could coexist in a single park.[56] Such is the essential problem of
multiple use, in that it tries to make nature all things to all people. But
the proposed park met with a surprising amount of resistance, both human and natural. Public opinion, scientific research, and physical limitations all dramatically modified the park in its formative stages.

Support for environmental issues tends to develop sporadically. People
are most likely to mobilize around either distinct natural features or familiar local landscapes where "the environment" is not abstract. Communities around the upper Great Lakes led the campaign to improve water
quality in the 1970s, for by this point the pollution in the Lakes clearly
affected their everyday lives.[57] The political tenor in the Thirty Thousand
Islands, however, is dominated by the voice of cottagers. Their vision for

Georgian Bay – how and why they wanted the landscape "protected" – was shaped by the new awareness of environmental issues, certainly, and intense feelings of place attachment; but it was also an idealized view of a recreational wilderness that reflected their urban, upper-class origins. The influence of a well-organized and socially powerful cottager population intent on protecting its island paradise was evident in the Blackstone area well before the creation of the park reserve. Cowper Township, south of Parry Sound, had been the site of copper operations at the turn of the nineteenth century. The Division of Mines refused to prohibit staking north of Moon Island, and only two-thirds of the area reserved for Blackstone Harbour under the Public Lands Act was also reserved from staking under the Mining Act.[58] With the spectre of renewed exploration in the late 1950s, a group of local landowners from Sans Souci decided to take action. Calling themselves the Cowper Conservation Authority (CCA), they purchased a large section of shoreline along Spider Bay, as well as its mining licences of occupation. In 1974 the CCA turned twenty-two lots in two concessions over to the province on the understanding that the land would be managed as wilderness, with no roads or campgrounds, for activities such as canoeing, hiking, and "casual environmental appreciation."[59] The proviso was necessary because the Provincial Parks Act did not preclude mineral exploration in existing claims or licences of occupation. Other branches of the MNR, however, did not appreciate the appropriation of such a large parcel of Crown land for the limited uses of a park. The Division of Mines was particularly persistent, arguing that the mineral potential was too substantial to ignore and challenging the legality of the CCA agreement.[60] Strong-armed in its own department and squeezed by budget constraints, the Parks Branch held on tenaciously, citing a "moral obligation," pressure "from influential 'Bay' interest groups," and a "commitment by the Minister to a community whose enmity he would not wish to invite." By the mid-1970s rising real estate prices were forcing the province to court private donations of this sort for parkland. The CCA deal represented a high-profile test case in land acquisition.[61]

Public participation shaped more than the park's boundaries: it helped reorient the park's raison d'être. Unregulated camping on both Crown and private land was certainly a problem; in 1969 the Parks Branch had to extend Killbear 600 feet into Georgian Bay and acquire three nearby islands in order to give park officials some authority over rowdy campers.[62] But many residents – permanent and seasonal alike – near the proposed Blackstone feared the damage that a provincial park could inflict more than the status quo. Their fervent opposition indicates just how suspect the designation of "park" had become by the 1970s. An opinion survey reported that residents feared "any such development will necessarily result in the ultimate destruction of everything that is valuable about the character of the area as it now exists." As one respondent from Crane Lake explained, "The attachment of the word 'park' to the area, no matter what category, will advertise, in most people's minds, that the area is a fully-serviced playground to be exploited." Residents insisted that any park or reserve protect what they saw as the area's wilderness character.[63] By the time budget cuts conveniently imposed a planning freeze in 1977, the MNR was already considering "a complete re-orientation of park objectives to cater to lower intensity uses and relatively experienced user groups." The next year, the deputy minister announced that 95 percent of the park would remain "natural, undeveloped and inaccessible" except by boat or on foot, to "preserve the environment values that have been ensured on this section of Georgian Bay by the past vigilance of permanent and seasonal residents in the area."[64] Meanwhile, the nascent French River Provincial Park was going through the same process. Residents objected to the idea of a waterway-class park on the grounds that it would attract users to the detriment of the river itself ("waterway park" does sound disconcertingly like a theme ride). Here too, the MNR backed off. It downsized the development plans and promised to not promote the park.[65]

By insisting that the parks recognize and protect the Bay's unique character, residents managed to some degree to adapt standard environmental policy to their locality. Was this their primary goal? Their defence of the place had more than a tinge of self-interest; their opposition to the park reflected the fear that it would "advertise" the area to such an extent that their prized island retreats would be overrun by city rabble. Repeated characterization of the archipelago as wilderness by its own users seems somewhat naïve, even disingenuous, and factually inaccurate. However, records indicate that the Parks Branch was becoming aware of the ecological value and limits of the archipelago. Proposals for 1,200 to 1,500 campsites were quickly scrapped,[66] and ultimately the park had only 135 water-access sites and seven mooring bays. The dramatically scaled-down version that prevailed was partly the result of budget cuts, but it was also an admission that extensive campground development was frankly impractical (if not impossible) in the convoluted landscape of inland lakes.

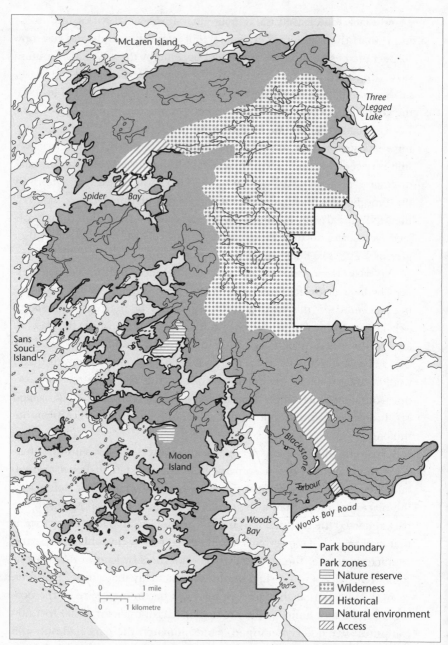

Map 11 Park zones in Massasauga Provincial Park. Reproduced from Ministry of Natural Resources, *The Massasauga Provincial Park Management Plan* (Toronto: Queen's Printer, 1993). (© Queen's Printer for Ontario, reproduced with permission.)

Blackstone's makeover reflected the infrastructure of policy and legislation that had been pieced together over its lifetime, making environmental assessment an obligatory and principal factor in park planning. The National Parks Act was amended in 1988 to make ecological integrity the priority consideration in park planning, and parks such as Georgian Bay Islands and Bruce Peninsula developed "ecosystem conservation plans" to assess the parks' health as part of a larger ecosystem.[67] A series of life science inventories in the late 1970s and early 1980s identified dozens of provincially significant species in the Blackstone reserve, as well as the habitats of several endangered species, notably the spotted turtle and the eastern massasauga rattlesnake. They also called attention to important geological formations and wetlands such as sphagnum bogs. When finally completed in 1993, the Management Plan for what was now The Massasauga Provincial Park declared that protection was "the highest priority among Park objectives, governing all Park planning, use and management."[68] Things had changed quite a bit in twenty years.

But older ideas about the nature of parks and park management have persisted. In 1999 the new master plan for Ontario's Crown land proposed a chain of expanded parks and a record number of conservation reserves between Gibson River and Killarney. In addition, it established a Great Lakes Heritage Coast from Severn Sound to the western end of Lake Superior. The inclusion of the archipelago, however, was partly the result of an old, familiar sentiment. As the executive director of the Georgian Bay Association remembered it, "the large lumbering companies and mining companies during the Lands for Life process were very quick to say, 'Sure, put your coastal thing in place, because we don't have an interest in it.'"[69] The Living Legacy plan also retained the tradition of multiple use: new park and conservation areas permit mineral exploration, commercial fishing and fur harvesting, and sport hunting and fishing.[70] And the recent privatization of provincial parks is a new spin on the old concept of making money from parks. The provincial government began contracting small parks out to the private sector in 1976; Sturgeon Bay was one of the first to be so managed.[71] In 1996 the government transferred responsibility for park management to an "entrepreneurial operating model" known as Ontario Parks. Now, parks resemble franchises. Park offices have sprouted gift shops; Massasauga markets "vernware," paraphernalia decorated with the park mascot, a cartoon rattlesnake named Vern.

Often, though, Vern is the least visible player on the scene. Environmental politics is essentially the "politics of turf," with different groups vying for access to a landscape. Conflicts over specific issues – such as development, zoning, and park creation – are localized expressions of a much larger struggle between competing agendas as to how people want nature in a particular place defined and managed. In Georgian Bay these conflicts are played out at both the community level and in a complex clutter of governmental jurisdictions. One of the greatest challenges facing archipelago municipalities is balancing the interests of permanent residents, who constitute less than 10 percent of the population, with those of the overwhelming seasonal majority. The two groups need and want different things from the Bay; as Katherine Govier has written, cottagers want to save the Bay, but locals just want to live there.[72] If cottagers lobby for strict zoning codes or limits on certain types of land use, in order to "protect" the environment, this restricts the options of year-round residents who need to earn a livelihood. Political involvement is framed by the same issues of class privilege that underwrite land ownership in the archipelago. The proposal to develop a ski hill in Killarney, for example, was supported by local developers because it promised spin-offs for the area's economy, an economy heavily dependent on outdoor recreation (and urban patrons) for close to a century. Year-round residents also have unique practical concerns which summer cottagers simply need not consider. For decades permanent residents of Sans Souci have used a snowmobile trail that runs through Spider Lake. This allows them to reach Parry Sound while avoiding the fast-flowing South Channel, which does not freeze as securely and is therefore much more dangerous. But Spider Lake is now part of The Massasauga Provincial Park's interior wilderness zone, a classification that prohibits motorized access. Gary Higgins, the park superintendent, explains his predicament: "To tell them, 'I'm sorry, this is a wilderness zone, put your float coats on and good luck on the South Channel,' isn't a very neighbourly thing to do." Yet cottagers were not pleased when, in 1994, the Shawanaga First Nation closed a road across its reserve to the public, forcing residents to find an alternative means of accessing their homes at Skerryvore.[73] These localized issues play into much larger battles over environmental authority, or who has rights to access and make decisions about the use of a place. The impression of an external constituency managing Georgian Bay from

Toronto echoes the longstanding struggle between a rural north and an urban south over control of Ontario's resources. Tourism is itself a "process of extraction" that perpetuates the characterization of northern Ontario as a hinterland for the south. Incidentally, one aim of the governmental reviews of municipal governments in the 1960s and 1970s was to provide the near north with a greater degree of autonomy ("home rule," as the MPP for Muskoka called it).[74] However, just who constitutes the "local" population is not entirely clear – the cottagers who carry the tax burden or the people who live there year-round?

If the local/seasonal split is seen as a contest between market and non-market (that is, aesthetic) values, it undermines the credibility of environmentalists who can appear romantic, impractical, and even selfish. Recreationists and preservationists are particularly vulnerable to charges of elitism. Cottage country environmentalism in the 1960s and 1970s was primarily a defensive reaction against accelerated subdivision development and overcrowding – appropriately referred to as "muskokafication." Limiting access to natural spaces *is* necessary to protect them from overuse, but cottagers have been accused of using the environment as an excuse to keep other people out of their recreational preserve. If cottagers argue that the landscape can't support greater numbers, it is easy to see their environmentalism as an attempt to maintain their community's privileged *social* exclusivity.[75] The owner of Camp Hurontario, Birnie Hodgetts, persuaded the MNR to establish a fish sanctuary adjacent to the camp; but apparently Hodgetts, an avid fisherman, was more interested in keeping other fishers out than in keeping any fish in. Former camp and MNR staffs admit the issue was manufactured largely to protect the camp's privacy. (Hodgetts also blocked proposals to blast Loon Portage into a boat channel, which would have substantially increased traffic through the camp bay.) More recently, the Georgian Bay Association (GBA) sought to prevent a marina operator in Wood's Bay from towing floating cottages into Massasauga; this kind of use contravenes the intention of "a wilderness park," but it also raises questions about rights to recreational anchorage and user density.[76] A park that is accessible only by boat may be better for the environment, but it also protects the people fortunate enough to already live there. The possessively protective character of such environmentalism may be increasingly effective given the trend toward private stewardship arrangements as a means of designating protected areas. Organizations such as the Georgian Bay Land Trust acquire "significant properties" of cultural, environmental, or historic value through donations and conservation easements (to date it has acquired five such properties). The GBA and Georgian Bay Land Trust, as well as non-local organizations such as the Nature Conservancy, are fortunate in that they can draw on a devoted, well-connected, and financially comfortable constituency as their base of support. Such support is easy to generate when they are "selling" a beautiful natural landscape with great sentimental value and symbolic potential for Central Canada.

Whatever their motives, their devotion to the Bay has made cottage communities in the archipelago prominent political players. They are vital sources of support for environmental research and heavily involved in publicizing environmental concerns, to a degree unusual even among cottagers associations. Recent interests of the GBA range from monitoring aquaculture operations and water quality to presentations to the International Joint Commission on water diversion.[77] Seasonal and permanent residents are not always at loggerheads over what is best for the Bay. Since the 1960s, planners have reported that seasonal and permanent residents have more in common than we might think. Regional studies and government documents repeatedly comment on the degree of grassroots participation on both sides and the overriding consensus as to the need to protect the environment. "There is a strong, and often passionate conviction in the region of the vital importance of the landscape, and of the environment itself, as a factor over-riding all considerations of development," concluded one report in 1969. "This concern for the quality of the environment is widespread and deeply felt." Even municipalities suffering from out-migration and economic decline preferred to maintain a "green" setting in the hope of attracting tourism rather than courting heavy industry. Outfitters have long been concerned with clear cutting, water pollution, and overfishing; as one outfitter pointed out, "the camp operator has to be a conservationist to stay in business."[78] It is not a matter of pitting livelihood against affection, or privileging one type of place attachment over another. Artist John Hartman lives on the Penetang Peninsula, summers at Norgate Inlet, and has wintered at Manitou. He offers this reflection:

> It's a slightly different culture, the two things, but the common thing that runs through it all is this very strong attachment to a place, to a sense of place that people have. Georgian Bay people are very – ferociously – protective about the Bay. They really don't want it to get screwed up, they don't want it to change, they feel very strongly about this. And that's true for the people who live full time on the Bay and also for the summer residents.[79]

Attributing the preservationist impulse to only middle-class cottagers ignores this convergence of interests; neither group wants their home to be "screwed up." But it also distracts us from the silent partner in all this: the carrying capacity of the ecosystem. Measures designed to limit human pressure on the Bay must be judged by how well they serve *this* constituency.

A basic problem is that it belongs to so many different political jurisdictions. This is the quandary that lies at the heart of environmental politics, which by definition marries a political dynamic of fixed domains and institutional control with an ecological dynamic of mobile, uncertain, biophysical elements. The political landscape and the natural landscape can be two very different things.[80] The Townships of Georgian Bay and the Archipelago were meant to resolve this discrepancy, for as shoreline municipalities they would be able to formulate site-specific environmental guidelines and represent the archipelago in regional planning. Even at the municipal level, though, it was difficult to establish clear lines of communication and cooperation between regional and local governance. The Muskoka and Parry Sound government reviews recognized the need for district-level government to oversee standardized planning and pollution control, a particularly vital concern in a region with so many interconnecting bodies of water. But the historic sense of difference felt by communities "up the shore" was bolstered by the neighbourhood movement of the 1960s, with its anti-bureaucratic and anti-consolidation feeling. Both reviews formally recognized the physical differences within each district and the "sense of community identity" and "community acceptability" these differences produced. In Georgian Bay Township, cottagers were unwilling to cast their lot with either the Muskoka Lakes or the town of MacTier because they feared the specific problems of the islands would be overlooked.[81] The reviews also confirmed the established position of the GBA as the archipelago's unofficial government when much of the shoreline had remained in unorganized townships. (Negotiating with the GBA, though, underlined another difference between permanent residents and cottagers, who had formed their own associations based largely on shared social interests, and who were often absent from fall municipal elections. One survey found 74 percent of residents placed "high confidence" in their local association, compared with 39 percent who had such confidence in township government and 13 percent in the Ministry of Natural Resources.)[82] While Georgian Bay would benefit from a single bioregional government along a shoreline corridor, an archipelago with scattered colonies will always be prone to fragmentation. Consequently, the GBA remains the closest thing there is to a regional authority in the archipelago: with twenty-four member associations running the full length of the northeast shore, it is the most reasonable political representation of a decentralized landscape. Currently it is also campaigning to have the archipelago, under the name of the Littoral, declared a United Nations biosphere reserve, though it is not clear how such a designation would affect the practical governance of day-to-day issues.

The confusion of NGOs, governmental agencies, municipalities, cottagers, and permanent residents makes the political landscape as complex as the natural landscape of the archipelago. But the problems of negotiating between local interests pale next to the complexities of policy making in a federal system and in international waters. In 1867 the British North America Act split what we today think of as "the environment" rather messily in two. Canada's constitution hands the provinces rights to natural resources, "all Lands, Mines, Minerals, and Royalties" and the "management and sale of the Public Lands"; to the federal government it gives powers over trade, fisheries, navigation, and Indian affairs. It takes a place like Georgian Bay – an inland marine environment, adjoining international boundary waters, with an archipelago where land and water intersect – to expose the full potential for friction and deadlock in such an arrangement. Technically, the province enjoys a more comprehensive jurisdiction thanks to its authority over public lands. But since 1970, amid growing concern over transboundary issues such as air and water pollution, Ottawa has asserted itself more forcefully in environmental politics, with the creation of a federal Department of the Environment and more expansive federal legislation, such as the Canada Water Act.[83] Competitive federalism, though, is not a recent phenomenon. "Environmental" issues have long been used by Ottawa and the provinces to gain political leverage. The fight for control over natural resources powered the provincial rights movement after Confederation. Ontario successfully challenged the federal government in a series of legal battles; one decision, for example, awarded the province licensing rights in the fishery. The provincial government also tried to retaliate against an American tariff by imposing its own restrictions on the export of lumber, even though international trade is technically a federal jurisdiction.[84] Ottawa and Queen's Park each tried to manoeuvre the Georgian Bay Ship Canal project into their respective jurisdictions. The federal government, seeking to exercise its authority over navigable waters, claimed it was first and foremost a canal; as a hydroelectric development (which it was by the 1910s), Ontario considered it a provincial responsibility.

Ottawa and the province also butted heads over Indian lands. This naturally chaotic, insufficiently mapped landscape had been parcelled out in several treaties over the course of the nineteenth century, and this guaranteed confusion and controversy. With "23,000 islands," it was difficult to know which islands were covered under which treaty. In 1876 Moose Deer Point was agreed on as the dividing line between federal and provincial jurisdictions, because it marked the northern limit of the territory surrendered in 1856 by the bands of Huron, Simcoe, and Couchiching (the corollary to the original surrender of Lake Huron's north shore, the 1850 Robinson-Huron Treaty). The islands south of Moose Deer Point, then, were managed by the Department of Indian Affairs (DIA); those to the north by the provincial Department of Lands and Forests. In reality, things were rarely so straightforward. After the partial surrender of Manitoulin in 1862, the federal government claimed this new treaty included islands "north of Collins Inlet and contiguous to Manitoulin Island," thereby giving it jurisdiction in the North Channel. Meanwhile, the province continued to insist that the 1850 treaty covered *all* the islands in "Lake Huron" and therefore they all belonged to its jurisdiction. When the Department of Indian Affairs issued a tender for island sales south of Moose Deer Point in 1897, the province claimed ownership of the islands and the right to their disposal, forcing the DIA to withdraw the tender until an agreement was reached in 1905 reaffirming Moose Deer Point as the dividing line. Unfortunately, the First Nations sometimes found themselves caught in the crossfire. The Wahta Mohawks (Gibson) land claim is a perfect example of how these tripartite negotiations can go awry. In 1884 the DIA purchased Crown land from the province of Ontario in the District of Muskoka, to create a reserve for part of the Oka Mohawk band from Lake of Two Mountains near Montreal. But in 1918 the province withdrew about 40 percent of the designated territory from the reserve agreement. The band launched a land claim in 1981, which entailed negotiating first with Ottawa and then with Queen's Park as well. In 1998 the provincial and federal governments agreed to extend the reserve by another 8,300 acres along with a cash settlement.[85]

A landscape like an archipelago makes it painfully obvious how overlapping jurisdictions and political rivalries hinder joint management. "Floating cottages" registered as vessels under the federal Canada Shipping Act were not bound to municipal building codes. Cleaning an oil spill simultaneously involves at least two different governmental agencies, and very likely agencies from different levels of government altogether. Then there is the relationship between the Crown and private landowners to consider. In 2000, Parks Canada noted its concern with habitat fragmentation in Georgian Bay Islands National Park, where the physical fragmentation of islands is compounded by the fact that property owned by the park is scattered between parcels of privately owned land. Wasaga Beach is vulnerable to storm damage, shoreline deposition, and erosion, but closing the beach or adopting a "loss-bearing" approach (as opposed to structural measures) is not popular with local landowners.[86] The need to respect other jurisdictions, rightly or wrongly, inhibits the formulation and enforcement of effective environmental regulation. Ottawa refused to adopt Ontario's 1945 policy of reserving small islands, saying the DIA had "no jurisdiction to deal with the islands in any way other than to sell them."[87] Recently the federal-provincial tango has hampered a coordinated policy on bulk water diversion in the Great Lakes. At the same time, the unusual situation sometimes forces parties to work together. Georgian Bay has pushed the envelope of park policy in Canada with a large number of parks created in substantially different marine environments and designated under both provincial and federal legislation, from the French River to Fathom Five National Marine Conservation Area, which was transferred to federal jurisdiction in 1987.

As the sixth Great Lake, Georgian Bay is also caught in a complicated transnational tangle. In the 1890s Canadian fishermen protested that closed seasons would be "giving the advantage to the enemy [a]s Georgian Bay is international waters, or the same thing." They feared that American fishermen would "take advantage of the restrictions laid upon us and fish all they like."[88] A century later, acidic or even "dead" lakes around Sudbury came to symbolize the damage wrought by acid rain deposition. Fish and pollutants pay no attention to political boundaries; but for eight states, one province, and two federal governments, the Great Lakes shoreline is a vital segment of their borders. The Lakes have been the target of some pioneering efforts in binational governance, from the Boundary Waters Treaty of 1909 and the International Joint Commission, to the Great Lakes Fishery Commission of 1955 and the Great Lakes Water Quality Agreement of 1972. But the chief obstacle to effective governance continues to be the sheer number of players involved, each of whom lacks the resources to effect change independently and yet is reluctant to concede any authority to anyone else. The result is dramatically uneven standards.[89] That another, supranational designation like a biosphere reserve would resolve the fundamental problem of jurisdictional fragmentation seems unlikely. But the experience of the Lakes has broadened our policy horizon. It is now a basic principle of environmental management that any single region must consider itself part of a larger whole.

This is true in a historical sense as well. Addressing environmental concerns requires an understanding of how a place has been used and managed in the past. When acid rain exacerbated the natural acidity of the exposed Shield around Sudbury, it drew new attention to the fact that the hills had been scoured by massive pollution from mining and smelting operations for much of this century. Environmental problems and remedies are both historical artifacts. They are artifacts of land use and industrial activity, of contests for political leverage, and of changing perceptions of nature. Designations such as parks and reserves are a way of imagining nature in political terms, or more to the point, trying to legislate cooperation from the natural world. A place like Georgian Bay, with a history littered with provincial forests, fishery reserves, parks, and a multitude of other rules, regulations, and zoning categories, reminds us that nature is not "preserved" by human planning, for these measures are temporary and localized, targeted to specific areas and for specific purposes. As well, it brings home the extent to which we have been involved in a particular ecosystem, our cumulative impact on it. The ways in which Georgian Bay has been imagined, managed, and redrawn have reflected the trajectory of environmental thought in North America over the past two hundred years. Rational, utilitarian conservation dominated political and popular discourse because it promised to accommodate multiple uses and replenish key resource stocks. Eventually, the preservationist wing elbowed its way into decision making as societal values shifted toward ecology and a preference for recreational wilderness. Ironically, this shift brought ever-increasing pressures on the localized areas – like parks, like the archipelago – constructed as "wilderness" in policy and in imagination. Georgian Bay has weathered intense use because it has borne the weight of numerous competing demands on nature.

But it has forced us to re-examine our expectations of environmental management. This kind of case study allows us to see how ideas formulated in changing historical climates actually fared in a particular place. The broad strokes of ideology are important in environmental history; the finer details of place (in both human and physical terms) equally so. Bureaucrats discovered that it was not easy to impose measures on resident communities or on a particularly intransigent landscape. Plans were abandoned or modified when they underestimated local opposition or the physical challenges of managing (or interfering with) such a complex and resilient environment. Residents (cottagers were the most influential) sought to have their sense of place – and their sense of *the* place – recognized in a political arena. Georgian Bay has imposed its own limits on our managerial ambitions, whether in terms of development or protection. (Indeed, the inaccessible, tempestuous, uncooperative nature of the Bay protected whitefish and white pine more effectively than man-made reserves or parks.) Its history reveals the enormous burden we place on environmental policy: to mediate between the demands of society and the constraints of geography; between local communities and larger political networks; and between ideas of protection and use. "What needs to be done?" asked Eric McIntyre of the Parry Sound MNR. "We need fewer people experiencing its wonders."[90] Yet the Bay is a wondrous place, physically beautiful, prominent in Canadian history, so it does not seem fair to deny people a right or opportunity to experience it. And how can we appreciate nature when it is alien or abstract? As one writer has asked, "How can you fight for something you have never known?"[91] Perhaps this is the role of the historian: to tell the story of such places, to explain what people have sought – and continue to seek – in different surroundings, and what structures our valuation of wilderness. Our contribution to the public sphere is our assessment of natural heritage. Showing why a place has been important to a community, or a nation, implicitly presents a case for its protection.

Listening to the Bay

You are obliged to pay attention, and talk and listen,
because it's powerful and you are small in it ...
You can't, no matter where you are, ignore the Bay when you're there ...
you know you're in a place where you need to pay attention,
because if you don't there will be some consequences.

– Peter O'Brian (1997), Nares Inlet resident and yachter

I have never been able to ignore the Bay. This may seem strange if only because I don't live there, and never really have; I lived in the Bay only for a handful of summers in my early childhood. I don't know it as intimately as others do, with each shoal or channel learned from years of experience. Yet, my feeling of place attachment is curiously real. I measure other landscapes against the place where I first became aware of the natural world. I respond to sensual cues I absorbed years ago from the islands: the smell of pine needles baking in the sun, of creosote from old dock planking or gasoline from a marina. I miss the feel of warm rock, the intense salmon pink of the granite, the sound of water, and the feeling of open space. Not surprisingly, I couldn't ignore the Bay as a scholar, either. Like many other historians, I wanted to write about something I knew, something close to home. I have never thought this a disadvantage – just the opposite. It was simply an added benefit to find and piece together for the first time such interesting material: coolly worded accounts by Henry Bayfield, faded field notes by surveyors, a love letter from A.Y. Jackson written by campfire.

My original aim in writing this book was to draw attention to a region I felt had been largely overlooked in Canadian history (on its own merit, and not just part of an Ontario-authored national story). As I delved into the academic literature on place, nature, and culture, however, I realized there were many questions in landscape and environmental history that should be applied to the Bay. What started out as an interest in place attachment – how people became "addicted to the wilderness"[1] – and the high-profile cultural exports of a particular place matured into a more complex undertaking (which included pointing out the problem with characterizing the Bay as "wilderness"). Writing amid a great deal of literature on the construction of landscape, I thought the history of Georgian Bay made an important statement about the role the physical environment plays in shaping human history, even as humans reinterpret that environment into different "landscapes." Cultural artifacts produced in response to the archipelago drew me into the relationship between landscape and regional culture, where a distinct and difficult environment became the source and symbol of a distinct regional experience. This is a particularly significant issue in Canada, where regional identities exist alongside – and, in the case of Georgian Bay, contribute to – national myths. A place study also adds to our understanding of environmental history, by exploring how ideas of nature (ideas about its uses, value, and meaning) played out on the ground in one particular place. Though subject to the demands typically made of wild nature, the archipelago frequently challenged these expectations, suggesting that environmental history can vary locally almost as much as the environment itself.

How was I to write such a history? It constantly moves between historical or cultural circumstances and environmental conditions, none of which is ever stable or uniform. There is no singular teleology of change, yet no aspect of the Bay remained unaffected. Over the past 400 years, the Bay has been occupied by different groups using it in different ways. Hurons and Algonquins, Champlain and Father Brébeuf, Black Pete Campbell and Sandy Gray, Tom Thomson and Arthur Lismer – all express the intellectual and cultural environment of their day. Each represents and evaluates nature in Georgian Bay accordingly, as valuable or desirable or avoidable. Their stories follow a basic arc of changing use, from resource industries to recreation, with the differences in values, technology, and settlement. But none of these users are able to ignore the Bay. The physical environment has been as important a factor in Georgian Bay history as any human agent. It presented limits to human knowledge and experience that continually forced people to respond to it. It has presented a constant challenge to human habitation, no matter how strong the attraction (resources or recreation) was. This is a continuity in itself, but there are also similarities in the types of problems (unmalleable bedrock, water access, isolation, disorientation), and to a lesser extent the types of adaptations (inside passages, seasonal rhythms, building sites). These form patterns of continuity through time and changing historical circumstances. Cumulative experiences and collective memory in turn form the bedrock for a sense of regional culture and community. The landscape has been modified, especially in the past 150 years, but it has always influenced, constrained, and directed human activity to some degree. Though it changes over time, the Georgian Bay of Huronia, of the *Waubuno*, and of *Night, Pine Island*, is always recognizable.

Now I imagine Georgian Bay as a kind of stage where the human and the non-human constantly confronted one another, a kind of middle ground. It was akin to an island shoreline, with waves washing against a rock and retreating. So I think of its history as a series of encounters with the archipelago. Why and how people came to the Bay; what they were prepared to find, and how they responded or adapted to the place itself – and what effect these encounters had on the Bay. To organize this book I took my cue from the different ways in which people related to the natural world. There are the ways they tried to conceptualize or understand it: mapping and measuring; incorporating it into histories and popular myth; and negotiating an artistic language and a regional vocabulary. These ways of seeing affected how they sought to use or manage it, whether in commercial exploitation and ownership or environmental and park policy. The outcome, though, was never predictable, for the story isn't only about people and what they brought to the Bay; it is also about what they found here.

European explorers, travellers, and surveyors told of their adventures attempting to map the myriad "Islands, Islets, and Rocks, of Lake Huron." The frank commentary in travel journals and survey notes belies the precise and detached language of cartography in their maps. They came to measure Georgian Bay for various purposes, seeing a useful strategic conduit, a stillborn agricultural frontier, or increasingly valuable cottaging space, depending on the expectations of the age. But they reached similar conclusions about their "work of great difficulty and labour"[2] in surveying a remote, hazardous, and largely unknown territory. Standard scientific training often left them unprepared for the obstreperous landscape, where anything that could go wrong often did. As the water level fluctuates, the lay of the land can change hourly and the line between land and water cannot be clearly or permanently drawn. To compensate for the limitations of cartography and hydrography, people "mapped" the Bay in other ways, using place names, for example, to identify landmarks and points of reference, as well as a way of claiming territorial possession simply by asserting the *right* to name the place. Because the archipelago resisted orderly and conventional settlement, it was invariably thought of as wilderness. But as the meaning of wilderness evolved from barren wilds to primeval retreat, the Bay was re-evaluated accordingly. What was seen for a long time as an unusable obstacle or a through route to elsewhere was "rediscovered" as an attractive destination and prized locale ("a beauty indeed ... A pretty little dot ... A perfect little gem"!)[3] Mapping, in turn, shaped the ways in which the Bay was imagined, as a geographical and a political entity. The ubiquitous survey grid determined how the archipelago was incorporated into the province of Ontario. Recent efforts to reconfigure municipalities along biophysical rather than geometric principles are a belated acknowledgement of the social, economic, and ecological consequences of carving up a place for political expediency.

If surveying the landscape was difficult, making use of it was even more so. For over three hundred years, a series of industries attempted to establish a foothold in the Bay. From the fur trade to the fishery and the timber trade, and finally cottaging and recreation, industry achieved a progressively greater presence in the landscape, and in the process raised the public profile of the Bay. After the 1850s, railways brought the region into the orbit of industrial North America, in a rhythm of exporting resources and importing people. State investigations into navigation, shipwrecks, forest stands, and fish stocks also brought a formerly remote hinterland into the public mind and onto the political agenda of the Canadas. Competition for resources could erupt into conflict, as tugs dragged log booms across gill nets or tourists complained about mills spewing smoke into the air. Industrial activity profoundly altered the Bay, as the fishers hauled away millions of pounds of fish, towns sprang up on shores cleared of pine, and the twin spectres of pollution and resource exhaustion permanently affected ecosystems on water and land.

But the usual image of nineteenth-century industry – a political and technological powerhouse, an unstoppable force of environmental manipulation and economic homogenization – does not reflect its peculiar fortunes in different places. Industries were forced to make concessions as well. The Bay promised a great deal: unsurpassed spawning grounds, open water routes for shipping, convenient sites for mills, and river access to rich timber stands in the interior. But it never really cooperated with those who sought to exploit or manage it. Its pine and minerals were deemed "inferior" or inaccessible, storms took an annual toll of ships, solid granite greeted proposed canals, and the islands proved a haven for poachers. Settlement remained shallow-rooted, a product of expediency:

company towns or shipping ports that rose and fell with the resource supply. Yet, a wilderness that resisted the mantra of utility and progress soon became an attractive resource in its own right. Even pragmatically-minded military observers at the end of the eighteenth century betrayed some of the Romantic love of wild nature that would propel the Bay's transformation into marketable cottage country a hundred years later. Recreation has secured the archipelago more thoroughly and more permanently than any other type of use, in part because it assigns value to private ownership, to being and staying in the Bay. Ironically, the group responsible for modernizing the Bay is the same one that prefers its nature in an unchanged, pre-industrial state. This almost schizophrenic dichotomy between celebrating progress and celebrating the primeval persists in Canadian attitudes toward the environment. By embracing a duality that sees nature in both functional and aesthetic terms, we try to have our outdoor cake and eat it too. This way a single landscape can be used simultaneously by industry and wilderness seekers: different types of consumption without segregation of physical space.

Despite its history of competing and often intensive uses, Georgian Bay's wilderness identity proved durable, in large part because of the Bay's associations with Canadian history. Its inhospitable frontier quality and the presence of First Nations were read as characteristics of a *terre sauvage*, from the Huron missionaries to twentieth-century tourists. The argument was circular and self-supporting: Indians represented the antithesis of civilization; their primitive way of life belonged to a wilderness environment; Georgian Bay was such a wilderness because the *sauvages* lived there. Wilderness and Aboriginals were judged on the same terms, sometimes hostile and treacherous, sometimes an uncorrupted example of the forest primeval, a model for urbanites seeking to get back to nature. History confined the Bay and its inhabitants to a specific era, a site of exploration and the fur trade: classic images of Canadiana safely distant from modernity. This identity was unusually vivid because romantic ideas of wilderness, and popular ideas of early Canadian history, merged smoothly with the physical environment (more smoothly than, for example, plans for a ship canal or homesteads). Precambrian rock scarred by glaciers, open water, an uninhabited shoreline, and wild, twisted pines made it easy to imagine the Bay as a fragment of an original New World, a return to an earlier age and to a wilderness life. It was viewed as an escape, from the turbulence of modernity – somewhere "to duck the atom bombs"[4] – or from the passage of time altogether. This image of the Bay tells us a great deal about what we look for in the past, about the uses of history as well as landscape. The history we assign to a place depends on what we want to use it for in the present; "the backwards view"[5] was more conducive to thinking of it as a place for canoeing like voyageurs or playing pioneer. Historical references were essential to constructing a mental landscape that preserved its increasingly desirable wilderness quality. The fixation on the era of exploration reflected the needs and values of early-twentieth-century Canada: the idealization of pioneering woodsmen, favouring a national historical narrative, and the longing of a young country for a sense of heritage and belonging in "this new world, ours." Georgian Bay offered a homeland, an ancestry, and an opportunity to participate in history. "And in the footsteps of Samuel de Champlain," as Adrienne Clarkson concluded in her inaugural speech as Governor General, "I am willing to follow."[6]

The Bay's wild nature was more than a romantic tourist construction. Its reluctance to be tamed or modernized was rooted in inescapable physical factors:

> unlike the other, more populated and civilized summer lakes, it also had rattlesnakes. As much as the weather, the cold water, the barren rocks, it was the reputation of rattlers that discouraged many visitors. While much impressed with the paintings they saw of the landscape, people who didn't care for poisonous snakes and wild storms tended to keep it at that. They stayed away. And the Waubuno Reaches stayed almost wilderness for the longest time.[7]

Life on *la mer douce* – this *un*civilized lake – meant negotiating with "the cold water, the barren rocks." Shield granite and water were the most fundamental physical factors shaping human activity in the Bay. The islands placed a kind of natural limit on concentrated development because they were harder to get to and harder to build on, and few people particularly wanted to live year-round in a harsh maritime environment. Communities that managed to persist "up the shore" developed pragmatic and aesthetic responses to the water. Boats as distinctive as the scoot and as indispensable as the supply steamer were built, adopted, or adapted to suit the rocky shoreline and open water. These conditions are also responsible for the danger inherent in the inland seas and the folklore that resulted "when the gales of November came slashing ... in the face of a hurricane west wind."[8] But subtle contradictions in descriptions of the Bay – awkward attempts to explain details of beauty and delicacy – suggest a savage garden within the stark framework of rock and water. It is impossible, in other words, to write environmental history without paying a great deal of attention to the environment in question; and it is irresponsible to imply that this environment does not have either an independent reality or a tangible effect on the course of human history.

But the ability to recognize, appreciate, and describe this complexity took time to develop. This became apparent when people viewed the Bay through the prisms of Romanticism and science, the most common languages of description of nature in the modern age. Both carried certain assumptions about what nature should look like and what should be done with it: lakelets and water lilies on the one hand, profitable intervention and management on the other. So people measured the Bay in existing terms of evaluation and expectation, and often found it wanting and "not quite paintable." Yet they also found themselves struggling to communicate the unique character of a "wave-worn, windblown" landscape, because their reactions were not entirely uniform, formulaic or contrived.[9] This struggle is a process that lies at the heart of the Canadian experience: the maturation of a regional perspective, the gradual definition of a regional identity. It is perhaps best illustrated by the arts, which evolved out of the quintessential colonial tension between imported paradigms and the specifics of "here," between inherited conventions and new experience. As people became more familiar with the Bay, they developed experiential relationships with it and an affective attachment to it. This propelled a shift in perspective from looking in at *un pays étranger* to describing a home place known and understood. They were learning to speak from within the landscape, and in the process cultivating a sense of locality. The Bay was increasingly thought of as a distinct place, separate even from adjacent sections of the Great Lakes and northern Ontario; the rivalry with Muskoka could only exist among people with a clear and local place attachment to the archipelago. That such cohesion would develop among seasonal and migratory populations including workers, cottagers, and campers is intriguing. However, the old link between metropolitan perception and local experience persists. Georgian Bay has been distilled into a recreational landscape and a series of signature images (from Group of Seven paintings to cut-off jeans and bare feet in cottage mythology) meant to support the cottagers' sense of their community. Moreover, place-specific images are exported from the Bay as generic symbols of the northland. Now the ubiquitous leaning pine is imposed on the rest of the country as part of a nationalist iconography.

External agendas and local realities also collided in the political realm. Legislation and policies that affected Georgian Bay were not written with the Bay specifically in mind; rather, they reflected how "nature" was generally understood and the demands made on this kind of landscape. Environmental management has been primarily concerned with reserving different parts of the Bay for different users. As a leading producer of fish and lumber, two of the most important industries in nineteenth-century Canada, the Bay attracted substantial attention once these valuable stocks were threatened. The spectre of resource exhaustion from the 1870s onward highlighted the effects of Crown ownership and the state's growing role in environmental regulation. Such regulation, designed to answer the needs of industry, measured the Bay in terms of supply and demand in natural resources. Provincial and federal officials designed spatial or geographical limits as well as restrictions on certain technologies and methods used in harvesting the resource. Their techniques reflected the principles of conservation as it was developed in late-nineteenth-century North America, especially in its faith that scientific management could achieve sustainable resource populations. After the Second World War, however, a growing demand for recreational as opposed to industrial space, coupled with wider support for environmental issues, shifted the emphasis to "preservation" in designated reserves, wilderness spaces, and parks. Parks, though, remained torn between the competing mandates of use and protection. The spirited politics of cottage country, and recreational landscapes as a whole, reflect the unresolved questions of preservation: what should be protected, for whom, and why? In all this, Georgian Bay is a typical illustration of the general currents in environmental thought and action in North America.

But how Georgian Bay was imagined in policy and how these policies actually fared in the archipelago could be two very different things. It was – and is – rather arrogant to think we could "manage" this landscape at all. Parks may have been intended "for the benefit, advantage and enjoyment of people," but they rarely anticipated the limits that the Bay's topography and marine environment could impose on plans for intense use. With no clear line between land and water (let alone between mainland, Indian lands, Crown land, and international waters), the archipelago was included in several overlapping jurisdictions and enmeshed in the rivalries between provincial and federal governments and between Canada and the United States. Residents added an additional consideration of "community acceptability," whether this was fishers opposed to limits imposed on their livelihood or cottagers demanding limits on building development or boat traffic. In fact, political or management issues are an important catalyst to the expression of a regional sensibility. Cottagers obviously stand to benefit if such limits protect not just the environment but their recreational space. However, it should be stressed that seasonal as well as permanent residents evince a land ethic, a sense of what is feasible in the archipelago, of what it can bear. Environmental management – as in the case of The Massasauga Provincial Park – has begun to incorporate a better understanding of the ecosystem and a greater sympathy to place difference.

The idea that the environment plays a role in human history is not new. For well over a century, North Americans have imagined the continent – in academic debate, popular history, and national myths – in the terms and geographies laid out by the frontier and the Laurentian shield. But recent scholarship has made these classic theses appear too straightforward. We know now that the environment does not wield a single, uniform influence over a nation; it is approached, experienced, and remembered differently by different groups. Nor is "the environment" a homogenous entity (like a single line of a hypothetical frontier); scholars now prefer to concentrate on identifying the distinctive patterns of individual landscapes, rather than the ideological meaning of "nation building" ones. We also have learned to question the *way* a landscape is represented, in different kinds of texts, to demonstrate how concepts such as wilderness are fluid and historic. Any number of variables contributed to the descriptions of a place we inherit – as mundane and immediate as the weather that day, as academic and removed as colonial policy. Our perspective on a place depends on how we are positioned relative to it, historically, culturally, and physically.

What of the other half of the equation? All records of a landscape, from paintings to poetry, are produced in a particular place. As John Wadland has said, "No cultural activity occurs exclusive of the environment required to sustain it."[10] The natural world exists independently of us, outside this kaleidoscope of cultural factors. An unusually intransigent landscape like the archipelago simply makes this more apparent. The Bay has never acted like mere scenery, so it has never been treated as such, a passive object of human contemplation or something designed for human consumption and enjoyment. People created a whole spectrum of perceptions, inventions, and practices in response to the specific contours of the archipelago, by "talking and listening" to it. Assumptions and ambitions were modified; political, industrial, and artistic agendas were rarely imposed intact. Patterns of travel, types of land use, literature and art, and political policy are all determined in large part by "what you actually encountered when you got here."[11] The danger with thinking of nature as a cultural "construct" is that it reduces the physical world to an ephemeral cluster of cultural intangibles; since we constructed it, it would seem to suggest we can de-struct it or reconstruct it in any way we please without actual consequences. But whether you see a stand of white pine as money in the bank or a cathedral of Nature, interfering with it has the same and very real ecological repercussions.

In short, the relationship between the human and the natural world is dialectical. People and nature continually respond to and redefine one another. We need to be able to distinguish where humans have imposed on the environment and where they adapted to it, and recognize that a landscape is a product of both dynamics. A map of Georgian Bay preserves imperial tributes to England's eighteenth-century naval heroes alongside sites from local memory of shoals and shipwrecks. Weather systems in the Lakes basin, wildlife populations, the post-glacial topography – all influenced human activity at the same time as human activity affected water quality, wildlife populations, and forest boundaries. Even industry alternated between harnessing nature and adjusting to it, sending skiffs and tugboats to seasonal fishing stations located near the spawning grounds of the outer shoals, or log booms through the islands to mills constructed where rivers emptied into the Bay. Place and culture have a symbiotic relationship. I would suggest, then, that to say nature or landscapes are "constructed" is not entirely accurate. Places are experienced. Things such as paintings and poems and boats and parks are constructed, as a way of both making sense of that experience and presenting an opinion about that place. Northrop Frye called this "imaginative digestion": studying a landscape through the arts in order to understand and eventually accept it.[12]

The Bay is a palimpsest of different cultural landscapes – a single setting reinhabited by different peoples yet always recognizable. Usually the term "cultural landscape" suggests a natural environment modified by human design, where nature has been overlaid or remade with patterns of settlement and distinctive architecture. But it should also include any concept or object, like art, created *in* that environment – the culture *of* that place – the imaginative landscape, if you will. Thus everything, from a sketch brought back by a military surveying party to the social relations of Ontario's cottage country – anything that reflects the contact and exchange between society and nature in a particular place – has contributed to how we think of Georgian Bay. This way we look for the connections between the artifacts as examples of encounters with a particular environment, rather than seeing them as isolated examples of different media (art, architecture, policy), different historical periods, or different users. A *terre sauvage* means something very different in the Jesuit *Relations* than in A.Y. Jackson's painting; but the inside passage and landmarks like the light at the *pointe au baril* appear in seventeenth-century descriptions of Huron trading, timetables for Victorian steamers, and cottagers' histories as they all navigate their way through the archipelago. The historical record is coloured by changing values, but people were not "making the landscapes up."

Such an interdisciplinary approach counters the parochial connotations of local or regional history. Of course, in one sense this *is* a local history, and it is easiest to see history as a dialogue with our surroundings at the local or regional level. Yet – and pardon the expression – no place is an island: it is impossible to look at a community or a landscape in isolation. There are always larger contexts to consider and comparative frameworks in which to operate: ecological (Great Lakes, Canadian Shield), functional (recreation, resource management), cultural (arts and architecture, or vernacular and high culture), or political (policy, jurisdictions, community mobilization). We cannot understand what happened in the Bay without knowing what was going on in the rest of Ontario, Canada, North America, and the North Atlantic world at the time. This might mean specific events, such as a world war; broad trends, such as industrialization; political frameworks, such as the expansion of empire or the friction of federalism; or ideological concepts, such as Romanticism or Progressivism. The 1945 fire at Depot Harbour, going full throttle in wartime production, lit the sky to Parry Sound; DDT bombs on Beausoleil and deer yards at Killbear reflected mid-century ideas about wildlife management. At the same time, Georgian Bay reached outward to affect events and thinking elsewhere. It anchored the country's inland fisheries, produced the Canadian Hydrographic Service, inspired the Group of Seven, and forced Ontario's park officials to reconsider the nature of wilderness parks.

Tracing the influence of one place illustrates the actual and symbolic power of individual landscapes in Canadian history. True, Georgian Bay has certain claims to fame – many people will recognize such luminaries as Champlain and the Group of Seven. But are there other landscapes (especially those outside the historical tradition of Central Canada) that have played a significant role in our national stories? A regional biography like this one can link familiar local experience with national narratives. This is one of the more useful aspects of this type of study in a country perennially insecure about its identity, its culture, and its prospects for unity. Communities from High River to Glace Bay have little in common. *except* the fact that their histories are integrally tied to their surroundings. Their economy, physical layout, population, and sense of identity can all be explained in part by the community's historical relationship with its environment. The arc we see in Georgian Bay – encountering a new landscape, adjusting imported ideas, negotiating with local conditions, and delineating local identity – has been repeated over and over across this country, making the development and expression of place identity one of the core themes of Canadian culture. Somewhat paradoxically, the more faithfully art and literature expresses someone's experience of (and loyalty to) a specific place, the more genuinely "Canadian" it is judged to be. Consider any survey of Canadian literature: L.M. Montgomery's Prince Edward Island, W.O. Mitchell's prairie, Mordecai Richler's Montreal, Alastair MacLeod's Cape Breton. In a country that holds natural landscape as a cultural totem, as our "secret ideology" and national passion,[13] we need to examine how and why particular landscapes become significant in both national myth and individual lives. A fourth-generation cottager associates Georgian Bay with family traditions and summer holidays but is also proudly aware of its role in Canada's history, keeping Group of Seven prints on the walls, Ojibwa quill boxes on the shelves, and a canoe at the dock.

This says a great deal about the place of history in popular culture. As we saw in Chapter 3, perceptions of the environment as wilderness played a large part in how the history of Georgian Bay – and Canadian history generally – was written. But history is also part of the fabric of everyday life. The elements we choose to emphasize reflect how we imagine our place in the landscape. History might mean a collection of practices assembled from decades of experience, like sheltered skidoo routes or shipwreck exhibits in local museums. Or it might be an imaginative, re-creative journey into the past, with provincial parks promising the opportunity to retrace the paths of the voyageurs in "Champlain country." But these different interpretations can and do coexist; one does not necessarily displace the other.[14] Both are legitimate; both are based on historical encounters with the physical qualities of the Bay. Different artifacts simply tend to emphasize one version or the other. If we think of history as C.W. Jefferys' illustrations of Champlain and Brûlé, we will think of the Bay as a wilderness from a past age; if we paddle past rusted anchor rings driven into the rock, we will see it as an inhabited workplace. The truth is we absorb history from a wide variety of sources. Cultural and natural heritage are indivisible; heritage, as Thomas Symons has said, is the total environment inherited from the past.[15]

Thinking of Georgian Bay as a historically inhabited space gives us better insight into its environmental health today. For a long time, the apparent resilience of the islands gave us a false confidence about its ability to remain in a natural state. We might tread more lightly on the land when we realize we are not the first to encounter it, that the Bay is not necessarily "pretty much as it had been when Champlain had passed through," and that our involvement in the ecosystem is neither erasable nor invisible. Current problems are often the historical residue of past use, such as when motorboats churn up deadhead logs and sawdust in bays where lumber mills were located a century ago. At the same time, a historical perspective forces us to acknowledge that a human presence makes change inevitable, no matter how much we might wish it otherwise. As one study of Muskoka concluded, "stopping the change is not as practical as planning for it, reducing it to a tolerable level, and managing the resource to keep it in the best condition that is possible given the circumstances."[16] William Cronon has argued that the concept of wilderness is often counterproductive, because it suggests such an unrealistic image of nature – unconfined, undefiled by humanity – that it doesn't help us relate to the landscapes we actually live in.[17] The archipelago is not a wilderness but a heritage landscape that comprises natural and human elements. Any plans to protect it must assume a human presence, past and present.

We are usually conscious of "the environment" only for brief moments and in specific situations: during vacations or smog alerts in summer, or headline-grabbing confrontations at places such as Burnt Church or Tofino when a resource we rely on appears threatened. On the other hand, it is difficult to embrace an abstract philosophy of bioethics with its idealistic "good for the planet" rhetoric, or even the causes adopted by mainstream environmental groups, when most of us have never seen an endangered species or an old-growth forest. We need to realize that the environment has a direct impact on our lives and communities. This is precisely what environmental history is all about. It explains how the world around us came to look the way it does, and why we act a particular way in it. It is easy to think the maple and beech forests at Killbear have been "preserved" in the park if we don't know that this is second-growth on a point logged a hundred years ago; it is easy to miss the crib of the old Manitou Dock if we don't know it is there, or why it is there; it is easy to mistake towns such as Port Severn and MacTier for cottage-country service centres until we think about where logging rivers flowed and where rail lines ran. Environmental history teaches us to scrutinize our surroundings and our behaviour in those surroundings, and to take nothing for granted.

The history of Georgian Bay is a usable past because it generates questions to ask about similar landscapes: nationally significant places such as Jasper, recreational spaces near urban regions, landscapes of transition between resource and recreational use, or other types of "threshold wilderness" under pressure. In each case, nature has been modified and even degraded by humans, but we are still powerfully affected by it. History provides the necessary background to present-day environmental politics and those high-profile confrontations. It identifies the actors who compete for a single landscape, their different interpretations of it, and their claims to its resources. Seeing how claims actually fared in the past helps us gauge the constraints and carrying capacity of the landscape before attempting to use or protect it. A historical perspective also explains the rationale for managerial designations, and challenges the notion that policy is a fixed and immutable bureaucratic creation. It shows how environmentalism, which we tend to associate with the radical 1960s, belongs to both an intellectual tradition of western ideas about nature and a political tradition of regulation and management. ("Aren't soap suds pollution?" asks Charles Gordon in his humorous memoir of cottage life. "Don't think so: we're always had soap suds. Pollution wasn't invented until the sixties, was it?")[18] In the same vein, a place history identifies sources of public support for environmental protection, in the grassroots attachment to distinct regional landscapes and what Samuel Hays called environments of personal meaning.[19] This is certainly the case in Georgian Bay. A place history can explain the reasons for that attachment, and the sources of that meaning.

Writing the history of a landscape like Georgian Bay is a unique type of public history. As Del Muise has argued, historians need to discover "communities of interest" between academia and the rest of society.[20] Landscapes are familiar, accessible, relevant, and meaningful for most people. Apart from actually being in the Bay, the most exciting and rewarding part of this project has been talking to people who know and love it. They are thrilled to discover that someone wants to talk to them about the Bay, mildly incredulous that it might warrant "serious" scholarly attention, and yet sometimes shy about voicing their emotions about, as one man said, "the power of nature and this yearning to understand and be part of it ...

this reverence for how beautiful the Bay could be, and the Bay of all places. This thing you hear in their voices, they don't talk about it like any other place."[21] History and landscape to them are very real. Their lives have all been shaped by the west wind. In this book I wanted to tell their stories and in the process, let Georgian Bay speak as well. And when I daydream, I can return to a flat, fractured, pink granite slab on a nameless island south of Killarney, surrounded by cold, impossibly turquoise water and pale straw-coloured grasses, sheltered by a stunted white pine, with the roar of wind "in our beloved Georgian Bay."

Notes

FOREWORD

1 Nicole Eaton and Hilary Weston, *In a Canadian Garden,* with Photographs by Freeman Patterson (Markham, ON: Penguin Books Canada, 1989), 9, for "contradiction," and pp. 104-7 for the garden of Nicole Eaton. One of Patterson's photographs of the Eaton garden is included in this volume (see Plate 14).

2 I think of many things, but specific reference is due Margaret Atwood, "The Planters," in *The Journals of Susanna Moodie* (Toronto: Oxford University Press, 1970), 16-17, and R. Cole Harris and Elizabeth Phillips, ed., *Letters from Windermere, 1912-1914* (Vancouver: UBC Press, 1984).

3 Douglas LePan, "Islands of Summer," and "Rough Sweet Land," in *Weathering It: Complete Poems 1948-97* (Toronto: McClelland and Stewart, 1987), 19 and 217-23; John Irving, *A Prayer for Owen Meany* (New York: Ballantine Books, 1989), 359, 505; Katherine Govier, *Angel Walk* (Toronto: Little Brown, 1996), 24 and 412.

4 Jonathan Bordo, "Jack Pine: Wilderness Sublime or the Erasure of the Aboriginal presence from the Landscape," *Journal of Canadian Studies* 27, 4 (1992-3): 98-128; Michael Ondaatje, *The English Patient* (Toronto: McClelland and Stewart, 1992), 296; Jill K. Conway, *True North* (New York: A.A. Knopf, 1994), 114-15; Claire Campbell, "'Our Dear North Country': Regional Identity and National Meaning in Ontario's Georgian Bay," *Journal of Canadian Studies* 37, 4 (Winter 2003): 68-91.

5 For a recent expression of this idea, see Roy MacGregor, *Escape: In Search of the Natural Soul of Canada* (Toronto: McClelland and Stewart, 2002).

6 Clifford Geertz, *The Interpretation of Cultures* (New York: Basic Books, 1973), 3-30.

7 LePan, "Canoe-Trip," in *Weathering It,* 77-8.

INTRODUCTION

1 Douglas LePan, "Islands of Summer," in *Weathering It: Complete Poems 1948-1987* (Toronto: McClelland and Stewart, 1987), 19.

2 W.F.W. Owen, "A Plan of the Straits from Lake Huron into the Manitoolin Lake from the Open Gut to Cabots Head from a Survey made 26th, 27th and 28th Sept. 1815," Library and Archives Canada, microfiche NMC 19403.

3 James Barry, *Georgian Bay: The Sixth Great Lake* (Toronto: Clarke; Irwin, 1968; 3rd ed., Erin, ON: Boston Mills Press, 1995), 3; Great Lakes Information Network 2002, <www.great-lakes.net>.

4 "Georgian Bay Littoral Biosphere Reserve," *GBA Update,* Summer 1999; Norman Pearson, *Planning for Eastern Georgian Bay* (London, ON: Tanfield Hall, 1996), 56.

5 Katherine Govier, *Angel Walk* (Toronto: Little Brown, 1996), 496-7.

6 Bill Mason, *Canoescapes* (Erin, ON: Boston Mills Press, 1995), 12.

7 David Wistow, *Landscapes of the Mind: Images of Ontario* (Toronto: Art Gallery of Ontario, 1986), 13; Northrop Frye, "Conclusion to *A Literary History of Canada*" (1965) in *Mythologizing Canada: Essays on the Canadian Literary Imagination*, ed. Branko Gorjup (Toronto: Legas, 1997), 71; Frank Underhill, "False Hair on the Chest," *Saturday Night*, 3 October 1936.

8 Frye, "Conclusion to *A Literary History of Canada*," 92.

9 See P.A. Buckner, "'Limited Identities' Revisited: Regionalism and Nationalism in Canadian History," *Acadiensis* 30, 1 (Autumn 2000): 4-15.

10 Email from Dr. Barry Gough, Wilfrid Laurier University, to Dr. Jonathan Vance, University of Western Ontario, 21 January 1999.

11 William Ratigan, *Great Lakes Shipwrecks and Survivals* (Grand Rapids, MI: Wm. B. Eerdmans, 1960), 125.

12 See Victoria Brehm, Introduction, in *"A Fully Accredited Ocean": Essays on the Great Lakes*, ed. Brehm (Ann Arbor, MI: University of Michigan Press, 1998).

13 Doug Aberley, "Interpreting Bioregionalism: A Story from Many Voices," 13, and Michael Vincent McGinnis, "A Rehearsal to Bioregionalism," in *Bioregionalism*, ed. M. McGinnis (New York: Routledge, 1999); Robert Gottleib, *Forcing the Spring: The Transformation of the American Environmental Movement* (Washington, DC: Island Press, 1993), 196; J.G. Nelson et al., "Overview of Protected Areas and the Regional Planning Imperative in North America: Integrating Nature Conservation and Sustainable Development," in *Protected Areas and the Regional Planning Imperative in North America*, ed. J.G. Nelson et al. (Calgary: University of Calgary Press; East Lansing, MI: Michigan State Press, 2002), 1-21; John H. Wadland, "Great Rivers, Small Boats: Landscape and Canadian Historical Culture," in *Changing Parks: The History, Future and Cultural Context of Parks and Heritage Landscapes*, ed. John Marsh and Bruce W. Hodgins (Toronto: Natural Heritage/Natural History, 1998), 20-1.

14 See Carl Berger, *The Writing of Canadian History: Aspects of English-Canadian Historical Writing Since 1900*, 2nd ed. (Toronto: University of Toronto Press, 1986); Matthew Evenden, "The Northern Vision of Harold Innis," *Journal of Canadian Studies* 34, 3 (Fall 1999): 162-86.

15 See William Morton, "Clio in Canada: The Interpretation of Canadian History," *University of Toronto Quarterly* 15, 3 (April 1946); J.M.S. Careless, "Limited Identities in Canada," *Canadian Historical Review* 50 (1969): 1-10; Cole Harris, "The Emotional Structure of Canadian Regionalism," Walter L. Gordon Lecture Series 1980-1, vol. 5, *The Challenges of Canada's Regional Diversity* (Toronto, 1981), 9-30; Ramsay Cook, "Canada: An Environment without a History?" unpublished paper, 1998, cited in Neil Forkey, *Shaping the Upper Canadian Frontier: Environment, Society and Culture in the Trent Valley* (Calgary: University of Calgary Press, 2003).

16 Arnold Berleant, *The Aesthetics of Environment* (Philadelphia: Temple University Press, 1992), 167.

17 See Deryck Holdsworth, "Landscape and Archives as Texts," in *Understanding Ordinary Landscapes*, ed. Paul Groth and Todd W. Bressi (New Haven, CT: Yale University Press, 1997); Alan R.H. Baker, "On Ideology and Landscape," in *Ideology and Landscape in Historical Perspective*, ed. Baker and Gideon Biger (Cambridge University Press, 1992).

18 See, for example, Bruce Trigger, *The Children of Aataentsic: A History of the Huron People to 1660* (Kingston and Montreal: McGill-Queen's University Press, 1987); Peter S. Schmalz, *The Ojibwa of Southern Ontario* (Toronto: University of Toronto Press, 1991);

Theresa A. Smith, *The Island of the Anishnaabeg: Thunderers and Water Monsters in the Traditional Ojibwe Life-World* (Moscow, ID: University of Idaho Press, 1995).

19 Donald Worster, *An Unsettled Country: Changing Landscapes of the American West* (Albuquerque: University of New Mexico Press, 1994), x-xii; see, for example, Michael E. Soulé and Gary Lease, eds., *Reinventing Nature: Responses to Postmodern Deconstruction* (Washington, DC: Island Press, 1995).

20 Carl Sauer, "The Morphology of Landscape," *University of California Publications in Geography* 2, 2 (1925): 19-54, cited in Michael Conzen, "The Historical Impulse in Geographical Writing," in *A Scholar's Guide to Geographical Writing on the American and Canadian Past,* ed. Michael P. Conzen et al. (Chicago: University of Chicago Press, 1993); John Brinckerhoff Jackson, *Discovering the Vernacular Landscape* (New Haven, CT: Yale University Press, 1984), 8.

21 Jonathan M. Smith, Andrew Light, and David Roberts, "Philosophies and Geographies of Place," in *Philosophies of Place,* ed. Andrew Light and Jonathan M. Smith (Lanham, MD: Rowman and Littlefield, 1998), 2; Paul Groth, "Understanding Ordinary Landscapes," in *Understanding Ordinary Landscapes,* ed. Groth and Todd W. Bressi, 7.

22 Simon Schama, *Landscape and Memory* (New York: Alfred A. Knopf, 1995), 15.

23 See, for example, *The Making of the American Landscape,* ed. Michael P. Conzen (Boston: Unwin Hyman, 1990), including the essay "French Landscapes in North America" by Cole Harris. Harris's work on the seigneurial system explores the extent to which the official plans for colonial settlement were thwarted and modified by the Canadian environment; explaining the "look of the land" as the product of negotiation and a certain disjunction between ideological and political expectation, social relations, and physical forces.

24 William Norton, *Explorations in the Understanding of Landscape: A Cultural Geography* (Westport, CT: Greenwood Press, 1989); Arnold R. Alanen and Robert Z. Melnick, "Why Cultural Landscape Preservation?" *Preserving Cultural Landscapes in America* (Baltimore: Johns Hopkins University Press, 2000), 1-21; Richard L. Nostrand and Lawrence E. Estaville, Introduction, in *Homelands: A Geography of Culture and Place across America,* ed. R.L. Nostrand and L.E. Estaville (Baltimore: Johns Hopkins University Press, 2001).

25 Graeme Wynn, "Geographical Writing on the Canadian Past," in *A Scholar's Guide to Geographical Writing,* ed. Michael P. Conzen et al.; Stephen Hornsby, *Nineteenth-Century Cape Breton: A Historical Geography* (Montreal and Kingston: McGill-Queen's University Press, 1992), xv; Conrad Heidenreich, *Huronia: A History and Geography of the Huron Indians, 1600-1650* (Toronto: McClelland and Stewart/Ontario Ministry of Natural Resources, 1971), 16. Wynn addresses the challenge of finding a cohesive interpretation for a region while recognizing its internal diversity, in "Ideology, Identity, Landscape and Society in the Lower Colonies of British North America 1840-1860," in *Ideology and Landscape in Historical Perspective,* ed. Alan R.H. Baker and Gideon Biger (Cambridge: Cambridge University Press, 1992), 197-229. Much of the recent writing from Atlantic Canada on "environmental" issues has been oriented toward issues of class conflict and political economy, as in Daniel Samson, ed., *Contested Countryside: Rural Workers and Modern Society in Atlantic Canada, 1800-1950* (Fredericton, NB: Acadiensis Press, 1994) and L. Anders Sandberg, ed., *Trouble in the Woods: Forest Policy and Social Conflict in Nova Scotia and New Brunswick* (Fredericton, NB: Acadiensis Press, 1992).

26 See, for example, Richard L. Nostrand and Lawrence E. Estaville, eds., *Homelands: A Geography of Culture and Place Across America* (Baltimore: Johns Hopkins University Press, 2001).

27 Rupert Brooke, 27 July 1913, in Greg Gatenby, ed., *The Wild Is Always There: Canada Through the Eyes of Foreign Writers* (Toronto: Vintage Books, 1994), 205. On oral history, compare, for example, the essays in *Using Wilderness: Essays on the Evolution of Youth Camping in Ontario*, ed. Bruce Hodgins and Bernandine Dodge (Peterborough, ON: Frost Centre, 1992) with those in "Shaped by the West Wind: Memories of Camp Hurontario, 1947-1997," by Claire Campbell (unpublished collection, 1997). Oral history also figures prominently in the community histories of Go Home Bay, Sans Souci, Pointe au Baril, and Cognashene.

28 On heritage and tourism, see D.A. Muise, "Who Owns History Anyway? Reinventing Atlantic Canada for Pleasure and Profit," *Acadiensis* 27, 2 (Spring 1998); James Overton, *Making a World of Difference: Essays on Tourism, Culture and Development in Newfoundland* (St. John's, NF: Institute of Social and Economic Research, Memorial University, 1996); Ian McKay, *The Quest of the Folk: Antimodernism and Cultural Selection in Twentieth-Century Nova Scotia* (Montreal and Kingston: McGill-Queen's University Press, 1994).

29 See, for example, J.B. Harley, "Deconstructing the Map," in *Writing Worlds: Discourse, Text and Metaphor in the Representation of Landscape*, ed. Trevor J. Barnes and James S. Duncan (London: Routledge, 1992), 231-47; Edmund C. Penning-Rowsell, "Themes, Speculations and an Agenda for Landscape Research," in *Landscape Meanings and Values*, ed. Penning-Rowsell and David Lowenthal (London: Allen and Union, 1986), 114; Martin Locock, "Meaningful Architecture," in *Meaningful Architecture: Social Interpretations of Buildings*, ed. M. Locock (Avebury: Ashgate Publishing, 1994), 6; Yi-Fu Tuan, *Topophilia: A Study of Environmental Perception, Attitudes, and Values* (Englewoood Cliffs, NJ: Prentice-Hall, 1974).

30 W.H. New, *Land Sliding: Imagining Space, Presence, and Power in Canadian Writing* (Toronto: University of Toronto Press, 1997), 117; Greg Halseth, *Cottage Country in Transition: A Social Geography of Change and Contention in the Rural-Recreational Countryside* (Montreal and Kingston: McGill-Queen's University Press, 1998).

31 Ronald Bordessa, "Moral Frames for Landscape in Canadian Literature," in *A Few Acres of Snow: Literary and Artistic Images of Canada*, ed. Paul Simpson-Housley and Glen Norcliffe (Toronto: Dundurn Press, 1992), 62, 67.

32 A.Y. Jackson, "Interview with A.Y. Jackson," by Lawrence Sabbath, *Canadian Art*, July 1960.

33 Alexander Henry, *Travels and Adventures in Canada and the Indian Territories Between the Years 1760 and 1776*, ed. James Bain (Edmonton: M.G. Hurtig, 1969), 169-70; LePan, "Red Rock Light," in *Weathering It*, 202-3.

34 Barker Fairley, "The Group of Seven," *Canadian Forum*, February 1925.

35 Ted Steinberg, *Down to Earth: Nature's Role in American History* (New York: Oxford University Press, 2002), 284.

36 I owe the idea of vantage point and refracting history to Stephen Pyne, *How the Canyon Became Grand* (New York: Viking, 1998), xv.

CHAPTER 1: WHAT WORD OF THIS CURIOUS COUNTRY?

1 Northrop Frye coined this famously Canadian phrase in his Conclusion to *A Literary History of Canada* (1965), reprinted in *Mythologizing Canada*, ed. B. Gorjup, 68. See also Margaret Atwood's comments about explorers in *Survival* (Toronto: House of Anansi Press, 1972), 17-8.

2 J.B. Harley, "Texts and Contexts in the Interpretation of Early Maps," in *From Sea Charts to Satellite Images: Interpreting North American History through Maps*, ed. David Buisseret (Chicago: University of Chicago Press, 1990), 10.

3 Conrad Heidenreich, "Mapping the Great Lakes: The Period of Exploration, 1603-1700," *Cartographica* 17, 3 (1980), and "Mapping the Great Lakes: The Period of Imperial Rivalries, 1700-1760," *Cartographica* 18, 3 (1981); Lt. Henry W. Bayfield, Penetanguishene, to Commander Robert Barrie, Kingston, 11 May 1821, Library and Archives Canada [LAC], Manuscript Group [MG] 12, adm. 1, vol. 3444.

4 Jeff Malpas, "Finding Place: Spatiality, Locality and Subjectivity," in *Philosophies of Place*, ed. Andrew Light and Jonathan Smith, 37.

5 A.F. Hunter, "Shore Lines between Georgian Bay and the Ottawa River," *Summary Report of the Geological Survey of Canada 1907*, Canada Sessional Papers (hereafter abbreviated as SP) no. 26 (Ottawa, 1908).

6 Peter Whitfield, *New Found Lands: Maps in the History of Exploration* (London: British Library, 1998), 141.

7 Fr. Gabriel Sagard, *Sagard's Long Journey to the Country of the Hurons*, ed. George M. Wrong, trans. H.H. Langton (Toronto: Champlain Society, 1939), 66, 185-9.

8 Gary Foster et al., *The Archaeological Investigations on Beausoleil Island, Georgian Bay Islands National Park, 1985* (Canada Parks Service, 1987), 14, 37-8; Heidenreich, *Huronia*, 231-2, 241.

9 Georges Sioui, *Huron-Wendat: The Heritage of the Circle*, rev. ed. (Vancouver: UBC Press, 1997), 62 and 176, 90, 94; Heidenreich, *Huronia*, 22. Sioui characterizes the etymology of the term "Huron" as disparaging (p. 1, n. 3), and while I respect this, I also feel obliged to retain the terms used in the historical documents.

10 Trigger, *Children of Aataentsic*, 779, 788; Donald B. Smith, "The Dispossession of the Mississauga Indians: A Missing Chapter in the Early History of Upper Canada," *Ontario History* 73, 2 (1981): 70. Arthur Ray comments on the traditional seasonal fishery of the "northern Ojibwa" on the north shore of Lake Huron in "'Ould Betsy and Her Daughter': Fur Trade Fisheries in Northern Ontario," in *Fishing Places, Fishing People: Traditions and Issues in Canadian Small-Scale Fisheries*, ed. Dianne Newell and Rosemary E. Ommer (Toronto: University of Toronto Press, 1998).

11 Peter A. Fritzell, "Changing Concepts of the Great Lakes Forest: Jacques Cartier to Sigurd Olsen," in *The Great Lakes Forest: An Environmental and Social History*, ed. Susan L. Flader (Minneapolis: University of Minnesota Press, 1983), 278; Denis Cosgrove, ed., *Mappings* (London: Reaktion Books, 1999), 2.

12 G.M. Lewis, "Mapping the Great Lakes between 1755 and 1795," *Cartographica* 17, 1 (1980): 7.

13 Samuel de Champlain, *Works of Samuel de Champlain*, vol. 4, ed. H.P. Biggar (Toronto: Algonquin Historical Society, 1932), 237; Sagard, *Long Journey*, 69; Trigger, *Children of Aataentsic*, 578-86.

14 Henry, *Travels and Adventures*, 32-3.

15 Fred Landon, *Lake Huron* (Indianapolis: Bobbs-Merrill, 1944), 289.

16 Alexander Macdonnell, "Diary of Lt. Governor Simcoe's Journey from Humber Bay to Matchedash Bay in 1793," *The Correspondence of Lieut. Governor John Graves Simcoe*, vol. 2, ed. E.A. Cruikshank (Toronto: Ontario Historical Society, 1924), 73-4.

17 Now known simply as St. Joseph Island. See Pierrett Désy and Frédéric Castel, "Native Reserves of Eastern Canada to 1900," in *Historical Atlas of Canada Vol. II: 1800-91: The Land Transformed*, ed. R. Louis Gentilcore (Toronto: University of Toronto Press, 1993); Canada, *Indian Treaties and Surrenders, from 1680-1890* (Ottawa, 1891); Helen Hornbeck Tanner, ed., *Atlas of Great Lakes Indian History* (Norman, OK: University of Oklahoma Press, 1987); *Map of Part of the Province of Upper Canada Showing Districts and Counties, 1816* (University of Western Ontario Map Library).

18 These sketches traditionally have been attributed to Elizabeth Simcoe, but research suggests they are in fact Pilkington's originals, which Simcoe kept in her collection. Public Archives Canada, *Elizabeth Simcoe*, microfiche 9 (Ottawa: 1978). James Angus summarizes the opinions of military observers before 1820 in *A History of the Trent-Severn Waterway, 1833-1920* (Kingston and Montreal: McGill-Queen's University Press, 1988), 359-61.

19 William Jones to Isaac Chauncey, 27 January 1813, reprinted in William S. Dudley, ed., *The Naval War of 1812: A Documentary History*, vol. 2: 1813 (Washington, DC: Naval Historical Center, 1992), 419-20.

20 Lt. Colonel Robert McDougall, Michilimackinac, to Lt. General Gordon Drummond, Kingston, 9 September 1814, in William Wood, ed., *Select British Documents of the Canadian War of 1812*, vol. 3 (Toronto: Champlain Society, 1926), 279. While the 1813 log of the *Nancy* has been reprinted as *Leaves from the War Log of the Nancy* (Toronto: Rous and Mann Press, 1936; republished by Huronia Historical Development Council/Ontario Department of Tourism and Information, 1968), apparently no log covering its 1814 stint in the Bay survives. For a detailed account of the naval war see Barry Gough, *Fighting Sail on Lake Huron and Georgian Bay: The War of 1812 and Its Aftermath* (Annapolis, MD: Naval Institute Press, 2002).

21 George Head, *Forest Scenes and Incidents in the Wilds of North America ... During Four Months' Residence in the Woods on the Borders of Lakes Huron and Simcoe* (London, UK: John Murray, Albemarle Street, 1829), 195.

22 Lt. Miller Worsley, Michilimackinac, to Sir James Yeo, Kingston, 15 September 1814, LAC, adm. 1, vol. 2378, reel B-2942.

23 Committee on Ottawa and Georgian Bay Territory, *Report*, 15 June 1864, *Journals of the Legislative Assembly of the Province of Canada* 1863-4, appendix no. 8.

24 Letter addressed to W.D. Thomas, author unknown, 6 August 1814, intercepted by Captain Sinclair at Nottawasaga, reprinted in *Niles' Weekly Register*, vol. 7 supplement, 132-3; Sinclair to Jones, USS *Niagara*, 9 August 1814, reprinted in *Niles* supplement, 130-1.

25 William Dunlop, *Recollections of the War of 1812* (Toronto: Historical Publishing, 1908), 91-9.

26 Head, *Forest Scenes and Incidents*, 177, 187, 207.

27 Sir George Prevost to Lord Bathurst, cited in E. Cruikshank, "An Episode of the War of 1812: The Story of the Schooner 'Nancy,'" *Ontario Historical Society Papers and Records*, vol. 9 (1907), 79.

28 Benedict Anderson, *Imagined Communities: Reflections on the Origin and Spread of Nationalism*, rev. ed. (London: Verso, 1991), 164, 172-3.

29 James Carmichael Smyth, *Copy of a Report to His Grace the Duke of Wellington*, Halifax, 9 September 1825, in *Muskoka and Haliburton 1615-1875: A Collection of Documents*, ed. Florence B. Murray (Toronto: Champlain Society/University of Toronto Press, 1963), 33.

30 Bayfield, Quebec, to Capt. Francis Beaufort, hydrographer, 9 May 1832, LAC, MG 12, adm. 1, Commander H.W. Bayfield, St. Lawrence Survey Correspondence 1828-35.

31 Bayfield, Penetanguishene, to Robert Barrie, Kingston, 20 October 1820, LAC, MG 12, adm. 1, vol. 3444.

32 See L.F.S. Upton, "The Origins of Canadian Indian Policy," *Journal of Canadian Studies* 8, 4 (November 1973): 51-61.

33 Committee on Ottawa and Georgian Bay Territory (n.p., 1864). Treaties and reserves are discussed further in Chapter 3.

34 Bayfield to Beaufort, 9 May 1832. He also complained of battling ague, "intermittent fever," and even scurvy.

35 5 August 1837 in David Thompson's "Journal of Occurrences from Lake Huron to the Ottawa River," cited in C.E.S. Franks, "David Thompson's Explorations of the Muskoka and Madawaska Rivers," in *Nastawgan*, ed. Bruce Hodgins and Margaret Hobbs (Toronto: Betelgeuse Books, 1985), 30; also David Thompson Papers, Archives of Ontario [AO] vol. 28, no. 66, in *Muskoka and Haliburton 1615-1875*, ed. F. Murray, 86.

36 J.W. Bridgland, *Report, Field Notes and Diary of Exploring Lines from the Eldon Portage to the Mouth of the River Muskoka*, 31 January 1853, Department of Crown Lands Field Notes, no. 1897, 1853, book 3.

37 Survey of Foley, Report of the Commissioner of Crown Lands of Canada 1866-7, Appendix 29 (Ontario SP no. 6).

38 A.G. Ardagh, "Survey of Islands in Georgian Bay, in front of Townships of Harrison and Shawanaga, in the District of Parry Sound," Report of the Department of Lands, Forests and Mines 1910-11, Appendix 32 (Ontario SP no. 3, 1911).

39 W.H.C. Napier, Toronto, 1 December 1856, to R.T. Pennefather, Superintendent General, Indian Affairs. Indian Land Records Canada West, Inspection Returns 1835-65, LAC, Record Group [RG] 10, vol. 727, 28-31.

40 Report of the Commissioner of Crown Lands of the Province of Ontario 1877 (SP no. 17, 1878), 38; Report of the Commissioner of Crown Lands 1876 (SP no. 1, 1877), 16; Bridgland, Department of Crown Lands Field Notes, no. 1897 (1853), 154-5.

41 Thomas McMurray, *The Free Grant Lands of Canada, from Practical Experience of Bush Farming in the Free Grant Districts of Muskoka and Parry Sound* (Bracebridge, ON: 1871), 125; *Guidebook and Atlas of Muskoka and Parry Sound Districts* (Toronto: H.R. Page, 1879); Alexander Kirkwood and J.J. Murphy, *The Undeveloped Lands in Northern and Western Ontario* (Toronto: Hunter Rose, 1878), 58; Committee on Ottawa and Georgian Bay Territory (1864).

42 Suzanne Zeller, "Mapping the Canadian Mind: Reports of the Geological Survey of Canada, 1842-1863," *Canadian Literature* 131 (Winter 1991): 165-6.

43 Alexander Murray, "Report, for the Year 1857" (1858), 10.

44 La Cloche Report on District (B109/e/1 Hudson's Bay Company Archives, 1828-35); John H. Peters, "Commercial Fishing in Lake Huron, 1800 to 1915: The Exploitation and Decline of Whitefish and Lake Trout" (MA thesis, University of Western Ontario, 1981), 24.

45 Calvin Colton, *Tour of the American Lakes, and Among the Indians of the Northwest Territories, in 1830*, vol. 1 (London, 1833), 90; David Thompson, "Report of the Route between Lake Huron and the Ottawa River," submitted to John Macaulay, Surveyor General of Upper Canada, 30 May 1837, LAC, R-2655-0-2-E, Macaulay Papers 1781-1921.

46 Ironically, the catalyst for the survey, the *Asia*, sank in a November storm, a fate which had nothing to do with inadequate charts. Robert Higgins, *The Wreck of the Asia: Ships, Shoals, Storms and a Great Lakes Survey* (Waterloo, ON: Escart Press, 1995), 65.

47 *Georgian Bay and North Channel Pilot*, 1899 ed. (Ottawa: Minister of Marine and Fisheries), 266.

48 Department of Public Works, *Georgian Bay Ship Canal: Report Upon Survey, with Plans and Estimates of Cost, 1908* (SP no. 19a, 1909); Robert Legget, *Canals of Canada* (Vancouver: Douglas, David and Charles, 1976), 123.

49 Report of the Commissioner of Crown Lands of Canada 1866-7, Appendix 29 (SP no. 6), 185-6, 193-4; George Stewart, Field Notes of Survey of Foley, 1866, book 1191.

50 McMurray, *Free Grant Lands of Canada*, 129.

51 Marion Thayer MacMillan, *Reflections: The Story of Water Pictures* (New York: Greenberg, 1936), 24.

52 J.G. Sing, "Survey of Islands, Georgian Bay," Department of Crown Lands, *Annual Report*, Appendix 27 (SP no. 3, 1900); Sing, Ontario Land Survey, 1899, LAC, RG 10, vol. 2852, file 176, 296-1B.

53 J. Satterly, "Mineral Occurrences in Parry Sound District," Department of Mines *Annual Report* 1942, vol. 51, pt. 2 (Toronto, 1943), 4; S. Bray to J.D. McLean, 18 September 1900, LAC, RG 10, vol. 2851, file 176, 296-1A.

54 See Department of Indian Affairs correspondence regarding island sales, LAC, RG 10, vol. 2853, file 176, 296-1H to 1R.

55 Ella V. Carmichael to Department of Indian Affairs, 13 May 1922, LAC, RG 10, vol. 2869, file 176, 296-157.

56 Memo to J.M. Wardle, Director of Surveys and Engineering Branch, Department of Mines and Resources, from Parks Branch, 8 June 1937, LAC, RG 84 A-2-a, vol. 1150, file GB16-8.

57 Letter to N.K. Ogden, Mines and Resources Branch, Department of Indian Affairs, from Gowling, Mactavish, Watt, Osborne & Henderson, Barristers & Solicitors, Ottawa, 15 April 1947, LAC, RG 10, vol. 2851, file 176, 296-1.

58 J.T. Coltham, Ontario Land Survey, Field Notes for Survey of Township of McDougall, book 1557, 1934; Carling, book 2763, 1953; J.H. Burd, "Survey of Islands in Georgian Bay, District of Parry Sound," Department of Lands and Forests, *Annual Report* 1910-11, Appendix 31 (SP no. 3, 1911).

59 "Romantic White Boats Cash for Georgian Bay," *Financial Post*, 18 July 1959.

60 Kenneth McNeill Wells, *Cruising Georgian Bay* (Toronto: Kingswood House, 1958), 85. He also complained about the sparse and inaccurate buoyage on the Trent-Severn in *Cruising the Trent-Severn Waterway* (1959).

61 Interview with John Hughes (Toronto: 1 June 1997).

62 Department of Fisheries and Oceans, *Sailing Directions Georgian Bay CEN 306* (Ottawa, 1998), 1-2; Stanley Fillmore and R.W. Sandilands, *The Chartmakers: The History of Nautical Surveying in Canada* (Toronto: NC Press, 1983), 62.

63 Wilbur Zelinksy, "The Imprint of Central Authority," in *Making of the American Landscape*, ed. Michael P. Conzen, 313.

64 Telephone conversation with Chris Baines, Past President, Georgian Bay Land Trust (Toronto: 16 August 1999); interview with John Birnbaum, Executive Director, Georgian Bay Association (GBA) (Toronto: 18 September 1999).

65 District of Parry Sound Local Government Study, *Final Report and Recommendations* (Toronto: Ontario Ministry of Treasury, Economics and Intergovernmental Affairs, 1976), 199.

66 See, for example, Ontario Department of Lands and Forests, *Forest Industry Opportunities in Georgian Bay Area* (1963); Georgian Bay Regional Development Council, *Regional Plan 1968-1972* (Guelph, ON: University of Guelph Centre for Resources Development, 1969).

67 Pearson, *Planning for Eastern Georgian Bay*, 28. Submissions universally in support of a shoreline township are summarized in Donald M. Paterson, *Muskoka District Local Government Review: Final Report and Recommendations* (1969), 73.

68 Parry Sound Local Government Study, 176.

69 W. Darcy McKeough, Treasurer of Ontario and Minister of Economic and Intergovernmental Affairs, 30 January 1978, "Proposals for the Improvement and Strengthening of Local Government in the District of Parry Sound," AO, [MU] 3830.

70 W.J. Dodd, Clerk-Treasurer for Medora and Wood Townships, to Darcy McKeough, 12 December 1969, AO, RG 19-6-1 26-6, box 48.

71 Letter from C.W. King, President, GBA, to McKeough, 31 January 1977; Federation of the Maganatawan Valley Conservation Associations, *For Better Governance and Planning: A Response to the District Parry Sound Local Government Study* (1977), 1, AO, MU 3830.

72 Paterson, *Muskoka District Local Government Review*, 73; Norman Pearson, *Environment Control, Planning and Local Government in Georgian Bay Archipelago* (Sans Souci and Copperhead Association, 1975), 41; Ontario Legislature *Debates*, 18 June 1979, 3012-6.

73 *Georgian Bay and North Channel Pilot*, 313.

74 Lt. Colonel George Croghan, USS *Niagara*, to the Secretary of War, 9 August 14 1814, reprinted in *Niles' Weekly Register*, vol. 7, 10 September 1814, 4; Capt. Arthur Sinclair, USS *Niagara*, to William Jones, Secretary of the Navy, 3 September 1815, reprinted in *Niles' Weekly Register*, vol. 7, supplement, 131-2.

75 S.J. Chapleau, "Detailed Description of Route and Project – Nipissing District," in *Georgian Bay Ship Canal*, by Department of Public Works, 121.

76 William Milawski, Burlington to T.E. Lee, Director, Park Planning Branch, Ontario Ministry of Natural Resources [MNR], 14 January 1974, AO, RG 1, [IB] 4, 26-14-7.

77 J.J. Bigsby, *The Shoe and Canoe, Or, Pictures of Travel in the Canadas* (London: Chapman and Hall, 1850), 96-7; Scott A. McLean, ed., *From Lochnaw to Manitoulin: A Highland Soldier's Tour through Upper Canada* (Toronto: Natural Heritage Books, 1999), 26.

78 *Georgian Bay and North Channel Pilot*, 309.

79 Cited in Charles Long, "Taxi's Waiting," *Cottage Life*, Spring 2000.

80 Herbert L. Harley, "A Cruise to Georgian Bay in 1898," *Rudder*, 1899, pt. 2, p. 35.

81 MacMahon, *Island Odyssey: A History of the San Souci Area of Georgian Bay*. (Toronto: San Souci and Copperhead Association, 1990), 106.

82 LePan, *Weathering It*, 217.

83 Kent Ryden's *Mapping the Invisible Landscape: Folklore, Writing, and the Sense of Place* (Iowa City: University of Iowa Press, 1993) discusses extensively the nature and role of cartography and different types of mapping.

84 Bigsby, *Shoe and Canoe*, 107.

85 Ronald Rees, *Land of Earth and Sky: Landscape Painting of Western Canada* (Saskatoon: Western Producer Prairie Books, 1984), 5, 10; R.H. Hubbard, "Landscape Painting in Canada," in *Canadian Landscape Painting 1670-1930* (Madison, WI: University of Wisconsin Press, 1973), 6; Northrop Frye, "Sharing the Continent," reprinted in *A Passion for Identity: An Introduction to Canadian Studies*, ed. Eli Mandel and David Taras (Toronto: Methuen, 1987), 208; Janet Clark, "William Armstrong: Watercolour Drawings of New Ontario – from Georgian Bay to Rat Portage," in *Watercolour Drawings of New Ontario – from Georgian Bay to Rat Portage*, by William Armstrong (Thunder Bay, ON: Thunder Bay Art Gallery, 1996), 35.

86 Maria Tippett, *Stormy Weather: F.H. Varley, A Biography* (Toronto: McClelland and Stewart, 1998), 127.

87 John Hartman, interviewed by Dick Gordon on "This Morning," Canadian Broadcasting Corporation (Toronto: 10 September 1999).

88 "Canoe Trip" and "Rough Sweet Land," LePan, *Weathering It*, 77 and 217.

89 Steven Tudor, "Aird Island," "Depth Sounder," and "Mathematics," in *Hangdog Reef: Poems Sailing the Great Lakes* (Detroit, MI: Wayne State University Press, 1989).

90 Evan Solomon, *Crossing the Distance* (Toronto: McClelland and Stewart, 1999), 194.

91 New, *Land Sliding*, 25-7; Zelinksy, "The Imprint of Central Authority," in *Making of the American Landscape*, ed. Michael P. Conzen.

92 James White, "Place Names in Georgian Bay," Ontario Historical Society *Papers and Records,* vol. 11 (1913).
93 Interview with Mike Chellew (Lakefield, ON: 23 June 1997). My father salvaged the signs from islands 220 and 226.
94 See Karl W. Butzer, "The Indian Legacy in the American Landscape," and Cole Harris, "French Landscapes in North America," in *Making of the American Landscape,* ed. Michael P. Conzen.
95 "On the Georgian," in *Wahsoune Guests Have Rights* (Hamilton, ON: John Gordon Gauld, 1933), 22.
96 Heidenreich, "Mapping the Great Lakes: The Period of Imperial Rivalries, 1700-1760," *Cartographica* 18, 3 (1981): 100-1; *Jean-Baptiste Perrault: marchand voyageur parti de Montreal le 28e de mai 1783,* ed. Louis-P. Cormier (Montreal: Boreal Express, 1978), 143-5.

CHAPTER 2: A REGION OF IMPORTANCE

1 Michael Williams, "The Clearing of the Forests," in *Making of the American Landscape,* ed. Michael P. Conzen, 159; Barbara Moon, *The Canadian Shield* (Toronto: Natural Science of Canada, 1970), 133.
2 Trigger, *Children of Aataentsic,* 357-8; Russell Floren et al., *Ghosts of the Bay: A Guide to the History of Georgian Bay* (Toronto: Lynx Images, 1994), 170; B.W. Hodgins and Jamie Benidickson, *The Temagami Experience: Recreation, Resources, and Aboriginal Rights in the Northern Ontario Wilderness* (Toronto: University of Toronto Press, 1989), 28; Hornbeck Tanner, *Atlas of Great Lakes Indian History,* 127.
3 Frances Simpson, diary entry, 10 May 1830, in "Journey for Frances," *Beaver,* March 1954, 13; Hudson's Bay Company Archives, microfilm copy (LAC, Report on La Cloche District, B109/e/1 to 7, 1827 to 1835).
4 Commissioner of Crown Land Reports, Province of Canada and Ontario sessional papers; Dominion Land Survey, Field Notes of Survey (for Parry Sound road, J. O'Hanley, book 2226, reel 80); R. Louis Gentilcore and Kate Donkin, *Land Surveys of Southern Ontario* (Toronto: York University, 1973); K.H. Tops et al., *Early Settlement Patterns in Selected Townships in the North Bay Area* (North Bay, ON: Department of Geography, Nipissing University College, 1981); Fleetwood K. McKean, "Early Parry Sound and the Beatty Family," *Ontario History* 56 (1964): 3; Pearson, *Planning for Eastern Georgian Bay,* 4.
5 In fact, Ontario's characteristic regional differences date to the 1820s, when townships to the northeast and the pioneer fringe along the Shield showed slow growth and population decline compared with townships in the southern peninsula between Lakes Ontario, Erie, and Huron. J. David Wood, *Making Ontario: Agricultural Colonization and Landscape Re-creation before the Railway* (Montreal and Kingston: McGill-Queen's University Press, 2000), 6-11, and 32; Tony Fuller, "Changing Agricultural, Economic and Social Patterns in the Ontario Countryside," in *The Countryside in Ontario: Evolution, Current Challenges and Future Directions,* ed. Michael Troughton and J. Gordon Nelson (Waterloo, ON: Heritage Resources Centre, University of Waterloo, 1998), 10. See also Geoffrey Wall, "Nineteenth-Century Land Use and Settlement on the Canadian Shield Frontier," in *The Frontier: Comparative Studies,* vol. 1, ed. David Harry Miller and Jerome O. Steffen (Norman, OK: University of Oklahoma Press, 1977), 227-41; Florence Murray, "Agricultural Settlement on the Canadian Shield: Ottawa River to Georgian Bay," *Profiles of a Province* (Toronto: Ontario Historical Society, 1967), 178-86.
6 Kirkwood and Murphy, *The Undeveloped Lands*; Report of the Commissioner of Crown Lands, Ontario SP, 1867-8, Appendix 25.

7   Herbert L. Harley, "A Cruise to Georgian Bay in 1898," iii, 110.

8   Forkey, *Shaping the Upper Canadian Frontier,* 83-92; Thomas R. Roach, "The Pulpwood Trade and the Settlers of New Ontario, 1919-1938," *Journal of Canadian Studies* 22, 3 (1987): 78-88.

9   R.J. Burgar, "Forest Land-Use Evolution in Ontario's Upper Great Lakes Basin," in *The Great Lakes Forest,* ed. Susan L. Flader, 191; Gerald Killan, *Protected Places: A History of Ontario's Provincial Park System* (Toronto: Dundurn Press, 1993); Brief from Blackstone Harbour and Wood's Bay Cottage and Resident Association, presented at the Public Hearings of the Ontario Provincial Parks Council, 5 February 1977 (AO, RG, 1-IB 3, box 37).

10   J.D. McLean, Secretary, Memo to the Minister, 17 August 1897, LAC, RG 10, vol. 2851, file 176, 296-1A; W.A. ORR, Lands and Timber Branch, to D. Breithaupt, 14 May 1920, LAC, RG 10, vol. 2871, file 176-296-192; J.C. Robertson, "Dominion over Rock and Pine," *Madawaska Club, Go-Home Bay, 1898-1923* (Madawaska Club, reprinted 1972).

11   See, for example, Norman Pearson's studies on the Georgian Bay region, notably *Environment Control, Planning and Local Government in the Georgian Bay Archipelago* (1975) and *Planning for Eastern Georgian Bay* (1996).

12   Canada, Special Commission to Investigate Indian Affairs in Canada, *Report* (1858); Bruce Bowden et al., *A History of Christian Island and the Beausoleil Band,* vol. 3, rev. ed. (London, ON: Department of History, University of Western Ontario, 1990); Edward S. Rogers and Donald B. Smith, *Aboriginal Ontario: Historical Perspectives on the First Nations* (Toronto: Dundurn Press, 1994), 131-2.

13   Thomas McIlwraith, *Looking for Old Ontario: Two Centuries of Landscape Change* (Toronto: University of Toronto Press, 1997), 41, 309, 311; William Cronon, "Telling Tales on Canvas: Landscapes of Frontier Change," in *Discovered Lands, Invented Pasts: Transforming Visions of the American West,* by Jules David Prown et al. (New Haven, CT: Yale University Press, 1992).

14   Ian Radforth, *Bushworkers and Bosses: Logging in Northern Ontario, 1900-1980* (Toronto: University of Toronto Press, 1987), 28.

15   *Guidebook and Atlas,* 2-3; McMurray, *Free Grant Lands of Canada,* 18-19.

16   Douglas Leighton, "The Historical Significance of the Robinson Treaties of 1850," paper presented to the Annual Meeting of the Canadian Historical Association, Ottawa, June 1982.

17   See Dianne Newell, *Technology on the Frontier: Mining in Old Ontario* (Vancouver: University of British Columbia Press, 1986).

18   William Norton, *Explorations in the Understanding of Landscape: A Cultural Geography* (Westport, CT: Greenwood Press, 1989), 79.

19   D.F. Hewitt, Geology and Mineral Deposits of the Parry Sound-Huntsville Area Geological Report 52 (Ontario Department of Mines, 1967); MNR provincial park files, AO, IB-4, 1974-6, [PS] 19, box 3.

20   Alexander Murray, *Geological Survey of Canada, Report of Progress* 1847-8 (Montreal: Lovell and Gibson, 1849), 122; and 1848-9 (Toronto: Lowell and Gibson, 1850), 45-6; A.P. Coleman, "Copper in the Parry Sound District," *Report* of the Bureau of Mines 1899, 260-2; Department of Public Works, *Georgian Bay Ship Canal,* 18; Satterly, "Mineral Occurrences," 21, 3; J.E. Thomson, Geological Branch, to G.T. Stevens, Mining Lands Branch, 16 January 1969, in MNR Lands File 171517 (Proposed Provincial Parks, District of Parry Sound); letter from Jim Trusler, former Regional Geologist, 14 June 2000.

21   James Williamson, *The Inland Seas of North America* (Kingston, ON: J. Duff, 1854), 43; Committee on Ottawa and Georgian Bay Territory (n.p., 1864).

22 Alan Morantz, *Where Is Here? Canada's Maps and the Stories They Tell* (Toronto: Penguin Canada, 2002), 153.

23 Christopher Andreae, *Lines of Country: An Atlas of Railway and Waterway History in Canada* (Erin, ON: Boston Mills Press, 1997); Niall Mackay, *Over the Hills to Georgian Bay: A Pictorial History of the Ottawa, Arnprior and Parry Sound Railway* (Erin, ON: Boston Mills Press, 1981); Keith Fleming, "Owen Sound and the CPR Great Lakes Fleet: The Rise of a Port, 1840-1912," *Ontario History* 76, 1 (March 1984): 21.

24 "Canada and the Tourist," *Canadian Magazine,* May 1900, 4.

25 Eric Jarvis, "Georgian Bay Ship Canal: A Study of the Second Canal Age, 1850-1915," *Ontario History* 69, 2 (June 1977); Thomas Keefer, *Report of Thomas Keefer of Survey of Georgian Bay Canal Route to Lake Ontario by Way of Lake Scugog* (Whitby, ON: W.H. Higgins, 1863); William Kingsford, *The Canadian Canals: Their History and Cost* (Toronto: Rollo and Adam, 1865).

26 Angus, *History of the Trent-Severn Waterway.*

27 Department of Railways and Canals, *Annual Report* 1920 (Canada SP no. 20, 1921), 5; Angus, *History of the Trent-Severn Waterway,* 410.

28 George W. Spragge, "Colonization Roads in Canada West, 1850-1867," *Ontario History* 49, 1 (1957); Crown Land Reports, Journal of Legislative Assembly/Ontario sessional papers, 1857-64; "Report of a Survey between the Ottawa and French Rivers," Journal of Legislative Assembly 1858, Appendix 15.

29 T.L. Church (Toronto Northwest), House of Commons *Debates,* 11 March 1927, 1138.

30 Department of Public Works, *Georgian Bay Ship Canal,* 506.

31 Church, *Debates,* 14 March 1927, 1160; *Globe,* "The Old Fight Is on Again," 7 March 1927; Thompson, "Report of the Route between Lake Huron and the Ottawa River," 30 May 1837, LAC, R-2655-0-2-E, Macaulay Papers 1781-1921.

32 Sagard, *Long Journey,* 185, 189.

33 Head, *Forest Scenes and Incidents,* 281, 300, 315. Paul Kane painted this in "Fishing by Torch Light" (1845).

34 Margaret Beattie Brogue, *Fishing the Great Lakes: An Environmental History, 1783-1933* (Madison, WI: University of Wisconsin Press, 2000); Peters, "Commercial Fishing in Lake Huron," 53.

35 John Birnie et al., *Report and Recommendations of the Dominion Fisheries Commission Appointed to Enquire into the Fisheries of Georgian Bay and Adjacent Waters* (Ottawa: Government Printing Bureau, 1908), also known as Georgian Bay Fisheries Commission and hereafter referred to as the GBFC, 1. The numbers are an approximation only: there was some confusion between the three types of fish in the whitefish family, and fish are of course a highly mobile resource; poaching undoubtedly resulted in low estimates. The 1908 Royal Commission complained of "grossly defective" information gathering and "erroneous totals" (9). DLF Annual Reports; Norman S. Baldwin et al., *Commercial Fish Production in the Great Lakes 1867-1977,* Technical Report no. 3 (Ann Arbor, MI: Great Lakes Fishery Commission, 1979), 91-7; J.R. McNeill, *Something New Under the Sun: An Environmental History of the Twentieth-Century World* (New York: W.W. Norton, 2000), 258.

36 Cited in *Report of the Dominion Fishery Commission on the Fisheries of the Province of Ontario 1893-4* (Canada SP no. 10c, 1893), also known as the Ontario Fishery Commission and hereafter the OFC, 222. For individual species profiles see W.B. Scott and E.J. Crossman, *Freshwater Fishes of Canada,* 2nd ed. (Oakville, ON: Galt House, 1998).

37 Robert B. Townsend, ed., *Tales from the Great Lakes, Based on C.H.J. Snider's "Schooner Days"* (Toronto: Dundurn Press, 1995), 165-7; James P. Barry, *Ships of the Great Lakes:*

300 Years of Navigation (Berkeley, CA: Howell-North Books, 1973), 114; A.B. McCullough, *The Commercial Fishery of the Canadian Great Lakes*, Studies in Archaeology, Architecture and History: Canadian Parks Service (Ottawa: Minister of Supply and Services Canada, 1989), 41, 68, 80; Peters, "Commercial Fishing in Lake Huron," 74, 112; Hawk Tolson, "The Boats That Were My Friends: The Fishing Craft of Isle Royale," in *"A Fully Accredited Ocean,"* ed. V. Brehm. Regarding their recreational use, see "A Sailing Dinghy," *Forest and Stream*, 15 April 1899; Donald T. Fraser, "The Sailing Dinghy and the Early Races," in *Madawaska Club at Go-Home Bay, 1898-1923*, Midland: Madawaska Club, 1923, 48.

38 Raymond S. Spears, *A Trip on the Great Lakes* (Columbus, OH: A.R. Harding, 1913), 179-82; Edward S. Warner, "Towing with Steam Tugs: An Aspect of the Great Lakes Commercial Trade," in *"A Fully Accredited Ocean,"* ed. V. Brehm.

39 Aldophus Martin, cited in OFC, BB; Norman Sanders, cited in OFC, 204.

40 GBFC, 12.

41 George Cuthbertson, *Freshwater: A History and a Narrative of the Great Lakes* (Toronto: Macmillan, 1931), 129, 226, 233, 241-2; Townsend, *Tales from the Great Lakes*, 22-3; Landon, *Lake Huron*, 305; James McCannel, "Shipping out of Collingwood," *Ontario Historical Society Papers and Records*, vol. 28 (1932).

42 Muskoka and Nipissing Navigation Co., *Guide to Muskoka Lakes, Upper Magnetawan, and Inside Channel of Georgian Bay* (Toronto, 1888); LAC, RG 84 A-2-a, vol. 47, GB 109.

43 See James Barry, *Georgian Bay: The Sixth Great Lake*, 3rd ed. (Erin, ON: Boston Mills Press, 1995).

44 Andrea Gutsche et al., *Alone in the Night: Lighthouses of Georgian Bay, Manitoulin Island and the North Channel* (Toronto: Lynx Images, 1996).

45 Douglas LePan, "A Northern River, with Figures," *Far Voyages* (Toronto: McClelland and Stewart, 1990), 15.

46 The Crown reserved to itself ownership of timber and minerals on Crown land, and had done so since the Napoleonic wars at the turn of the nineteenth century. Generally, however, it profited from selling licences to private interests, giving them the rights to certain limits on Crown land. A.R.M. Lower, *The North American Assault on the Canadian Forest* (1938; reprint, New York: Greenwood Press, 1968); Richard Lambert and Paul Pross, *Renewing Nature's Wealth: A Centennial History of the Public Management of Lands, Forests and Wildlife in Ontario, 1763-1967* (Department of Lands and Forests, 1967), 118; also H.V. Nelles, *The Politics of Development: Forests, Mines and Hydro-Electric Power in Ontario, 1849-1941* (Toronto: Macmillan, 1974).

47 Though, interestingly, a seminal study in ecoregionalization by G.A. Hills in 1959, which divided the province into regions based on biomass productivity, identified the archipelago as distinct from the interior. In later studies, however, it is again subsumed into the single unit of the near north. See Ajith H. Perera and David J.B. Baldwin, "Spatial Patterns in the Managed Forest Landscape of Ontario," in *Ecology of a Managed Terrestrial Landscape: Patterns and Processes of Forest Landscapes in Ontario*, ed. Ajith Perera et al. (Vancouver: University of British Columbia Press, 2000), 75-7.

48 Commissioner of Crown Lands Report, SP 1873 no. 14; "A Statement on Timber Berths on North Shore of Lake Huron," SP 1873 no. 11; Lambert and Pross, *Renewing Nature's Wealth*, 118-20; James T. Angus, *A Deo Victoria: The Story of Georgian Bay Lumber Company, 1871-1942* (Thunder Bay: Severn Publications, 1990), 38; Burgar, "Forest Land-Use Evolution," in *The Great Lakes Forest*, ed. Susan L. Flader, 182; K.W. Horton and W.G.E. Brown, *Ecology of White and Red Pine in the Great Lakes-St. Lawrence Forest Region*, Forest Research Division, Technical Note 88 (Department of Northern Affairs and National Resources, 1960).

49 "The Teams at Wanapitei," ed. Edith Fowke, *Lumbering Songs from the Northern Woods* (Austin: University of Texas Press/American Folklore Society, 1970), 79-80; telephone interview with Frank King (London, ON: 31 January 1998).

50 A. Ernest Epp, "Ontario Forests and Forest Policy Before the Era of Sustainable Forestry," in *Ecology of a Managed Terrestrial Landscape*, ed. Ajith Perera et al., 237-75; Mark Kulberg, "'We are the Pioneers in This Business': Spanish River's Forestry Initiatives after the First World War," *Ontario History* 93, 2 (Autumn 2001): 150.

51 Department of Lands and Forests, *Forest Industry Opportunities in Georgian Bay Area* (Toronto, 1963), 9-10.

52 Memo from W.W. Orr, Lands and Timber Branch, DIA, 24 October 1906, LAC, RG 10, vol. 2852, file 176, 296-1C.

53 Angus, *A Deo Victoria*, 25.

54 "The Falling of the Pine," which Franz Rickaby attributes to Georgian Bay in *Ballads and Songs of the Shanty-Boy* (Cambridge, MA: Harvard University Press, 1926), 82-4.

55 Bridgland, Field Notes, no. 1897 (1853); Napier to Pennefather, 1856 (NAC, RG 10, vol. 727); Alexander Murray, "Report, for the Year 1857," 9.

56 Barry, *An Illustrated History of the Georgian Bay* (Erin, ON: Boston Mills Press, 1992) 173.

57 James Cleland Hamilton, *Georgian Bay: An Account of Its Position, Inhabitants, Mineral Interests, Fish, Timber and Other Resources, with Map and Illustrations* (Toronto: James Bain and Son, 1893), 155.

58 George Warecki, for example, sees the same unresolved contradictions with today's preservationists. They enjoy an improved standard of living owed largely to the consumption of natural resources yet oppose utilitarian management of those resources. *Protecting Ontario's Wilderness: A History of Changing Ideas and Preservation Politics, 1927-1973* (New York: Peter Lang, 2000), 313.

59 *Picturesque Parry Sound on Georgian Bay*, Parry Sound Yachting Fleet, n.d., AO Pamphlets no. 7; *Four Years on Georgian Bay: Life Among the Rocks* (Toronto: Copp Clark, 188?), 64; Angus, *History of the Trent-Severn Waterway*, 380.

60 J.C. Robertson, "Dominion over Rock and Pine," *Madwaska Club 1898-1923* (Madwaska Club, 1923).

61 Minnicognashene Summer Resort pamphlet, n.d. (Regional Collection, University of Western Ontario); Laurence C. Walker, *The North American Forests: Geography, Ecology, and Silviculture* (Boca Raton: CRC Press, 1999); Chao Li, "Fire Regimes and Their Simulation with Reference to Ontario," in *Ecology of a Managed Terrestrial Landscape*, ed. Ajith Perera et al., 118-9.

62 *Four Years on Georgian Bay*, 77.

63 Ardagh, "Survey of Islands in Georgian Bay," emphasis mine. He noted that even cutting timber for island numbering posts was not allowed – something that would have contributed to the surveyors' confusion.

64 The DLF permitted selected harvesting in 1939 for purposes of fire prevention, but this was obviously not exhaustive. I am grateful to Dr. Gerald Killan (King's College, University of Western Ontario) for sharing his information on Franklin Island Provincial Park (Land and Forest park files, Land Records files 16089); Ontario's Living Legacy: Land Use Strategy (Ontario MNR, 1999), 79.

65 DLF Annual Report 1929 (Ontario SP no. 7, 1930), 151; Hodgins and Benidickson, *Temagami Experience*, 107, 175. They attribute this to several factors, from limited railway access around the Temagami reserve to the small size of licences in the reserve.

66 William Bradshaw, "Georgian Bay Archipelago," *Canadian Magazine*, May 1900, 23; "Little Current: A Holiday Trip to the Commercial Capital of the Manitoulin Island,"

Mer Douce, June 1923, 16-17; *Georgian Bay and Thirty Thousand Islands: Illustrated Souvenir* (Owen Sound, ON: Fleming Publishing, 191-), Toronto Library Canadiana Collection.

67 Grand Trunk Railway and Muskoka Navigation Co., *Picturesque Muskoka: To the Highland Lakes of Northern Ontario* (1898); Georgian Bay Fisheries Commission, *Interim Report of Georgian Bay Fisheries Commission*, 17 January 1907, Appendix A in GBFC Report, 35; "Captain Mac," *The Muskoka Islands and Georgian Bay* (England: W.J. Welch, 1884), 55-8.

68 Alexander Fraser, "Field of Opportunity for the Algonquin Historical Society," *Mer Douce*, October-November 1922, 26-7; Marlow A. Shaw, *The Happy Islands: Stories and Sketches of Georgian Bay* (Toronto: McClelland and Stewart, 1926), 61; Lewis R. Freeman, *By Waterways to Gotham* (New York: Dodd, Mead, 1926), 71-2; Stella E. Asling-Riis, *The Great Fresh Sea* (New York, 1931), 1.

69 Percy Robinson, *Georgian Bay* (Toronto: privately printed, 1966), 11; Ella J. Reynolds, "The Pine Tree," in *Wahsoune Guests*, 33; interview with Margaret Rossiter (London, ON: 28 October 1997).

70 Virgil Martin, *Changing Landscapes of Southern Ontario* (Erin, ON: Boston Mills Press, 1988), 16.

71 LePan, "A Northern River," *Far Voyages*.

72 "The Ballad of Georgian Bay," in *Souvenir: The Story of the Georgian Bay*, by W.H. Adams (Toronto: Fristbrook Box, 1911).

73 Cameron Taylor, *Enchanted Summers: The Grand Hotels of Muskoka* (Toronto: Lynx Images, 1997); Barbaranne Boyer, *Muskoka's Grand Hotels* (Erin, ON: Boston Mills Press, 1987); Geoffrey Wall and John S. Marsh, *Recreational Land Use: Perspectives on Its Evolution in Canada* (Ottawa: Carleton University Press, 1982).

74 Barry, *An Illustrated History of the Georgian Bay* (Erin, ON: Boston Mills Press, 1992) 164.

75 "Comfort in Camp," *Harper's Weekly*, 3 October 1896. Two good surveys of nineteenth-century vacationing are Patricia Jasen, *Wild Things: Nature, Culture, and Tourism in Ontario, 1790-1914* (Toronto: University of Toronto Press, 1995) and Cindy S. Aron, *Working at Play: A History of Vacations in the United States* (New York: Oxford University Press, 1999).

76 Elmes Henderson, "Some Notes on a Visit to Penetanguishene and Georgian Bay in 1856," Ontario Historical Society *Papers and Records*, vol. 28 (1932).

77 "Statement Showing Islands on the East Shore of Georgian Bay, South of Moose Deer Point, Sold by This Department under Surrender of 1856," 19 May 1905, LAC, RG 10, vol. 2851, file 176, 296-1A; MacMahon, *Island Odyssey*, 234.

78 Richard B. Wright, *The Age of Longing* (Toronto: HarperCollins, 1995), 230.

79 Theodore D. Wakefield, "Cruising Georgian Bay in the Grand Manner," *Inland Seas* 47, 4 (Winter 1991); Michael Morris, "The Contrasting Social Environments of a Vernacular Building Tradition: A Study of the Inter-War Weekend Cabins in Cheshire," in *Meaningful Architecture*, ed. Martin Locock, 272; Alan Gowans, *The Comfortable House: North American Suburban Architecture 1890-1930* (Cambridge, MA: MIT Press, 1986), 76.

80 Freeman, *By Waterways to Gotham*, 242; *Globe*, "Spinner," "An Angler's Paradise," 7 September 1889.

81 OFC, 246; Minnicognashene Summer Resort, n.d.; Harry Symons, *Ojibway Melody* (Ambassador Books/Copp Clark, 1946), 45-6; "A Thrilling Expedition Up North," in *Guide to Muskoka Lakes*, by Muskoka and Nipissing Navigation Co., 37. Emphasis mine.

82 Amy Willard Cross, *The Summer House: A Tradition of Leisure* (Toronto: HarperCollins, 1992), 190. See also Joe LeMoine, Island F45, *Wind, Water Rock and Sky*, vol. 2 (1997), 31.

83 Henry Hubbard and Theodora Kimball, *An Introduction to the Study of Landscape Design,* rev. ed. (New York: Macmillan, 1929), 189-90; Harold Kalman, *A History of Canadian Architecture,* vol. 2 (Toronto: Oxford University Press, 1994), 724; George T. Kapelos, *Interpretations of Nature: Contemporary Canadian Architecture, Landscape and Urbanism* (Kleinburg, ON: McMichael Canadian Art Collection, 1994), 23; Gowans, *The Comfortable House;* William Comstock, *Bungalows, Camps and Mountain Houses* (New York: William T. Comstock, 1924), 10; Linda Flint McClelland, *Building the National Parks: Historic Landscape Design and Construction* (Baltimore: Johns Hopkins University Press, 1998).

84 Mary Byers, *Longuissa* (Erin, ON: Boston Mills Press, 1988); CPR Collection, "French River Hotels," AO, box 869 C310-1-0-7.

85 Angus, *A Deo Victoria,* 9.

86 L.M. Montgomery, *The Blue Castle* (Toronto: McClelland and Stewart, 1926), 188; Ardagh, "Survey of Islands in Georgian Bay"; letter from Clive Powsey (Bowmanville, ON: 27 January 1998).

87 Richard Walker, "Unseen and Disbelieved: A Political Economist among Cultural Geographers," in *Understanding Ordinary Landscapes,* ed. Paul Groth and Todd W. Bressi, 165.

88 LePan, "Islands of Summer," *Weathering It,* 19.

89 Interview with Gillian Michel (London, ON: 5 March 1998).

90 F.M. de la Fosse, "Early Days in Muskoka," *Ontario Historical Society Papers and Records,* vol. 34 (1942), 105; *100 Years Go-Home Bay, 1898-1998,* 15-16.

91 Telephone interview with Chris Cavanaugh (London, ON: 31 January 1998); Wells, *Cruising Georgian Bay,* 2, 73; *Globe and Mail,* "Captains Outrageous," August 1986, Toronto supplement.

92 Letter from Eric McIntyre, MNR Fisheries biologist, to the author (Parry Sound, ON: 9 February 1998), emphasis in original.

CHAPTER 3: A VIVID REMINDER OF A VANISHED ERA

1 Her Excellency, Rt. Honourable Adrienne Clarkson, the Senate, Ottawa, 7 October 1999.

2 Trigger, *Children of Aataentsic,* 779, 788; Donald Smith emphasizes the enduring hostility between the Ojibwa and the Iroquois in "The Dispossession of the Mississauga Indians: A Missing Chapter in the Early History of Upper Canada," *Ontario History* 73, 2 (1981): 67-87; Diamond Jenness, *The Ojibwa Indians of Parry Island, Their Social and Religious Life* (Ottawa: Department of Mines/National Museum of Canada, 1935), 1.

3 Fr. Paul Ragueneau, "Account of the Mission of the Holy Ghost," *Jesuit Relations and Allied Documents,* 1649-50, vol. 35, ed. Reuben Gold Thwaites (Cleveland: Burrows, 1898), 179.

4 Thompson in *Muskoka and Haliburton 1615-1875,* ed. Florence B. Murray, 86.

5 Brébeuf, "Relation of what occurred among the Hurons in the year 1635," letter to Fr. Paul le Jeune, *Relations,* 1634-6, vol. 8, 81.

6 Cormier, *Jean-Baptiste Perrault,* 144.

7 Written 1 March 1649 at St. Marie, *Relations,* vol. 35, 257-8.

8 Anna Jameson, *Winter Studies and Summer Rambles in Canada* (Toronto: McClelland and Stewart, 1923), 356.

9 Bond Head to Lord Glenelg, 20 August 1836, "Communications and Despatches Relating to Recent Negotiations with the Indians," Appendix to *Journals of Legislative Assembly,* 1838.

10 James A. Clifton, *A Place of Refuge for All Time: Migration of the American Potawatomi into Upper Canada, 1830-1850* (Canadian Ethnology Service Paper no. 26. Ottawa: National Museums of Canada, 1975), 96.

11 *Guidebook and Atlas of Muskoka and Parry Sound Districts*, 16.

12 Canada, *Indian Treaties and Surrenders*, 149, 204-5; David Shanahan, "The Manitoulin Treaties, 1836 and 1862: The Indian Department and Indian Destiny," *Ontario History* 86, 1 (March 1994): 16-7, 22; W.R Wightman, *Forever on the Fringe: Six Studies in the Development of the Manitoulin Island* (Toronto: University of Toronto Press, 1982), 13-15.

13 On the Jesuit reaction, see James Morrison, "Upper Great Lakes Settlement: The Anishinabe-Jesuit Record," *Ontario History* 86, 1 (March 1994): 58-9, 62; Theresa A. Smith, *Island of the Anishnaabeg*, 104; Ragueneau, *Relations*, 1648, vol. 33, 155. On the opposition to Bond Head, Memorial to Lord Durham by Committee of Aborigines Protection Society, 3 April 1838, APS *Annual Reports* 1838-46 (London, UK), 26, at the D.B. Weldon Special Collection, University of Western Ontario; "Report of the Special Commission ... to Investigate Indian Affairs in Canada," *Journal of the Legislative Assembly of the Province of Canada*, 1858, Appendix 21.

14 McLean, *From Lochnaw to Manitoulin*, 34.

15 9 August 1793 at York. *Mrs. Simcoe's Diary*, ed. Mary Q. Innis (Toronto: Macmillan, 1965), 103.

16 Schmalz, *The Ojibwa of Southern Ontario*, 89.

17 See Tuan, *Topophilia*, 125; Carl Berger, *Science, God and Nature in Victorian Canada* (Toronto: University of Toronto Press, 1983). On the interest in Native culture in the new field of ethnography, see Morrison, "Upper Great Lakes Settlement," 65-6; Worster, *Nature's Economy*, 172; Carole Henderson Carpenter, *Many Voices: A Study of Folklore Activities in Canada and Their Role in Canadian Culture*, Canadian Centre for Folk Culture Studies, Paper no. 26 (Ottawa: National Museums of Canada, 1979), 270.

18 "A Boy Scout," *Mer Douce*, November-December 1921, 29.

19 Henry, *Travels and Adventures*, 169-70.

20 Allan Smith, "Farms, Forests and Cities: The Image of the Land and the Rise of the Metropolis in Ontario, 1860-1914," in *Old Ontario: Essays in Honour of J.M.S. Careless*, ed. David Keane and Colin Read (Toronto: Dundurn Press, 1990), 74-5.

21 MacMillan, *Reflections*, 26.

22 See, for example, Asling-Riis, *The Great Fresh Sea* and the serial "The Cave of the Spirit: A Traditional Legend of Indian War and Love and Hate," which ran in *Mer Douce* in 1921.

23 *Guidebook and Atlas of Muskoka and Parry Sound Districts*, 2; *Picturesque Parry Sound on Georgian Bay*, 10.

24 See an example of such a deed at LAC, RG 10, vol. 6831, file 506-2-1.

25 "Annals of Go-Home," *Madawaska Club 1898-1923*, 27.

26 DIA *Annual Report* 1896 (Canada SP no. 14, 1897), xxviii; Bowden et al., *A History of Christian Island*, 66.

27 Francis Parkman, *France and England in North America, Part Second: The Jesuits in North America* (Boston: Little, Brown, 1906), 141; Sioui, *Huron-Wendat*, has argued that non-Native historians have overemphasized both the role and warlike nature of the Iroquois (75).

28 OFC, 183-5.

29 Interviews with Keith Townley (Toronto: 22 May 1997) and Neil Campbell (Toronto: 19 May 1997).

30 Franz Mischaquod Koennecke, "Once Upon a Time There Was a Railway on Wasauksing First Nation Territory," in *Co-Existence? Studies in Ontario-First Nations Relations,* ed. Bruce W. Hodgins et al. (Peterborough, ON: Frost Centre for Canadian Heritage and Development Studies, 1992); *Journal of the Legislative Assembly of the Province of Canada,* 1858; James T. Angus, "How the Dokis Protected Their Timber," *Ontario History* 81, 3 (September 1989): 181-200.

31 I borrow this term from Ian McKay's *The Quest of the Folk: Antimodernism and Cultural Selection in Twentieth-Century Nova Scotia* (1994).

32 Symons, *Ojibway Melody,* 4.

33 John Bentley Mays, *Arrivals: Stories from the History of Ontario* (Toronto: Penguin Canada, 2002), 201-2. Mays draws heavily on a 1901 account of the Penetang francophones: A.C. Osborne. "The Migration of *Voyageurs* from Drummond Island to Penetanguishene in 1828," *Ontario Historical Society Papers and Records,* vol. 3 (1901), 123-66.

34 Adelaide Leitch, "The Island Galaxy of Georgian Bay," *Canadian Geographical Journal,* August 1955, 123; Interview with Margaret Hankinson (London, ON: 9 February 1998). Regarding the Varley painting, apparently Robert McMichael assigned it the title *Indians Crossing Georgian Bay* when it was donated to the gallery in 1969. The title was changed in 1982 by the gallery's curator (email from Linda Morita, Librarian/ Archivist, McMichael Gallery, 2 January 2003).

35 Asling-Riis, *The Great Fresh Sea,* 91-2.

36 Ronald H. Perry, *Canoe Trip Camping* (Toronto: J.M. Dent and Sons, 1943), 2.

37 *Globe,* "An Angler's Paradise," 7 September 1889.

38 Wilfred Campbell, *The Canadian Lake Region* (Toronto: Musson Book Company, 1910), 113.

39 Spears, *A Trip on the Great Lakes,* 65. How those of the "Indian faculties" are supposed to conquer *and* rejoice in the wilderness at the same time, he doesn't explain.

40 John Newlove, "The Pride" (1968), cited in *Land Sliding,* New, 39.

41 The only other creation story for the islands I have encountered is recorded by Jenness, *Ojibwa Indians of Parry Island,* as told by Jonas King: "Nanibush was hunting the giant beaver, *wabnick.* He drove it from lake Superior to Georgian bay, where the beaver, thoroughly exhausted, crawled half-way out of the water and turned to stone. Nanibush, seeking its hiding-place, smote the land, with his club, and shattered it into the maze of islands that exist today" (38).

42 Daniel Francis, *National Dreams: Myth, Memory and Canadian History* (Vancouver: Arsenal Pulp Press, 1997), 147-8; Margaret Atwood, *Strange Things: The Malevolent North in Canadian Literature* (Oxford: Clarendon Press, 1995), 47, 60.

43 See, for example, Eric Kaufmann, "'Naturalizing the Nation': The Rise of Naturalistic Nationalism in the United States and Canada," *Comparative Studies in Society and History* 40, 4 (October 1998): 666-95; Cara Aitchison et al., "Heritage Landscapes: Merging Past and Present," *Leisure and Tourism Landscapes: Social and Cultural Geographies* (London: Routledge, 2000), 94-109; William Cronon, "Telling Tales on Canvas: Landscapes of Frontier Change," 81, and Brian W. Dippie, "The Moving Finger Writes: Western Art and the Dynamics of Change," 106, in *Discovered Lands, Invented Pasts,* by Jules David Prown et al.; W.L. Morton, "The North in Canadian Historiography," *Transactions, Royal Society of Canada* 4, 8 (1970): 31-40; Susan Wood, "The Land in Canadian Prose, 1845-1945" (PhD thesis, Carleton University, 1988), 82-3, 102. Kaufmann traces this aesthetic of "naturalizing the nation" to Switzerland and Scandinavia.

44 For a useful international comparison of wilderness ideology and national park creation, see C. Michael Hall and Stephen J. Page, "Tourism and Recreation in the Pleasure

Periphery: Wilderness and National Parks," *The Geography of Tourism and Recreation: Environment, Place and Space* (London: Routledge, 2002); Steinberg, *Down to Earth,* 148-52. At Yellowstone, for example, efforts were made to remove resident First Nations because they did not suit expectations of a "wilderness" park.

45 "On the Georgian," in *Wahsoune Guests,* 21; E.J. Pratt, *Brébeuf and His Brethren* (Toronto: Macmillan, 1940), 47.

46 Paul Wallace, "A Bit of Rock, River and Legend," *Canadian Forum,* September 1922.

47 See Wilfrid Jury and Elsie McLeod Jury's account of their 1948-51 excavation in *Sainte-Marie Among the Hurons* (Toronto: Oxford University Press, 1954), which followed an initial excavation by the Royal Ontario Museum between 1940-3. The site was adopted by the province in 1964. On Beausoleil, see "Georgian Bay Islands National Park: History and Establishment," LAC, RG 84 A-2-a, vol. 487, file GB2 (U325-9-6).

48 Cormier, Introduction, *Jean-Baptiste Perrault,* 21; Allan Greer, *The People of New France* (Toronto: University of Toronto Press, 1997), 79, 84-5.

49 Interview with Monte Hummel (Toronto: 29 May 1997). Hummel, a founding member of Pollution Probe, is currently the president of World Wildlife Fund Canada.

50 Anna G. Young, "The Rough Diamond," *Inland Seas* 18, 2 (Summer 1962); Floren et al., *Ghosts of the Bay,* 106; Ian MacLaren, "Cultured Wilderness in Jasper National Park," *Journal of Canadian Studies,* Fall 1999, 13.

51 Herbert L. Harley, "A Cruise to Georgian Bay in 1898," iii, 111.

52 Interview with Dave Hanna (Toronto, 22 May 1997).

53 Symons, *Ojibway Melody,* 192-3.

54 Henry R. Schoolcraft, *Personal Memoirs of a Residence of Thirty Years with the Indian Tribes on the American Frontiers* (Philadelphia: Lippincott, Grambo, 1851. Reprint, New York: Arno Press, 1975), 587; Sarah M.F. Ossoli, *Summer on the Lakes, in 1843* (New York, 1844), 246.

55 Hector Charlesworthy, "The Canadian Girl: An Appreciative Medley," *Canadian Magazine,* May 1893, 187-8; J.D. Logan, "Canadian Womanhood and Beauty," *Canadian Magazine,* January 1913, 253.

56 Interview with Tom Lawson (Port Hope, ON: 2 June 1997). For correspondence on island sales, see, for example, LAC, RG 10, vol. 2851, file 176, 296-1A. In 1931, 125 girls attended the Y camps on Beausoleil, compared with 350 boys (Superintendent Report, 20 November 1931, LAC, RG 84, vol. 2381, C-1445-3, pt 1; also Lloyd to Harkin, 26 January 1929).

57 Ruth H. McCuaig, *Our Pointe au Baril* (Hamilton, ON: R.H. McCuiag, 1984), 224.

58 Interview with Tom Lawson (Nares Inlet, ON: 12 August 1999).

59 At Go Home, for example, debates over hydro lasted for twenty years, until lines were installed in 1952; Sans Souci followed in 1955. "Annals of the Club," *Madawaska Club 1898-1973* and *100 Years,* 1999; MacMahon, *Island Odyssey,* 163.

60 Susan Perrin, cited in Eaton and Weston, 1995; Ross Skoggard, "A Summer on Georgian Bay," *City and Country Home* (June 1988), describing the structure on Deer Island.

61 R. Bruce Hull, "Image Congruity, Place Attachment and Community Design," *Journal of Architectural and Planning Research* 9, 3 (Autumn 1992).

62 LePan, "Astrolabe," *Weathering It,* 173. Regarding the murals, see Dennis Reid, *The MacCallum Bequest* (Ottawa: National Gallery of Canada, 1969) and Pierre Landry, *The MacCallum-Jackman Cottage Mural Paintings* (Ottawa: National Gallery of Canada, 1990).

63 LePan, "Rough Sweet Land," *Weathering It,* 217-23.

64 Interim Report of Georgian Bay Fisheries Commission, 17 January 1907, Appendix A in GBFC Report, 35.

65 Northrop Frye, "Conclusion," *A Literary History of Canada,* 18; William Morton, "The Relevance of Canadian History," in *The Canadian Identity* (Madison, WI: University of Wisconsin Press, 1961), 93; Atwood, *Survival,* 30; Roald Nasgaard, *The Mystic North: Symbolist Landscape Painting in Northern Europe and North America, 1890-1940* (Toronto: Art Gallery of Ontario/University of Toronto Press, 1984), 163.

66 Shaw, *Happy Islands,* 207. Emphasis in original.

67 Robinson, *Georgian Bay,* 9.

68 Conversation with Clive Powsey (Ashton/Evicta Gallery, Toronto: 8 April 2000).

69 Aron, *Working at Play,* 143, 157; George Stankey, "Beyond the Campfire's Light: Historical Roots of the Wilderness Concept," *Natural Resources Journal* 29 (Winter 1989).

70 Govier, *Angel Walk,* 31.

71 Maria Tippett, *Art at the Service of War: Canada, Art and the Great War* (Toronto: University of Toronto Press, 1984), 59, 108-9; Susan Butlin, "Landscape as Memorial: A.Y. Jackson and the Landscape of the Western Front, 1917-1918," *Canadian Military Journal* 5, 2 (Autumn 1996): 62-70.

72 The Queen Elizabeth Sea Cadet Camp opened on Beausoleil in 1942, and the Princess Alice Cadet Camp at Minnicog in 1943. On the *Penetang 66,* see Neil Campbell interview (Toronto: 19 May 1997).

73 Arthur Lismer, "To Georgian Bay," *Canadian Forum,* May 1921, 240-2.

74 Jameson, *Winter Studies,* 391; Wright, *Age of Longing,* 236-7; Campbell interview (Toronto, 25 June 1997).

75 M.W. Bowman to D.J. Allan, Superintendent of Reserves and Trusts, DIA, 1 June 1947, LAC, RG 10, vol. 2851, file 176, 296-1; interview with Wendy Willoughby (Toronto: 27 May 1997).

76 Kathleen Coburn, *In Pursuit of Coleridge* (Toronto: Clarke, Irwin, 1977), 184.

77 Michael Ondaatje, *The English Patient* (Toronto: McClelland and Stewart, 1992), 86-7, 296.

78 Hummel interview (29 May 1997).

79 David Macfarlane, *Summer Gone* (Toronto: Alfred A. Knopf Canada, 1999), 122-4; interviewed by Eleanor Wachtel, "The Arts Today," CBC Radio (12 October 1999).

CHAPTER 4: ROCKS AND REEFS

1 A.Y. Jackson, *A Painter's Country* (Toronto: Clarke, Irwin, 1964), 31.

2 Richard West Sellars argues that people are likely to see any natural landscape and unoccupied space as unimpaired, in *Preserving Nature in the National Parks: A History* (New Haven, CT: Yale University Press, 1997), 287.

3 Govier, *Angel Walk,* 558. Emphasis in original.

4 Interview with Mary Lee Otto (London, ON: 7 February 1998).

5 Smith, *Island of the Anishnaabeg,* 47; Sioui, *Huron-Wendat,* 16.

6 John Hartman, *Georgian Bay: Drawings by John Hartman* (Peterborough, ON: Broadview Press, 1989).

7 Simpson, "Journey for Frances," 13; LePan, "A Stream of Images," *Far Voyages,* 21-3.

8 J. Gray Sweeney, *Great Lakes Marine Painting of the Nineteenth Century* (Muskegon, MI: Muskegon Museum of Art, 1983); Roger B. Stein, *Seascape and the American Imagination* (New York: Clarkson N. Potter/Whitney Museum of American Art, 1975).

9 W.W. Campbell, "Sunset, Lake Huron," *Lake Lyrics and Other Poems* (St. John, NB: J. and A. McMillan, 1889), 55-6.

10 Colton, *Tour of the American Lakes,* 66.

11 Freeman, *By Waterways to Gotham,* 10.

12 Department of Marine and Fisheries, *Annual Report 1894* (SP no. 11, 1895), 10.
13 Worsley to his father, Nottawasaga, 6 October 1814, reprinted in the 1968 edition of *Leaves from the War Log of the Nancy*, xlviii.
14 Interview with Andy and Jean Hamelin (Midland, ON: 16 June 1997).
15 Daniel Marchildon estimates that the scoot population reached between 60 and 100 in the mid-1950s. *Flying Low: A History of the Georgian Bay Scoot* (Midland, ON: Huronia Museum, 1994), 6, 34; GBINP records, LAC, RG 84, vol. 2831, C-1445-3, pt. 4.
16 Peter Labor, "The Canot du Maître – Master of the Inland Seas," in *The Canoe in Canadian Cultures*, ed. John Jennings et al. (Toronto: Natural Heritage/Natural History, 1999).
17 Hughes interview (Toronto: 1 June 1997). In 1888 the Department of Marine and Fisheries sent the *Cruiser*, the first steam-powered vessel to patrol the Canadian Great Lakes, to inspect Lake Huron and Georgian Bay; it remained on the Bay through 1891. Brogue, *Fishing the Great Lakes*, 217-8.
18 Jack Legault, Honey Harbour, cited in Long, "Taxi's Waiting"; this article includes good photographs of local Bay boats. On the Georgian Bay launch, see *100 Years: Go Home Bay 1898-1998* (Madawaska Club at Go Home Bay, 1999), 42; interviews with Neil Campbell (Toronto: 19 and 25 May, 15 June 1997; 18 September 1999; 23 August 2000).
19 Michael Mitchell, "Papa Duck's Tale," *City and Country Home*, June 1988, 64.
20 Sagard, *Long Journey*, 246.
21 Bayfield, Penetanguishene, to Barrie, Kingston, 20 October 1820, LAC, MG 12, adm. 1, vol. 3444.
22 Letter on the "Ottawa Timber Trade," author unknown, ca. 1830, LAC, MG 55/24, no. 292.
23 MacMahon, *Island Odyssey*, 129-30. On fluctuations in water levels, see *Shoreline Resource Management and Ontario's Provincial Parks* (Occasional Paper 8. Heritage Resources Centre, University of Waterloo, 1998).
24 Report of Colonization Road Operations during 1866, *Report of the Commissioner of Crown Lands*, 1867-8, Appendix 25.
25 Official Ontario Road Maps, Department of Highways, 1941-70 (Serge Sauer Map Library, University of Western Ontario); interview with David Hodgetts (Hamilton, ON: 26 June 1997).
26 Gould, "A Sixth Great Lake," (1967) 125.
27 See Chapter 6 regarding nature/wilderness reserves at Moon Island, McCrae, and Go Home Lakes.
28 Jameson, *Winter Studies*, 156.
29 Taped comments submitted by Dr. Ross Hodgetts (Spruce Grove, Alberta, received by the author 6 August 1997). On the effect of this perspective on ecological sympathies, see D.M.R. Bentley, *The Gay/Grey Moose: Essays on the Ecologies and Mythologies of Canadian Poetry, 1660-1990* (University of Ottawa Press, 1992), 94, 206, 215; Jamie Benidickson, *Idleness, Water and a Canoe: Reflections on Paddling for Pleasure* (Toronto: University of Toronto Press, 1997), 228.
30 Macfarlane, *Summer Gone*, 104; LePan, "Astrolabe," *Weathering It*, 173.
31 Milne cited in David P. Silcox, *Painting Place: The Life and Work of David B. Milne* (Toronto: University of Toronto Press, 1996), 241; Peter Crossley, "The Simple Life," *City and Country Home*, May 1989; *Viewpoints: One Hundred Years of Architecture in Ontario, 1889-1989* (Ontario Association of Architects/Agnes Etherington Art Centre, 1989), 15; on maritime architecture see Brendan Gill and Dudley Witney, *Summer Places* (Toronto: McClelland and Stewart, 1978), 151; and Peter Ennals and Deryck

Holdsworth, "Vernacular Architecture and the Cultural Landscape of the Maritime Provinces: A Reconnaissance," *Acadiensis* 10, 2 (1981): 188.

32 Landry, *MacCallum-Jackman Cottage Mural Paintings*, 12, 17; "Honey Harbour History," *Honey Harbour Hoots*, June 1933, LAC, RG 84 A-2-a, vol. 47, file GB 109; Sing, Ontario Land Survey, 1898, LAC, RG 10, vol. 2852, file 176, 296-1B. The environmental and political controversy of floating cottages is discussed further in Chapter 6.

33 Murray, *Geological Survey of Canada, Report of Progress*, 1857 (1858), 10; Freeman, *By Waterways to Gotham*, 71.

34 Alan Andrew MacEachern, "In Search of Eastern Beauty: Creating National Parks in Atlantic Canada, 1935-1970" (PhD thesis, Queen's University, 1997), 50.

35 Jameson, *Winter Studies*, 160; Bigsby, *Shoe and Canoe*, 87; "All Around Trip, by Dolly," in *Guide to Muskoka Lakes*, by Muskoka and Nipissing Navigation Co., 61.

36 Govier, *Angel Walk*, 496-7.

37 Metaphors used to describe the shoals include a flotilla, Govier, "The Orange Kite," in *The Immaculate Conception Photography Gallery and Other Stories* (Toronto: Little, Brown, 1994), 85; lurking submarines, Symons, *Ojibway Melody*, 94; and basking whales, Margaret Atwood, "True Trash," in *Wilderness Tips* (Toronto: McClelland and Stewart, 1997), 15.

38 For concise explanations of the Bay's geology, see Philip S.G. Kor, "Rhythm in Rock: Georgian Bay," in *Islands of Hope: Ontario's Parks and Wilderness*, ed. Lori Labatt and Bruce Litteljohn (Willowdale, ON: Firefly Books, 1992), 155, and John B. Theberge, *Legacy: The Natural History of Ontario* (Toronto: McClelland and Stewart, 1989), 38-40.

39 Doris McCarthy, "The Pre-Cambrian Shield," in *Islands*, ed. Alan Stein (Parry Sound, ON: Church St. Press, 1994).

40 Interview with Peter O'Brian (Toronto: 10 June 1997); Susan Toth, *England As You Like It* (New York: Ballantine Books, 1995), 137-8.

41 A.J.M. Smith, "The Lonely Land," reprinted in *Canadian Anthology*, 1st ed., ed. Carl Klinck (Toronto: Gage, 1966), 276-7; Dennis Lee, "Civil Elegies," cited in W.J. Keith, *Literary Images of Ontario* (Toronto: University of Toronto Press, 1992), 9.

42 Stankey, "Beyond the Campfire's Light," 18; J.B. Jackson, *Discovering the Vernacular Landscape*, 61-2. Jules David Prown et al., *Discovered Lands, Invented Past*, is a useful introduction to American visions of the "ideal" landscape in art.

43 The McBrien Island home was designed by architect C. Blakeway Millar; ironically, the owners of Little Jane are the brother and sister-in-law of the owner of McBrien. Nicole Eaton and Hilary Weston, *In a Canadian Garden* (Markham, ON: Viking Studio Books, 1989), 106. Thank you to Graeme Wynn for his comments on this subject.

44 Thomas Schlereth, "Regional Culture Studies and American Culture Studies," in *Sense of Place: American Regional Cultures*, ed. Barbara Allen and Thomas J. Schlereth (Lexington, KY: University of Kentucky, 1990), 179.

45 In J.J. Talman, ed., *Basic Documents in Canadian History* (Princeton, NJ: D. Van Nostrand, 1959), 25-7. *Ghosts of the Bay* identifies two wrecks that may be the *Griffon*, one off Manitoulin, the other near Tobermory (264-6).

46 Madeleine Béland, *Chansons des Voyageurs, Coureurs de Bois et Forestiers* (Quebec: Presses de L'Université Laval, 1982), 141-5, 260; William M. Doerflinger, *Shantymen and Shantyboys: Songs of the Sailor and Lumberman* (New York: Macmillan, 1951), 221. Ryden, in *Mapping the Invisible Landscape*, sees this same negative image of the woods as "an antagonistic force" in logging folklore (82).

47 Wilfred Campbell, *Canada* (Toronto: Macmillan, 1906), 192; on the storm of 1913, see Landon, *Lake Huron*, 325; Ratigan, *Great Lakes Shipwrecks and Survivals*, 125-8, and Schlereth, "Regional Culture Studies," 180.

48 Department of Public Works, *Georgian Bay Ship Canal,* 20, 148.
49 Soaphy Anderson, "The Journey of the First White Settlement across Georgian Bay," *Mer Douce,* September-October 1921, 8. Anderson was the daughter of Capt. Thomas G. Anderson, who was responsible for relocating the Coldwater Band to Manitoulin.
50 Merrill Cook, "Georgian Bay at Indian Summer," *Shore Lines and Sand Songs* (Boston: Beacon Press, n.d.), 14.
51 Interview with Keith Townley (Toronto: 22 May 1997).
52 Donald Worster, *Nature's Economy: A History of Ecological Ideas* (1994 [1977]), 178-9; Carolyn Merchant, *The Death of Nature: Women, Ecology and the Scientific Revolution* (San Francisco: Harper and Row, 1980), 8-9; Atwood, *Survival.* Karen Dubinsky has argued that provocative imagery of danger and pleasure transformed Niagara Falls into a pseudo-sexual experience for the male viewer; see "'The Pleasure Is Exquisite but Violent': The Imaginary Geography of Niagara Falls in the Nineteenth Century," in *Gender and History in Canada,* ed. Joy Parr and Mark Rosenfeld (Toronto: Copp Clark, 1996).
53 Solomon, *Crossing the Distance,* 153.
54 Govier, "Aliens," 1994, 139.
55 Hamelin interview (16 June 1997).
56 Smith, *Island of the Anishnaabeg,* 121, 146.
57 J.E.H. MacDonald, *The Tangled Garden,* text by Paul Duval (Toronto, ON: Cerebrus/Prentice-Hall, 1978), 45; Kenneth M. Dewar, "Perceptions of the Canadian Wilderness: Literary and Visual Responses to the North Shore of Lake Superior, 1663-1926" (MA thesis, York University, 1983), 113-14.
58 Gaile McGregor, *The Wacousta Syndrome: Explorations in the Canadian Landscape* (Toronto: University of Toronto Press, 1985), 54-5.
59 Robinson, *Georgian Bay,* 19; Solomon, *Crossing the Distance,* 153; Kor, "Rhythm in Rock," in *Islands of Hope,* ed. Lori Labatt and Bruce Litteljohn, 156. For the shoal metaphors, see Katherine Govier, "The Orange Kite," in *Immaculate Conception,* 85, and Symons, *Ojibway Melody,* 94.
60 Symons, *Ojibway Melody,* 226; Atwood identifies ice, water, and rock as the favourite tools for dispatching protagonists in Canadian literature in *Survival* (55).
61 Freeman, *By Waterways to Gotham,* 9.
62 Berleant, *The Aesthetics of Environment,* 167.
63 LePan, "Red Rock Light," *Weathering It,* 202-3.
64 Interview with John Hartman (LaFontaine, ON: 16 June 1997).
65 Interview with Ulla Elliott (2 June 1997).
66 Interview with John Lord (Toronto: 28 May 1997).
67 O'Brian interview (10 June 1997).
68 *Relations,* 1649-50, vol. 33, 79-83; Pearson, *Planning for Eastern Georgian Bay,* 26; Smith, *Island of the Anishnaabeg,* 4; Schmalz, *The Ojibwa of Southern Ontario,* 20.
69 Muskoka and Nipissing Navigation Co., *Guide to Muskoka Lakes,* 13.
70 Solomon, *Crossing the Distance,* 153.
71 LePan, "Emblem," *Far Voyages,* 179; "Astrolabe," *Weathering It,* 173.
72 Bigsby, *Shoe and Canoe,* 98-100.
73 Wright, *Age of Longing,* 230; Govier, *Angel Walk,* 556. Histories of Canadian art single out the clear northern light as a key factor in the evolution of a distinct Canadian landscape art, from the late nineteenth century through to the Group of Seven.
74 Jay Appleton argues that our definition of beauty in landscape evolved from innate survival instincts: what we find beautiful or appealing is that which affords us the

ability to see (the prospect) without being seen (the refuge). *The Experience of Land-scape*, rev. ed. (London: John Wiley and Sons, 1996), 200-1, 222.

75 Email from Rob Thomson (Charlottetown, PEI: 12 July 1997).

76 Paul Simpson-Housley and Glen Norcliffe, "No Vacant Eden," in *A Few Acres of Snow*, ed. Simpson-Housley and Norcliffe, 2.

77 Govier, *Angel Walk*, 155.

78 Herbert L. Harley, "A Cruise to Georgian Bay in 1898," iii, 112; *Blackstone Harbour Provincial Park Public Participation Phase 1* (MNR, 1975), 9; "Briefing on GBA Water and Air Quality Initiatives in Georgian Bay Area," to Tony Clement, Minister of the Environment, 9 September 1999, Birnbaum files.

79 Hodgetts (tape received 6 August 1997).

80 Macfarlane, *Summer Gone*, 94; Freeman, *By Waterways to Gotham*, 71; *Massasauga Provincial Park Management Plan* (MNR, 1993), 3.

CHAPTER 5: OUR DEAR NORTH COUNTRY

1 Jackson cited in "An Apprehended Vision: The Philosophy of the Group of Seven," by Ann Davis (PhD thesis, York University, 1973), 155; Charles C. Hill, *Art for a Nation* (Ottawa: National Gallery of Canada, 1995), 50.

2 Atwood, *Survival*, 50-1, 62.

3 See Cara Aitchison et al., *Leisure and Tourism Landscapes*.

4 Keith, *Literary Images of Ontario*, 195; Jameson, *Winter Studies*, 159.

5 Schoolcraft, 16 August 1837, *Personal Memoirs*, 566.

6 Bradshaw, "Georgian Bay Archipelago," 19.

7 Such descriptions are ubiquitous; see, for example, "Georgian Bay to Parry Sound" and "Holiday in Parry Sound," in *Guide to Muskoka Lakes*, by Muskoka and Nipissing Navigation Co.; *Georgian Bay and Thirty Thousand Islands: Illustrated Souvenir* (191-); *Canadian Magazine*, "The Captain," "A Honeymoon in a Sailing Dinghy," March 1901; Herbert L. Harley, "A Cruise to Georgian Bay in 1898," iii, 111.

8 New, *Land Sliding*, 68; *Globe*, "Spinner," 7 September 1889.

9 Gatenby, Introduction, in *The Wild Is Always There*, ed. Gatenby, xiv; *Picturesque Parry Sound on Georgian Bay*, 2.

10 "A Trip to Parry Sound and Back, by Althea," in *Guide to Muskoka Lakes*, by Muskoka and Nipissing Navigation Co., 78; Colton, *Tour of the American Lakes*, 69.

11 Both Oelschlaeger, *The Idea of Wilderness from Prehistory to the Age of Ecology* (New Haven: Yale University Press, 1991), and Worster, *Nature's Economy* (1994), structure their comprehensive histories of ecological thought around the fundamental dialectic between these two schools, the "imperial" and the "arcadian." For more on Thoreau, Emerson, and other influential American writers see Daniel Payne, *Voices in the Wilderness: American Nature Writing and Environmental Politics* (Hanover, NH: University Press of New England, 1996).

12 Jasen, *Wild Things*, 26; "By An Old Camper," in *Guide to Muskoka Lakes*, by Muskoka and Nipissing Navigation Co., 49.

13 Coburn, *In Pursuit of Coleridge*, 184-5; Robinson, *Georgian Bay*, 19.

14 Jehanne Bietry Salinger, "Elizabeth Wyn Wood," *Canadian Forum*, May 1931; Salem Bland, "Preferences," *Canadian Forum*, December 1929. Jonathan F. Vance sees this same continuity in interwar literature, in *Death So Noble: Memory, Meaning, and the First World War* (Vancouver: University of British Columbia Press, 1997), 174.

15 Waubuno Reaches is an alias for Georgian Bay in *Summer Gone* (Macfarlane, 93); LePan, "Canoe-Trip," *Weathering It*, 77.

16   New, *Land Sliding*, 96; Atwood, *Survival*, 62; W.H. New, *A History of Canadian Literature* (London: Macmillan Education, 1989), 90.

17   Keith, *Literary Images of Ontario*, 103; *Globe*, "Spinner," 1889, citing "Dawn in the Island Camp," W.W. Campbell, *Lake Lyrics*, 27.

18   Asling-Riis, *The Great Fresh Sea*; Hugh Cowan, *La Cloche: The Story of Hector MacLeod and His Misadventures in Georgian Bay and La Cloche Districts* (Toronto: Algonquin Historical Society), 1928; Robinson, *Georgian Bay*, 7, 9.

19   Regarding the Quebec and Rocky Mountain schools, which have been called the foundation of a Canadian "national" art, see Brian S. Osborne, "The Iconography of Nationhood in Canadian Art," in *The Iconography of Landscape*, ed. Stephen Daniels and Denis Cosgrove (Cambridge: Cambridge University Press, 1988), 166; Dennis Reid, *"Our Own Country Canada": Being an Account of the National Aspirations of the Principal Landscape Artists in Montreal and Toronto, 1860-1890* (Ottawa: National Gallery of Canada, 1979), 6. For early interest in Ontario, see Reid, 1988; Robert Stacey, "The Myth – and Truth – of the True North," in *The True North: Canadian Landscape Painting, 1896-1939*, ed. Michael Tooby (London: Lund Humphries Publishers, 1991). On "regionalism" as a sign of cultural independence, see Nancy K. Anderson, "'Curious Historical Artistic Data': Art History and Western American Art," in *Discovered Lands, Invented Pasts*, by Jules David Prown et al., 24.

20   On impressionism see Carol Lowrey, "Into Line with the Progress of Art: The Impressionist Tradition in Canadian Painting, 1885-1920," and Robert Stacey, "The Sensations Produced by (Their Own) Landscape: Some Art History Paths to – and Away from – Canadian Impressionism," in Lowrey et al., *Visions of Light and Air: Canadian Impressionism, 1885-1920* (New York: Americas Society Art Gallery, 1995). On MacDonald, lecture given at the Art Gallery of Toronto, 17 April 1932, cited in Paul Duval, *The Tangled Garden* (Scarborough, ON: Cerebrus/Prentice-Hall, 1978), 48; Nasgaard, *The Mystic North*, discusses this connection between the Group of Seven and Scandinavian art.

21   J. Russell Harper, *Painting in Canada: A History* (Toronto: University of Toronto Press, 1977), 276-7.

22   Roald Nasgaard calls this the "national Romantic" style in his study of landscape painting in Scandinavia and North America between 1890 and 1940 (166-8).

23   Bentley Mays, *Arrivals*, 372-3.

24   See, for example, Maria Tippett, *Making Culture: English-Canadian Institutions and the Arts before the Massey Commission* (Toronto: University of Toronto Press, 1990) and Joyce Zemans, "Establishing the Canon: Nationhood, Identity and the National Gallery's First Reproduction Program of Canadian Art," *Journal of Canadian Art History* 16, 2 (1995).

25   Douglas Cole, "Artists, Patrons and Public: An Enquiry into the Success of the Group of Seven," *Journal of Canadian Studies* 13, 2 (Summer 1978): 69.

26   Atwood, *Strange Things*, 8.

27   "Imagined communities" was coined by Benedict Anderson in *Imagined Communities*.

28   William Westfall, "On the Concept of Region in Canadian History and Literature," *Journal of Canadian Studies* 15, 2 (Summer 1980); Peter Ennals and Deryck Holdsworth, *Homeplace: The Making of the Canadian Dwelling over Three Centuries* (Toronto: University of Toronto Press, 1998), 9.

29   Monique Taylor, "'This Our Dwelling': The Landscape Experience of the Jesuit Missionaries to the Huron, 1626-1650," *Journal of Canadian Studies* 33, 2 (Summer 1998): 86-8, 93; Pratt, *Brébeuf and His Brethren*, 9.

30 Reproduced in Landon, *Lake Huron*, 281-2. Floren et al., *Ghosts of the Bay*, is one source of the story of Sandy Gray (74).

31 E.A. LaPierre (Nipissing), *Debates*, 7 March 1927, 953.

32 Paterson, *Muskoka District Local Government Review*, 117.

33 Champlain, *Works of Samuel de Champlain*, 19.

34 Freeman, *By Waterways to Gotham*, 246. Penetang Peninsula is surveyed to the Severn by 1822.

35 See, for example, Polly Stewart, "Regional Consciousness as a Shaper of Local History: Examples from the Eastern Shore," 77, and Thomas Schlereth, "Regional Culture Studies and American Culture Studies," 173, in *Sense of Place*, ed. Barbara Allen and Thomas J. Schlereth.

36 Jasen, *Wild Things*, 98; Allan Smith, "Farms, Forests and Cities," in *Old Ontario*, ed. David Keane and Colin Read, 79-80.

37 "By an Old Camper," in *Guide to Muskoka Lakes*, by Muskoka and Nipissing Navigation Co., 47; Bradshaw, "Georgian Bay Archipelago," 16.

38 Interview with Tim Stinson (Toronto: 5 June 1997); *Muskoka: Naturally Beautiful, 1999 Guide to Muskoka* (N.p.: Muskoka Tourism, 1999), 32. Stinson has a foot in both camps: an alumni of Hurontario, his parents own a cottage on Parry Island, and his in-laws own a cottage in Muskoka.

39 Douglas Lee, "Building at Go-Home Bay, 1948-1973," in *Madawaska Club: 1898-1973, Go Home Bay* (Madawaska Club, 1973), 81; also J.C. Robertson, "The Second Summer and the Building of the First Cottages," in *Madawaska Club: 1898-1948, Go Home Bay* (Madawaska Club, 1948).

40 Berleant, *The Aesthetics of Environment*, 147.

41 "The New Tourists Guide, 1912," LAC, RG 10, vol. 2870, file 176, 296-172; Hamelin interview (16 June 1997).

42 Otto interview (7 February 1998); interview with Gregor Beck (Toronto: 7 June 1997); Michel interview (5 March 1998); Michael Hough, *Out of Place: Restoring Identity to the Regional Landscape* (New Haven, CT: Yale University Press, 1990), 2, 58, 19.

43 Richard Perren, cited in Eaton and Weston, *Canadian Garden*, 50.

44 LePan, "Islands of Summer," *Weathering It*, 19.

45 In fact, the merit of the provincial or local was something Frye accepted rather late in his career. See Frye, "Culture and Society in Ontario, 1784-1984" (1984) and "Levels of Cultural Identity" (1989) in *Mythologizing Canada*, ed. B. Gorjup; George Woodcock, "There Are No Universal Landscapes," in *Documents in Canadian Art*, ed. Douglas Fetherling (Peterborough, ON: Broadview Press, 1987).

46 Lawren Harris, "The Group of Seven in Canadian History," *Canadian Historical Association Report* (Toronto, 1948), 36.

47 T.C. Keefer, *Report of the Committee on the Ottawa and Georgian Bay Territory* (1864); Wilfred Campbell, *The Canadian Lake Region*, 113-14. On the nationalist symbolism of the Canadian Shield, see Susan J. Wood, "The Land in Canadian Prose, 1845-1945" (PhD thesis, Carleton University, 1988), 108; D.M.R. Bentley, "Charles G.D. Roberts and William Wilfred Campbell as Canadian Tour Guides," *Journal of Canadian Studies* 32, 2 (Summer 1997); Eric Kaufmann, "'Naturalizing the Nation': The Rise of Naturalistic Nationalism in the United States and Canada," *Comparative Studies in Society and History* 40, 4 (October 1998).

48 Francis, *National Dreams*, 151.

49 Jill Ker Conway, *True North* (Toronto: Vintage Canada, 1995), 114-15; F.R. Scott, "National Identity," in *The Blasted Pine*, ed. F.R. Scott and A.J.M. Smith, rev. ed. (Toronto:

Macmillan, 1967), 8. On environmental determinism in the field of geography, see Norton, *Explorations in the Understanding of Landscape*, in Canadian culture, Carl Berger, "True North Strong and Free: The Myth of the North in Canadian History," in *Nationalism in Canada*, ed. Peter Russell (Toronto: McGraw-Hill, 1966). Berger also discusses its impact on Canadian historical writing, notably in the work of Arthur Lower, in *The Writing of Canadian History: Aspects of English-Canadian Historical Writing Since 1900*, 2nd ed. (Toronto: University of Toronto Press, 1986).

50 Shelagh Grant, "Myths of the North in the Canadian Ethos," *Northern Review* 3/4 (1989).

51 Interview with Thomas Hockin (Toronto: 31 May 1997).

52 Doris McCarthy, *A Fool in Paradise: An Artist's Early Life* (Toronto: Macfarlane, Walter and Ross, 1990), 196.

53 LePan, "A Rough Sweet Land," *Weathering It*, 220; see also his "The Canadian Dialectic," reprinted in *Notes for a Native Land: A New Encounter with Canada*, ed. Andy Wainwright (Toronto: Oberon Press, 1969), 59.

54 Lambert and Pross, *Renewing Nature's Wealth*, 147; on the controversial Dingley Tariff of 1898 and Empire Ontario protectionism, see Nelles, *Politics of Development*, chap. 2.

55 GBFC, 21-4, 35; Ontario Game and Fish Commission *Report* (Ontario SP no. 79, 1892); Ontario Department of Tourism files, for example, AO, RG 5-35, MB 113, series E-5: Advertising Recommendations – Department of Tourism and Information 1965, file 5-8. Americans still make up approximately 20 percent of the cottage population. "GBA requests Support from American Members in Clean Air Efforts," *GBA Update*, Summer 1999, 5.

56 John Irving, *A Prayer for Owen Meany* (New York: Ballantine Books, 1989), 558, 505, 359. Emphasis in original.

57 Ryden, *Mapping the Invisible Landscape*, 38-40, 76.

CHAPTER 6: SOME PROPER RULE

1 Appleton, *The Experience of Landscape*, 3-4.

2 Michael Williams, "Thinking about the Forest: A Comparative View from Three Continents," in *The Great Lakes Forest*, ed. Susan L. Flader; Warecki, *Protecting Ontario's Wilderness*, 4.

3 See Nelles's conclusion, especially 491; Epp discusses the holdings and influence of the "newsprint cartel" in "Ontario Forests and Forest Policy," in *Ecology of a Managed Terrestrial Landscape*, ed. Ajith Perera et al., 252.

4 B.W. Clapp, *An Environmental History of Britain since the Industrial Revolution* (London: Longman, 1994), 105.

5 Richard White, *The Organic Machine: The Remaking of the Columbia River* (New York: Hill and Wang, 1995), 34-5.

6 M. Bates, cited in 1950 Summary of Proceedings, file 1-4, Ontario Tourist Conference Reports, AO, RG 5-35, TB, MB 64, series B-13, p. 16.

7 Ontario Game and Fish Commission, 189.

8 Clayton R. Koppes, "Efficiency, Equity, Esthetics: Shifting Themes in American Conservation," in *The Ends of the Earth: Perspectives on Modern Environmental History*, ed. Donald Worster (Cambridge: Cambridge University Press, 1988); Robert Craig Brown, "The Doctrine of Usefulness: Natural Resources and National Park Policy in Canada, 1887-1914," in *Canadian National Parks: Today and Tomorrow*, vol. 1, ed. J.G. Nelson and R.C. Scace (Calgary: University of Calgary Press, 1968). A good source for the flurry of Progressive-style studies and commissions between 1885 and 1921, culminating in the

Commission of Conservation, is Janet Foster, *Working for Wildlife: The Beginning of Preservation in Canada,* 2nd ed. (Toronto: University of Toronto Press, 1998).

10 Edward Harris, OFC, 186.

11 Lambert and Pross, *Renewing Nature's Wealth,* 112-13, point out that as long as the government profited from timber dues, it had limited interest in interfering with loggers' productivity; Margaret Beattie Bogue, *Fishing the Great Lakes: An Environmental History, 1783-1933* (Madison, WI: University of Wisconsin Press, 2000), 147; Joseph Gough, *Fisheries Management in Canada, 1880-1910* (Halifax: Department of Fisheries and Oceans, 1991); McCullough, *The Commercial Fishery.*

12 Clapp, *Environmental History of Britain,* 72-8.

13 GBFC, 12.

14 OFC, x-y.

15 On the Noble brothers, see Bogue, *Fishing the Great Lakes,* 233-7. On the Wikwemikong controversy and Native fishing, see Douglas Leighton, "The Manitoulin Incident of 1863: An Indian-White Confrontation in the Province of Canada," *Ontario History* 69, 2 (June 1977); Wightman, *Forever on the Fringe,* 43-8; Andrea Gutsche and Cindy Bisaillon, *Mysterious Islands: Forgotten Tales of the Great Lakes* (Toronto: Lynx Images, 1999), 156-7; Morrison, "Upper Great Lakes Settlement," 64. David T. McNab, "Making a Circle of Time: The Treaty-Making Process and Aboriginal Land Rights in Ontario," in *Co-Existence?* ed. Bruce W. Hodgins et al., discusses the historic Wikwemikong claim to the adjacent lakebed and islands. On the growing influence of the sports lobby, see "Interim Report: Proposed Georgian Bay Fish Preserve" (17 January 1907) in GBFC, 12-13, and Capt. W. Whartman, cited in OFC, 238; R. Peter Gillis, "Rivers of Sawdust: The Battle over Industrial Pollution in Canada, 1865-1903," *Journal of Canadian Studies* 21, 1 (Spring 1986); J. Michael Toms, "An Ojibwa Community, American Sportsmen, and the Ontario Government in the Early Management of the Nipigon River Fishery," in *Fishing Places, Fishing People: Traditions and Issues in Canadian Small-Scale Fisheries,* ed. Dianne Newell and Rosemary E. Ommer (Toronto: University of Toronto Press, 1998), 170-92.

16 Ontario Fisheries Branch, *Annual Report 1899* (SP no. 11, 1900), 47; Peters, "Commercial Fishing in Lake Huron," 128; McCullough, *The Commercial Fishery,* 105-6; P.C. 324, 16 February 1909, and P.C. 2577, 20 September 1912, LAC, RG 23, vol. 371, file 3244.

17 On the Biological Board of Canada and the Go Home Bay station, see Brogue, *Fishing the Great Lakes,* 303-4.

18 McNeill, *Something New Under the Sun,* 258; OFC, ix; Lee Emery, *Review of Fish Species Introduced into the Great Lakes, 1819-1974* (Ann Arbor, MI: Great Lakes Fishery Commission, 1985), 4; Scott and Crossman, *Freshwater Fishes of Canada,* 71.

19 Worster, *Nature's Economy,* 315; MNR Fisheries Management Fact sheet, "Great Lakes Stocking Program," February 2001; Emery, *Review of Fish Species Introduced into the Great Lakes.* Other critics of the concept of stewardship include John A. Livingston, "Thoughts for the Future," in *Islands of Hope,* ed. Lori Labatt and Bruce Litteljohn, 240; and Stan Rowe, *Home Place: Essays on Ecology* (Edmonton: NeWest Press, 1990), 75.

20 James McLean, Provincial Land Survey, Field Notes for the Survey of Township of Baxter, book 931, 1878.

21 "An Act to Preserve the Forests from Destruction by Fire" (1878) in "A History of Crown Timber Regulations," in *Annual Report of the Clerk of Forestry for the Province of Ontario, 1898-9* (SP no. 36, 1899), 62-3.

22 Radforth, *Bushworkers and Bosses,* 14.

23 DLF Annual Report 1928 (SP no. 3, 1929); Michael Howlett and Jeremy Rayner, "Opening Up the Woods? The Origins and Future of Contemporary Canadian Forest Policy Conflicts," *National History* 1, 1 (1997); Monique Ross, *A History of Forest Legislation in Canada, 1867-1996* (Calgary: Canadian Institute of Resources Law, University of Calgary, 1997).

24 DLF Annual Report 1959, 19; MNR Lands File 41946 (Sturgeon Bay); Bruce Hodgins, R. Peter Gillis, and Jamie Benidickson, "The Ontario Experiments in Forest Reserves," in *Changing Parks*, ed. John Marsh and Bruce W. Hodgins. Hodgins and Benidickson's *The Temagami Experience* provides a critical look at the policy of forest reserves.

25 DLF, *Forestry Industry Opportunities* (1963), 11.

26 This shift is discussed by Samuel Hays, *Beauty, Health, and Permanence: Environmental Politics in the United States, 1955-1985* (Cambridge: Cambridge University Press, 1987); White, *Organic Machine*, discusses how we no longer measure the landscape in terms of work and labour.

27 George Altmeyer, "Three Ideas of Nature in Canada, 1893-1914," *Journal of Canadian Studies* 11, 3 (August 1976).

28 Cronon, "The Trouble with Wilderness; or, Getting Back to the Wrong Nature," in *Uncommon Ground: Toward Reinventing Nature*, ed. Cronon (New York: W.W. Norton, 1995), 78-80.

29 This phrase occurs in American and Canadian legislation, including that responsible for the creation of Banff/Rocky Mountain (1885) and Algonquin (1893) parks. On parks and national park legislation, see Steinberg, *Down to Earth*, 148; Sellars, *Preserving Nature*; E.J. Hart, *The Selling of Canada: The CPR and the Beginnings of Canadian Tourism* (Banff: Altitude Publishing, 1983); Sid Marty, *A Grand and Fabulous Notion: The First Century of Canada's Parks* (Toronto: NC Press, 1984); Leslie Bella, *Parks for Profit* (Montreal: Harvest House, 1987); J.A. Kraulis and Kevin McNamee, *The National Parks of Canada* (Toronto: Key Porter, 1998).

30 "Resolution passed by Georgian Bay Association at the Annual General Meeting," Toronto, 1 April 1922, in MNR Lands File 16089 (Franklin Island Provincial Park); R.B. Orr to J.B. Harkin, 1 September 1921, LAC, RG 84 A-2-a, vol. 487, file GB2 (U325-9-6). I am indebted to Dr. Gerald Killan for sharing his material on Franklin Island with me.

31 F.H.H. Williamson, Assistant Commissioner of Parks to W.W. Cory, Deputy Minister of Interior, 19 October 1928, LAC, RG 84 A-2-a, vol. 487, GB2 (1924-30).

32 Stead to Harkin, 12 September 1921, LAC, RG 84 A-2-a, vol. 487, file GB2 (U325-9-6).

33 R.B. Orr to A.A. Pinard, 22 September 1920; undated memo "Beausoleil Island," LAC, RG 84 A-2-a, vol. 487, file GB2 (U325-9-6). Excavation in the 1970s revealed extensive archaeological sites in what is now Awenda Provincial Park (originally known as Methodist Point) at the south shore of the Bay.

34 C.J. Taylor, *Negotiating the Past: The Making of Canada's National Historic Parks and Site* (Montreal and Kingston: McGill-Queen's University Press, 1990); Bruce J. Noble, "At Historical Parks: Balancing a Multitude of Interests," in *Public History: Essays from the Field*, ed. James B. Gardner and Peter S. LaPaglia (Melbourne, FL: Krieger, 1999).

35 LAC Georgian Bay Islands National Park files, including Camps, RG 84 A-2-a, vol. 1149 and 1150, GB 16-1 to 16-9; Superintendents' Reports 1931-69, RG 84, vol. 2381 C-1445-3, pts. 1-4; Wardens' Reports, vol. 2381 C-1445-5; Campground Statistics 1959-61, RG 84, series B-3-e, vol. 2360, 8373-1; wildlife B-3-e, vol. 9810-1, 9875-1.

36 J.R.B. Coleman, Chief of National Parks and Historic Sites Services, to J.C. Browne, GBINP Superintendent, 12 February 1951). See Georgian Bay National Park files, LAC, RG 84 A-2-a, vol. 144, files GB 181 and GB296-12, RG 84, series B-3-e, vol. 2360, files

9875-1 and 9810-1; DLF Annual Reports 1928, 1951, 1958-9, and 1963-4; Killbear Park files, AO, RG 1, 1A7-27-1403.
37 Statement of Visitors, Superintendents' Report 1931, LAC, RG 84, vol. 2381 C-1445-3, pt. 1; Annual Report 1962, LAC, RG 84, vol. 2381 C-1445-3, pt. 4, 1961-9.
38 R.A. Gibson, Director, Lands, Parks and Forests Branch, draft letter to L.O. Breithaupt, 23 January 1947; also memo from Parks Branch, 4 June 1940; J.C. Browne, Park Superintendent, to J. Smart, Parks Branch, 13 September 1947. All in LAC, RG 84 A-2-a, vol. 144, file GB 28, "GBINP/Development – General 1930-50."
39 Franklin Island Park files, Lands Records file 16089; Killan, *Protected Places*, 29.
40 MNR Lands File 41946 (Sturgeon Bay).
41 R.S.O. "An Act to Amend the Public Lands Act," 1962, c. 117, s. 1; DLF Annual Report 1963-4.
42 DLF, *North Georgian Bay Recreational Reserve: A Summary Report* (1971), 11, 22, emphasis in original; Norman Pearson, *Planning for the North Georgian Bay Recreational Reserve* (University of Waterloo/Department of Lands and Forests, 1965); Lambert and Pross, *Renewing Nature's Wealth*, 496-9.
43 M.A. Adamson, District Forester to the Parks Branch, 21 June 1962; DLF Report to Ontario Parks Integration Board, 44; Frank Nowak to W.R. Peck, District Manager, MNR Parry Sound (5 October 1977), AO, RG 1, IB 3, box 37.
45 Interview with Gerry Bird, former biology instructor (Lakefield, ON: 23 June 1997; email, 31 August 2000).
46 See Payne, *Voices in the Wilderness*, 137-8; Warecki, *Protecting Ontario's Wilderness*, 213; Hays, *Beauty, Health, and Permanence*; and Worster, *Nature's Economy*, 351.
47 DLF Annual Report 1945 (1946), 57.
48 AO, RG 1, IB 4, file RG 1 26-14 NR (Proposed Parks and Reserves Files 1954-71); MNR Lands File 171517 (Proposed Provincial Parks, Parry Sound District); DLF Annual Reports.
49 In fact, McCrae Lake and McDonald Bay were first closed to disposition and fishing in 1950, and incorporated into a "no development" area in 1954. A.W. Leman, District Forester, Leman to F.A. MacDougall, Deputy Minister, 20 July 1954, and to F.W. Beatty, Surveyor General, 5 August 1954; MNR Lands File 171517. M.A. Adamson, District Forester, to P. Addison, Chief, Parks Branch, 15 January 1969, RG 1, IB 3 25-14-100; Adamson to A.B. Wheatley, Chief, Parks Branch, 22 August 1961; Leman to Addison, Regional Forester, 25 November 1960, AO, RG 1, IB 4, 26-14-1; DLF Annual Report 1960, 21-22. On Moon Island, see Park Planning Branch, "Expansion of Moon Island Park Reserve, South Georgian Bay, Parry Sound District" (1 November 1972) AO, RG 1, IB 4, file 26-14-7.
50 Sellars, *Preserving Nature*, 207-14.
51 Roderick, Nash, *Wilderness and the American Mind*, 3rd ed. (New Haven, CT: Yale University Press, 1982), 326-8.
52 Primitive class parks were later renamed wilderness class. See, for example, the claim made in Algonquin Wildlands League, *Wilderness Now*, 3rd ed. (Toronto: Algonquin Wildlands League, 1980); Warecki, *Protecting Ontario's Wilderness*.
53 MNR Lands File 170316 (Killarney); Parks Established and Operating Files, Killarney, AO, RG1 1A-7 27-1903; *North Georgian Bay Recreational Reserve: A Summary Report* (1971), 3; Killan, *Protected Places*, 115. For the early history of the park, see Kevin McNamee, "Killarney – The Wilderness Edge," in *Islands of Hope*, ed. Lori Labatt and Bruce Litteljohn. A small park reserve was established around Trout (now Jackson) Lake in 1933, and the park was expanded to roughly its present boundaries in 1963. Today, at

48,500 hectares, Killarney is the second largest provincial park in Georgian Bay, after the French River at 51,740 hectares.

54 E.G. Pleva, "The Parks of Ontario," in *The Canadian National Parks: Today and Tomorrow*, eds. J.G. Nelson and R.C. Scace, proceedings of a conference organized by the National and Provincial Parks Association of Canada and the University of Calgary, Calgary, 1968, 446.

55 Davis, Ontario *Debates* (25 June 1971), 3219.

56 The boundaries of the Blackstone Harbour Park reserve were established in 1973, after which point hydro extensions and severances were prohibited, though the park was not formally opened until 1996. Submission to the Resource Policy Field Committee, "Establishment of the South Georgian Bay Provincial Park, Parry Sound District," 11 January 1973, AO, RG1, IB4 26-14-7; W.T. Foster, "Proposal re. Blackstone Harbour Park Reserve," 11 March 1976, AO, RG 1, IB 3, box 37.

57 A. Wilson, *The Culture of Nations*, 242-3; Hays, *Beauty, Health, and Permanence*, 36-8, 153-5.

58 The park reserve encompassed 6,070 hectares, but initially only 3,960 hectares were reserved from staking. MNR Lands File 171517 (Proposed Provincial Parks, Parry Sound 59. This figure is from the legal transfer documents; a later report ("Land Acquisition History"), perhaps done after the park's northern boundary was adjusted, put the figure of the CCA deal at 1,064 acres, still representing 61 percent of the private property to be acquired for a park. Both sets of documents put the price at about $150,000. At this time the MNR also acquired mineral rights to twenty lots for a nominal fee from a "Lake Boundary Project Ltd.," whose director was named as Fred Eaton, another Sans Souci cottager. Hewitt, 31-3; A.R. Sawh, Solicitor, to J.R. McGinn, Lands Administration Branch, 24 June 1974; J.S. Ball, Regional Director Algonquin Region to Park Planning Branch, 14 March 1975, MNR, file 10-5-1; McGinn to K.H. Stahl, Financial Management Branch, 4 October 1974, file PS 19/LO 463. All AO, RG 1, IB-4, PS 19, 1974-6, box 33. Also "Land Acquisition History: Blackstone Harbour," 2 November 1976, AO, RG 1, IB-4, PS 19, box 35, 1976-7; interviews with Neil Campbell.

60 According to Mines, the CCA's action represented an abuse of the licenses of occupation by commandeering them not for mining purposes but to hold territory in an undeveloped state. Provincial Parks Act R.S.O. 1970, c. 371, s. 18; email from Jim Trusler, former Regional Geologist, 14 June 2000; memo to J.K. Reynolds, Parks Branch, 28 June 1977, AO, RG 1, IB 3, box 37.

61 "Review of Material to be Presented to the Provincial Parks Council," 25 May 1976; R.A. Beatty to J.W. Keenan, Executive Director of Parks, 1976, draft; Park Planning Branch, "Response to W.T. Foster's Options and Recommendations for Blackstone Harbour Provincial Park," 1976; J.A. van der Meer, Regional Parks and Recreation Coordinator, to L.H. Eckel, Executive Director of Parks Division, 6 April 1977, AO, RG 1, IB 3, box 37.

62 C.R. Tilt, OPIB, to P. Addison, Chief Parks Branch, 24 January 1969, in MNR Lands File 163882, vol. 2 (Killbear).

63 John Wilson, *The Blackstone Wilderness: The People Speak. A Report on a Survey of Attitudes and Opinions in the South Archipelago* (Georgian Bay Association, 1987), 46, 55; MNR, *Blackstone Harbour Public Participation Phase 1 and 2* (MNR: 1975), 20.

64 J.K. Reynolds, Deputy Minister, MNR, to John Sweeney, MPP, 27 February 1978; Park Planning Branch, "Response to W.T. Foster's options ..." (1976) AO, RG 1, IB-3, box 37.

65 MNR, *French River Provincial Park Management Plan*, 6-8.
66 "Blackstone Harbour Provincial Park Proposal: Project Review Highlights" (n.d./1976), presented to Provincial Parks Council, June 1977; Park Planning Branch, "Response to W.T. Foster's options ..." (1976); "Blackstone Harbour Provincial Park Project Review," 13 October 1976, AO, RG 1, 1B-3, box 37.
67 Kevin McNamee, "Preserving Canada's Wilderness Legacy: A Perspective on Protected Areas," 25-44, and J.G. Nelson, Patrick Lawrence, and Heather Black, "Assessing Ecosystem Conservation Plans for Canadian National Parks," 113-32, in Nelson et al., *Protected Areas and the Regional Planning Imperative* (2002).
68 Telephone interview with Gary Higgins, Superintendent, Massasauga and Oastler Lake Provincial Parks (25 January 2000); *The Massasauga Provincial Park Management Plan* (1993), 3.
69 Birnbaum interview (18 September 1999).
70 MNR, *Ontario's Living Legacy: Land Use Strategy* (MNR: July 1999), 22-4; by February 2000 the MNR was investigating the possibility of allowing sport hunting in Killarney and three other wilderness parks, a move the Partnership for Public Lands feared would include motorized access. Email from Partnership for Public Lands, owner-landsforlife-l@list.web.net (9 February 2000); letter from John Snobolen, Minister of Natural Resources, to the author (7 April 2000).
71 Killan, *Protected Places*, 236.
72 Govier, *Immaculate Conception*, 130. Statistical sources from the 1960s consistently estimate the seasonal population at over 90 percent of the population of the archipelago townships. See Paterson, *Muskoka District Local Government Review*, 131-2, and District of Muskoka Planning Department, *Permanent and Seasonal Population Totals Through the District of Muskoka* (1999), which puts the figure at about 92 percent. In the Township of the Archipelago, numbers may be even higher: the 1987 opinion survey *Blackstone Wilderness: The People Speak* gives it as 94 percent. In 1999 cottages accounted for 90.6 percent of residences, but cottage assessment actually underestimates seasonal numbers because it does not include the transitional summer population: boaters, cottage guests, campers, and so on. In addition, the usual municipal data – assessment, building permits, hydro connections – are often an incomplete measure in the unconventional landscape of the archipelago; cottages still may not have hydro or phone lines. According to the township's assessment office, this makes it "impossible to gauge" the exact numbers of the seasonal population. Brenda Laughlin, Archipelago Assessment Office, telephone call to the author (25 November 1999); also Darcy McKeough, Minister of Economic and Intergovernmental Affairs, "Proposals for the Improvement and Strengthening of Local Government in the District of Parry Sound," 30 January 1978, AO, MU 3830. "Politics of turf" is an expression frequently used by Halseth, *Cottage Country in Transition.*
73 Higgins interview (25 January 2000). The Skerryvore Road was the subject of a sixteen-year legal wrangle as the band sought compensation for public use of a private road. Eventually cottagers paid a levy to construct an alternative access road. See *Globe and Mail*, "Band to Close Access Road to Press Demand for Money," 11 May 1981, 10; *Globe and Mail*, "Cottagers Lose Bid to Keep Road Open," 28 July 1994, A5; and *Toronto Star*, "Families Leave Homes as Indians Close Road," 26 April 1997, A6. Robert Gottlieb examines the class, race, and gender differences in environmental politics, and the differences in power manifested in the spatial segregation of different groups, in *Forcing the Spring*.

74 G.R. Weller, "Hinterland Politics: The Case of Northwestern Ontario," *Canadian Journal of Political Science* 10 (December 1977): 752; R.J. Boyer, Muskoka, Ontario *Debates* (28 May 1970), 3297.

75 Warecki, *Protecting Ontario's Wilderness*, discusses how Ontario environmentalists wrestled with charges of elitism; see also Raleigh Barlowe, "Changing Land Use and Policies: The Lake States," in *The Great Lakes Forest*, ed. Susan L. Flader, 174, and Oeschlaeger, 292; Halseth, *Cottage Country in Transition*, 146; "muskokafication" is from Hodgins and Benidickson, *Temagami Experience*, 251.

76 On Hodgetts, see interviews with David Hodgetts (Hamilton, ON: 26 June 1997); Gerry Bird and Mike Chellew (Lakefield, ON: 23 June 1997); Don Marston (Camp Hurontario, ON: 20 July 1997). On Wood's Bay, see Penny Pepperell, on Ontario Morning, CBC (30 September 1999); GBA to Andy Mitchell, MP Parry Sound/Muskoka, 8 July 1998, and Wood's Bay Community Association to Ernie Eves, MPP Parry Sound, 14 March 1999, Birnbaum/GBA files; *Toronto Star*, "Waterfront Living," 18 September 1999. In 1999 the Federation of Ontario Naturalists, the Weston Foundation, and the Nature Conservancy of Canada cooperated to purchase 7,000 hectares of an alvar limestone bedrock habitat on the south shore of Manitoulin.

77 David E. Meers, "The Use of a Communal Tourism/Recreation Resource: Case Study, The District of Muskoka" (MA thesis, University of Western Ontario, 1990), 75; "Briefing on GBA Water and Air Quality Initiatives in Georgian Bay Area," GBA memo to Tony Clement, Minister of the Environment (9 September 1999), Birnbaum files.

78 See, for example, Paterson, *Muskoka District Local Government Review*, 20; Pearson, *Environment Control*, 3; Parry Sound Local Government Study Group, *Research Report* (1976), 11. On attitudes toward the environment, see Georgian Bay Regional Development Council, *Regional Plan 1968-1972*, 14; Mike Bates, files 1-4, Ontario Tourist Conference Reports 1950, AO, RG 5-35, TB, MB 64, series B-13, p. 14.

79 Hartman interview (16 June 1997).

80 G. Bruce Doern and Thomas Conway, *The Greening of Canada: Federal Institutions and Decisions* (Toronto: University of Toronto Press, 1994).

81 Regarding cottagers and Georgian Bay Township see Paterson, *Muskoka District Local Government Review*, 150-4.

82 D. Rotenburg, Wilson Heights, Ontario *Debates* (18 June 1979), 3011; J. Wilson, *The Blackstone Wilderness: The People Speak*, 70-1.

83 See Duncan MacLellan, "Shifting from the Traditional to the New Political Agenda: The Changing Nature of Federal-Provincial Environmental Relations," *American Review of Canadian Studies* 25, 2/3 (1995). I would, however, challenge his suggestion that federal-provincial conflict has emerged only in the past few decades.

84 On the fishery decision, see McCullough, *The Commercial Fishery*, 95-6; Peters, "Commercial Fishing in Lake Huron," 130; Bogue, *Fishing the Great Lakes*, 305-6; the best examination of the controversy over the Dingley tariff in 1898 is Nelles, *Politics of Development*, chap. 2.

85 Memo to the Minister (Clifford Sifton), 17 August 1897, from J.D. McLean, Secretary, LAC, RG 10, vol. 2851, file 176, 296-1A; 296-1B; Indian Affairs, LAC, RG 10, vol. 2851, file 176, 296-1A. Sylvia Du Vernet, *An Indian Odyssey: Tribulations, Trials and Triumphs of Gibson Band of the Mohawk Tribe of the Iroquois Confederacy* (Islington, ON: Muskoka Publications, 1986); "Wahta Land Claim Agreement-in-Principle Signed," Canada, Indian Affairs, 13 May 1998; "Wahta Mohawks to Vote on Land Claim Agreement," *The Muskokan*, 10 January 2002; Rogers and Smith, *Aboriginal Ontario*, 268-9. The Moose Deer Point band is also seeking additional land around O'Donnell Point and Tadenac,

claiming that the Pottawami who moved north in the 1840s and 1850s never received the land which the British promised its Native allies from the War of 1812. Indian Claims Commission: Moose Deer Point First Nation Inquiry, Pottawatomi Rights (April 1999) and minister's response (29 March 2001); also "Commission Supports Historic Claim Launched by Moose Deer Point," *Parry Sound North Star,* 28 April 1999.

86 *"A Gathering of Our Georgian Bay Family": Conference Report of Georgian Bay Association's Eastern Georgian Bay Conference* (Honey Harbour, 5-6 October 1991), 13, 38; GBA memo to Mitchell, 8 July 1999, Birnbaum files; "Foresight Averts Problem of Floating Cottages," *The Muskokan,* 18 July 2002; Parks Canada Agency, *Unimpaired for Future Generations? Protecting Ecological Integrity in Canada's National Parks,* Report of the Panel on the Ecological Integrity in Canada's National Parks (Ottawa, 2000); Ken Cain et al., "Fluctuating Great Lakes Water Levels and Ontario Provincial Parks: A Call for Coastal Management," in *Shoreline Resource Management and Ontario's Provincial Parks* (1998), 54.

87 D.J. Allan, Superintendent, Reserves and Trusts, DIA to T.U. Fairlie, GBA, 27 December 1945, LAC, RG 10, vol. 6831, file 506-2-1.

88 Charles Little, in OFC, 194.

89 Barry Rabe, "The Politics of Ecosystem Management in the Great Lakes Basin," *American Review of Canadian Studies* 27, 3 (Autumn 1997), argues that the United States leads the two countries in both federal and interstate agreements because the Canadian government tends to devolve responsibility to the provinces. Jennifer Read discusses the management of boundary waters in "Addressing 'A Quiet Horror': The Evolution of Ontario Pollution Control Policy in the International Great Lakes, 1909-1972" (PhD thesis, University of Western Ontario, 1999).

90 McIntyre letter, 9 February 1998.

91 Janet Foster, "Mobilizing a Sense of Wonder," in *Islands of Hope,* ed. Lori Labatt and Bruce Litteljohn, 21.

CONCLUSION

1 Clarkson, installation speech, 1999.

2 Bayfield to Beaufort, 9 May 1832, LAC, MG 12, adm. 1, Commander H.W. Bayfield St. Lawrence Survey Correspondence 1828-35.

3 J.G. Sing, Ontario Land Survey 1899, LAC, RG 10, vol. 2852, file 176, 296-1B.

4 M.W. Bowman to D.J. Allan, Superintendent of Reserves and Trusts, DIA, 1 June 1947, LAC, RG 10, vol. 2851, file 176, 296-1.

5 Shaw, *Happy Islands,* 61.

6 LePan, "Astrolabe," *Weathering It,* 173; Clarkson, installation speech (Ottawa, 7 October 1999).

7 Macfarlane, *Summer Gone,* 93-4. To be fair, the eastern massasauga rattlesnake is dangerous only if provoked, and in fact is in more danger from humans, as it currently ranks as an endangered species. In Scott Young's novel *That Old Gang of Mine* (Toronto: McClelland and Stewart, 1982), a Liberal cabinet minister announces that her government ensures serum is kept at the Parry Sound hospital, whereupon the backbenchers behind her yell across the floor of the Commons, "'We'd save you guys too, if a snake ever got that hard up,' and, 'What we can't figure out is how to save the snake if it bites a New Democrat'" (9-10).

8 Gordon Lightfoot, "The Wreck of the Edmund Fitzgerald" (Moose Music, 1976).

9 A.Y. Jackson, quoted in Davis, "An Apprehended Vision," cited in Nasgaard, *The Mystic North,* 164; Robinson, *Georgian Bay,* 7.

10 Wadland, "Great Rivers, Small Boats: Landscape and Canadian Historical Culture," in *Changing Parks*, ed. John Marsh and Bruce W. Hodgins, 20-1.

11 Atwood, *Survival*, 50-1.

12 Northrop Frye, "Canadian and Colonial Painting," *Canadian Forum*, 1940, in *Documents in Canadian Art*, ed. Douglas Fetherling, 89.

13 Daniel Francis, *National Dreams*, 151.

14 Overton and MacKay, in contrast, argue that the version of Maritime heritage constructed for tourists has overshadowed the genuine, organic culture of local residents.

15 Thomas Symons, "Greening of Heritage: Historic Preservation and the Environmental Movement," in *Linking Cultural and Natural Heritage*, ed. John Marsh and Janice Fialkowski (Peterborough, ON: Frost Centre for Canadian Heritage, 1995), 9.

16 David E. Meers, "The Use of a Communal Tourism/Recreation Resource: Case Study, The District of Muskoka" (MA thesis, University of Western Ontario, 1990), 134; the Champlain comment is from Jackson, *A Painter's Country*, 31.

17 Cronon, "The Trouble with Wilderness," in *Uncommon Ground*, ed. Cronon.

18 Charles Gordon, *At the Cottage* (Toronto: McClelland and Stewart, 1989), 41.

19 Hays, *Beauty, Health, and Permanence*, 9, 36, 62.

20 D.A. Muise, "Organizing Historical Memory in the Maritimes: A Reconnaissance," *Acadiensis* 30, 1 (Autumn 2000): 60.

21 Hummel interview (29 May 1997).

Bibliography

PRIMARY SOURCES

Archival Collections

Archives of Ontario (AO)
RG 19 Department of Municipal Affairs
RG 5 Department of Tourism
RG 1 Department of Lands and Forests/Ministry of Natural Resources Pamphlets
  Collection
Photographs: Special Collections and CPR Collection

Library and Archives Canada (LAC)
MG 1 Série C11A France. Archives des Colonies. Correspondence Générale, Canada.
RG 10 Department of Indian Affairs
Macaulay Papers
MG 12 Adm. 1 Admiralty and Secretariat papers
Henry Wolsey Bayfield fonds
Hudson's Bay Company Archives
MG 55/24 #292 [Ottawa Timber Trade]
RG 23 Department of Marine and Fisheries
RG 84 Canadian Parks Service, Georgian Bay Islands National Park Files
Anne Savage Fonds: A.Y. Jackson Correspondence
Thomas Mower Martin Art

Ontario Ministry of Natural Resources (MNR), Peterborough
Lands Files

University of Western Ontario
J.J. Talman Regional Collection, Pamphlet and Postcard Collections
Serge A. Sauer Map Library, Historical Maps Upper Canada-Ontario, Ontario road
  maps, and Great Lakes Collections

Toronto Public Library
Canadiana Collection

Political and Government Sources
Aborigines Protection Society. *Report on the Indians of Upper Canada* 1839 and *Annual
  Reports*. London, 1838-46.

Algonquin Wildlands League. *Wilderness Now,* 3rd ed. Toronto: Algonquin Wildlands League, 1980.

Birnie, John, J.J. Noble, and E.E. Prince. Georgian Bay Fisheries Commission. *Report and Recommendations of the Dominion Fisheries Commission Appointed to Enquire into the Fisheries of Georgian Bay and Adjacent Waters.* Ottawa: Government Printing Bureau, 1908.

Canada. *Indian Treaties and Surrenders, From 1680-1890.* Ottawa, 1891.

–. Department of Indian Affairs. *Annual Reports.* Ottawa.

–. Department of Marine and Fisheries. *Annual Reports.* Ottawa.

–. Department of Public Works. *Georgian Bay Ship Canal Interim Report* 1907-8 (Sessional Paper no. 178) and *Report Upon Survey, with Plans and Estimates of Cost 1908* (Sessional Paper no. 19a 1909). Ottawa.

–. Department of Railways and Canals. *Annual Reports.* Ottawa.

–. *House of Commons Debates.* 25 February-28 March, 1927.

–. Indian Claims Commission. *Moose Deer Point First Nation Report on: Pottawatomi Rights Inquiry.* Ottawa, 1999.

–. Special Commission to Investigate Indian Affairs in Canada, *Report.* Ottawa, 1858.

Cognashene Cottager. 1970-1.

Coleman, A.P. "Copper in the Parry Sound District." *Report of the Bureau of Mines 1899.* Toronto: Bureau of Mines, 1899.

Committee on Ottawa and Georgian Bay Territory. *Report.* 15 June 1864. *Journals of the Legislative Assembly of the Province of Canada 1863-4.* Vol. 23, Appendix no. 8.

District of Parry Sound Local Government Study. *Parry Sound District Atlas, Research Report,* and *Final Report and Recommendations.* Toronto: Ministry of Treasury, Economics and Intergovernmental Affairs, 1976.

Dudley, William S., ed. *The Naval War of 1812: A Documentary History.* Vol. 2, 1813. Washington: Naval Historical Center, 1992.

Georgian Bay Association. *"A Gathering of Our Georgian Bay Family": Conference Report of the Georgian Bay Association's Eastern Georgian Bay Conference.* Honey Harbour, 5-6 October 1991.

Georgian Bay and North Channel Pilot. Ottawa: Minister of Marine and Fisheries, 1899.

Georgian Bay Association (GBA). Files of John Birnbaum, Executive Director. Toronto.

Georgian Bay Regional Development Council. *Regional Plan 1968-1972.* Guelph, ON: University of Guelph Centre for Resources Development, 1969.

Globe (Toronto). "The Old Fight Is on Again," 7 March 1927.

Globe and Mail. "Cottagers Access Road Blocked by Indian Band," 2 June 1994, A10.

–. "Cottagers Lose Bid to Keep Road Open," 28 July 1994, A5.

Head, Sir Francis Bond. *Communications and Despatches Relating to Recent Negotiations with the Indians.* Appendix to *Journals of the Legislative Assembly,* Canada 1838.

–. *Copies or extracts of despatches from Sir F. B. Head, Bart., K.C.H., on the subject of Canada, with copies or extracts of the answers from the Secretary of State.* London, 1839.

Hewitt, D.F. *Geology and Mineral Deposits of the Parry Sound-Huntsville Area.* Geological Report 52. Toronto: Ontario Department of Mines, 1967.

Horton, K.W., and W.G.E. Brown. *Ecology of White and Red Pine in the Great Lakes-St. Lawrence Forest Region.* Forest Research Division Technical Note 88. Department of Northern Affairs and National Resources, 1960.

Hunter, A.F. "Shore Lines Between Georgian Bay and the Ottawa River." *Summary Report of the Geological Survey of Canada.* Sessional Paper no. 26, 1908.

Jenness, Diamond. *The Ojibwa Indians of Parry Island, Their Social and Religious Life.* Ottawa: Department of Mines/National Museum of Canada, 1935.

London Free Press. "Province to Protect Manitoulin Shore," 20 October 1999.

–. "Province's Snubs Rile LEDC Head," 22 June 2000.

Murray, Alexander. *Geological Survey of Canada. Report of Progress 1847-8.* Montreal: Lowell and Gibson, 1849.

–. *Geological Survey of Canada. Report of Progress 1848-9.* Toronto: Lowell and Gibson, 1850.

–. *Geological Survey of Canada. Report of Progress 1857.* Toronto: Lowell and Gibson, 1858.

Niles' Weekly Register. Vol. 7, 1814-15.

Ontario. Department of Crown Lands/Lands and Forests/Ministry of Natural Resources. *Annual Reports.* Toronto, 1853 to 2002.

–. Department of Lands and Forests. *Forest Industry Opportunities in the Georgian Bay Area.* 1963.

–. *North Georgian Bay Recreational Reserve: A Summary Report.* 1971.

–. Department of Mines. *Annual Reports.* Toronto.

–. Game and Fish Commission. *Report.* Sessional Paper no. 79, 1892. Toronto.

–. Legislature. *Debates.* Toronto, 1969-82.

–. Ministry of Natural Resources. *Blackstone Harbour Provincial Park: Public Participation Phase 1* and *Phase 2.* Toronto, 1975.

–. *French River Provincial Park Management Plan* (1985) and *The Massasauga Provincial Park Management Plan* (1993). Toronto.

–. *Ontario's Living Legacy: Land Use Strategy.* Toronto, 1999.

–. Ministry of Natural Resources/Parks Canada. *French River Candidate Provincial Waterway Park Management Plan.* Toronto: July 1984.

Ontario Morning. "Floating Cottages." Canadian Broadcasting Corporation. 30 September 1999.

Parks Canada. *Unimpaired for Future Generations? Protecting Ecological Integrity in Canada's National Parks.* Report of the Panel on the Ecological Integrity in Canada's National Parks. Ottawa, 2000.

Paterson, Donald M. *Muskoka District Local Government Review: Final Report and Recommendations.* 1969.

Pearson, Norman. *Environment Control, Planning and Local Government in the Georgian Bay Archipelago.* Sans Souci and Copperhead Association, 1975.

–. *Planning for Eastern Georgian Bay.* London, ON: Tanfield Hall, 1996.

–. *Planning for the North Georgian Bay Recreational Reserve.* University of Waterloo/Ontario Department of Lands and Forests, 1965.

Provincial Land Survey/Ontario Land Survey. Field notes.

Satterly, J. "Mineral Occurrences in Parry Sound District." Department of Mines *Annual Report 1942.* Vol. 51, Pt. 2. Toronto, 1943.

Sharpe, J.F., and J.A. Brodie. *Forest Resources of Ontario 1930.* Ottawa: Department of Lands and Forests, 1931.

Shoreline Resource Management and Ontario's Provincial Parks. Occasional Paper 8. Waterloo, ON: Heritage Resources Centre, University of Waterloo, 1998.

Wilmot, Samuel, and Edward Harris. *Report of the Dominion Fishery Commission on the Fisheries of the Province of Ontario. 1893-4.* Ontario Fishery Commission. Sessional Paper no. 10c, 1893. Ottawa: Queen's Printer, 1894.

Wilson, John. *The Blackstone Wilderness: The People Speak. A Report on a Survey of Attitudes and Opinions in the South Archipelago.* Toronto/Township of the Archipelago: Georgian Bay Association, 1987.

Wood, William, ed. *Select British Documents of the Canadian War of 1812.* Vol. 3. Toronto: Champlain Society, 1926.

Promotional and Popular Literature

Adams, E. Herbert, ed. *Toronto and Adjacent Summer Resorts.* Toronto: Murray Printing Co., 1894.

Adams, W.H. *Souvenir: The Story of the Georgian Bay.* Toronto: Fristbrook Box, 1911.

Campbell, Wilfred. *Canada.* Toronto: Macmillan, 1906.

–. *The Canadian Lake Region.* Toronto: Musson Book Company, 1910.

Canadian Magazine. "Canada and the Tourist," May 1900, 16-23.

Charlesworthy, Hector W. "The Canadian Girl: An Appreciative Medley." *Canadian Magazine,* May 1893.

Cherry, Zena. "Man Who Named District Overlooked 30,000 Islands." *Globe and Mail,* 29 July 1965.

Financial Post. "Romantic White Boats Cash for Georgian Bay," 18 July 1959.

Fleming Publishing. *The Georgian Bay and Thirty Thousand Islands: Illustrated Souvenir.* Owen Sound: Fleming Publishing, n.d. [191-?]. Canadiana Collection, North York Public Library, Toronto.

Fulford, Robert. "Cottage Industry." *Toronto Life,* July 1999.

Globe and Mail. "Captains Outrageous," August 1986, Toronto supplement.

Gould, Filomena. "A Sixth Great Lake," *Inland Seas* 23, 2 (1967): 121-30.

Grand Trunk Railway and Muskoka Navigation Co. *Picturesque Muskoka to the Highlands and Lakes of Northern Ontario.* N.p.: Grand Trunk Railway and Muskoka Navigation, 1898.

Grant, George M. *Picturesque Spots of the North.* Chicago, 1899.

Guidebook and Atlas of Muskoka and Parry Sound Districts. Toronto: H.R. Page, 1879.

Harper's Weekly. "Comfort in Camp," 3 October 1896.

Hubbs, Carl L., and Karl F. Lagler. *Fishes of the Great Lakes Region.* Bloomfield Hills, MI: Cranbrook Institute of Science, 1947.

Keefer, Thomas. *Report of Thomas Keefer of Survey of Georgian Bay Canal Route to Lake Ontario by Way of Lake Scugog.* Whitby, ON: W.H. Higgins, 1863.

Kingsford, William. *The Canadian Canals: Their History and Cost.* Toronto: Rollo and Adams, 1865.

Kirkwood, Alexander, and J.J. Murphy. *The Undeveloped Lands in Northern and Western Ontario.* Toronto: Hunter Rose, 1878.

Klein, Naomi. "The Tourist Trap." *Saturday Night,* September 1999.

London Free Press. "Cruising Ontario's Lakes in Style," 7 August 1999.

Mer Douce. Magazine of the Georgian Bay/Algonquin Historical Society, 1921-3.

Muskoka and Nipissing Navigation Co. *Guide to Muskoka Lakes, Upper Magnetawan, and Inside Channel of the Georgian Bay.* Toronto, 1888.

National Post. "The Cottage," 18 May 2000, special section.

–. "Georgian Splendour," 5 August 2000.

Osborne, A.C. "The Migration of *Voyageurs* from Drummond Island to Penetanguishene in 1828." Ontario Historical Society *Papers and Records* 3, 1901. 123-66.

Parkman, Francis. *France and England in North America, Part First: Pioneers of France in the New World.* Boston: Little, Brown, 1906.

–. *Part Second: The Jesuits in North America in the Seventeenth Century* and *Part Third: La Salle and the Discovery of the Great West.* Toronto: George N. Morang, 1898.

Perry, Ronald H. *Canoe Trip Camping.* Toronto: J.M. Dent and Sons, 1943.

Ports: The Cruising Guides. Overleaf Design, 1992.

Snider, C.J.J. *The Story of the "Nancy" and Other Eighteen-Twelvers.* Toronto: McClelland and Stewart, 1926.

Toronto Star. "Picture Yourself in a Boat on a River ...," 15 August 1999.

–. "Realm of the Great Spirit," 17 July 1999.

Williamson, James. *The Inland Seas of North America.* Kingston, ON: J. Duff, 1854.

Memoirs and Travel Accounts

Aflalo, F.G. *A Fisherman's Summer in Canada.* Toronto: McClelland and Goodchild, 1911.

Bayfield, Henry W. *The St. Lawrence Survey Journals of Captain Henry Wolsey Bayfield, 1829-1853.* Ed. with an introduction by Ruth McKenzie. Toronto: Champlain Society, 1984-6.

Beaver, "Journey for Frances," March 1954.

Bigsby, J.J. *The Shoe and Canoe, Or, Pictures of Travel in the Canadas.* London: Chapman and Hall, 1850.

Bradshaw, William. "The Georgian Bay Archipelago." *Canadian Magazine,* May 1900.

By Canoe from Toronto to Fort Edmonton in 1872, Among the Iroquois and Ojibways with a Chapter on Winter in Canada, by an Anonymous Traveler. Toronto: Canadiana House, 1968.

Canadian Magazine, The Captain. "A Honeymoon in a Sailing Dinghy." March 1901.

"Captain Mac." *The Muskoka Lakes and the Georgian Bay.* Toronto: J.T. McAdam, 1884.

Coyne, J.H. "Across Georgian Bay in 1871." Ontario Historical Society *Papers and Records.* Vol. 28, 1932.

Carver, Jonathan. *Three Years' Travels Through the Interior Parts of North America, For More Than Five Thousand Miles.* Philadelphia, 1784.

Champlain, Samuel de. *Works of Samuel de Champlain.* Vol. 4. Ed. H.P. Biggar. Toronto: Algonquin Historical Society, 1929.

Clark, Greg. *A Supersonic Day: From the Packsack of Gregory Clark.* Ed. Hugh Shaw. Toronto: McClelland and Stewart, 1980.

Clarkson, Adrienne. Installation Speech as Governor General. Ottawa, 7 October 1999.

Coburn, Kathleen. *In Pursuit of Coleridge.* Toronto: Clarke, Irwin, 1977.

Colton, Calvin. *Tour of the American Lakes, and Among the Indians of the Northwest Territories in 1830.* Vol. 1. London, 1833.

Conway, Jill K. *True North.* New York: A.A. Knopf, 1994.

Cormier, Louis-P., ed. *Jean-Baptiste Perrault: marchand voyageur parti de Montréal le 28e de mai 1783.* Montreal: Boréal Express, 1978.

Cruikshank, E.A., ed. *The Correspondence of Lieut. Governor John Graves Simcoe.* Vol. 2. Toronto: Ontario Historical Society, 1924.

Dunlop, William. *Recollections of the American War.* Toronto: Historical Publishing, 1908.

de la Fosse, F.M. "Early Days in Muskoka." Ontario Historical Society *Papers and Records.* Vol. 34, 1942.

Four Years on the Georgian Bay: Life Among the Rocks. Toronto: Copp Clark, 188-.

Fray, Henry A. "Commerce and Physical Features of the Great Lakes." Presented to the Canadian Society of Civil Engineers. 1895. CIHN 06594.

Freeman, Lewis R. *By Waterways to Gotham.* New York: Dodd, Mead, 1926.

Globe (Toronto), Spinner. "An Angler's Paradise: Mr. Kedgery's Experiences on Georgian Bay," 7 September 1889.

Gordon, Charles. *At the Cottage.* Toronto: McClelland and Stewart, 1989.

Hamilton, James Cleland. *The Georgian Bay: An Account of Its Position, Inhabitants, Mineral Interests, Fish, Timber and Other Resources, with Map and Illustrations.* Toronto: James Bain and Son, 1893.

Harley, Herbert L. "A Cruise to Georgian Bay in 1898." *Rudder,* 1899.

Head, George. *Forest Scenes and Incidents in the Wilds of North America; Being a Diary of a Winter's Route from Halifax to the Canadas, and During Four Months' Residence in the Woods on the Borders of Lakes Huron and Simcoe.* London: John Murray, Albemarle Street, 1829.

Henderson, Elmes. "Some Notes on a Visit to Penetanguishene and the Georgian Bay in 1856." Ontario Historical Society *Papers and Records.* Vol. 28, 1932, 30-4.

Hennepin, Louis. "A Description of a Ship of Sixty Tuns ..." 1698. Reprinted in J.J. Talman, ed., *Basic Documents in Canadian History.* Princeton, NJ: D. Van Nostrand, 1959.

Henry, Alexander. *Travels and Adventures in Canada and the Indian Territories Between the Years 1760 and 1776.* Ed. James Bain. Edmonton: M.G. Hurtig, 1969.

Hutchinson, Bruce. *The Unknown Country: Canada and Her People.* Toronto: Longmans, Green, 1942.

Jackson, Alexander Y. *A Painter's Country: The Autobiography of A.Y. Jackson.* Toronto: Clarke, Irwin, 1964.

Jameson, Anna. *Winter Studies and Summer Rambles in Canada.* Toronto: McClelland and Stewart, 1923.

Jesuit Relations and Allied Documents. Ed. Reuben Gold Thwaites. Vols. 33 (1647-8) and 35 (1649-50). Cleveland: Burrows Brothers, 1898.

Kane, Paul. *Wanderings of an Artist Among the Indians of North America.* 1858. Reprint, Toronto: Radisson Society of Canada, 1925.

Leaves from the War Log of the Nancy, 1813. Toronto: Rous and Mann Press, 1936. Reprint, Toronto: Huronia Historical Development Council/Ontario Department of Tourism and Information, 1968.

Leitch, Adelaide. "The Island Galaxy of Georgian Bay." *Canadian Geographical Journal,* August 1955.

Macfarlane, David. "Paradise Lost." *Toronto Life,* July 1995.

MacMillan, Marion Thayer. *Reflections: The Story of Water Pictures.* New York: Greenberg, 1936.

"The Manitoulin Letters of the Reverend Charles Crosbie Brough." Transcribed by Rundall M. Lewis. *Ontario History* 48, 2 (1956): 63-80.

McCarthy, Doris. *A Fool in Paradise: An Artist's Early Life.* Toronto: Macfarlane, Walter and Ross, 1990.

McLean, Scott A., ed. *From Lochnaw to Manitoulin: A Highland Soldier's Tour Through Upper Canada.* Toronto: Natural Heritage Books, 1999.

McMurray, Thomas. *The Free Grant Lands of Canada, from Practical Experience of Bush Farming in the Free Grant Districts of Muskoka and Parry Sound.* Bracebridge, ON, 1871.

Ossoli, Sarah. *Summer on the Lakes, in 1843.* New York, 1844.

Quill, Greg. "Lakes Great for Fun and Profit" [Lynx Images]. *Toronto Star,* 23 January 1999.

Rourke, Juanita. *Up the Shore: A Timeless Story of Georgian Bay.* Midland: Up the Shore Enterprises, 1994.

Sagard, Gabriel Fr. *Sagard's Long Journey to the Country of the Hurons.* Ed. George M. Wrong, trans. H.H. Langton. Toronto: Champlain Society, 1939.

Sander-Lofft, Suzanne. "Roughing It with the Rich: I Married a Georgian Bay Island." *Saturday Night,* June 1978.

Schoolcraft, Henry R. *Personal Memoirs of a Residence of Thirty Years with the Indian Tribes on the American Frontiers.* Philadelphia: Lippincott, Grambo, 1851. Reprint, New York: Arno Press, 1975.

Shaw, Marlow A. *The Happy Islands: Stories and Sketches of the Georgian Bay.* Toronto: McClelland and Stewart, 1926.

Simcoe, Elizabeth. *Mrs. Simcoe's Diary.* Ed. Mary Q. Innis. Toronto: Macmillan, 1965.

Spears, Raymond S. *A Trip on the Great Lakes.* Columbus, OH: A.R. Harding, 1913.

Symons, Harry. *Ojibway Melody.* Toronto: The Author and Ambassador Books/Copp Clark, 1946.

Toth, Susan A. *England As You Like It.* New York: Ballantine Books, 1995.

Wainwright, Andy, ed. *Notes for a Native Land: A New Encounter with Canada.* Toronto: Oberon Press, 1969.

Wahsoune Guests Have Rights. Hamilton, ON: John Gordon Gauld, 1933.

Wallace, Paul. "A Bit of Rock, River and Legend." *Canadian Forum,* September 1922.

Wells, Kenneth McNeill. *Cruising the Georgian Bay.* Toronto: Kingswood House, 1958.

–. *Cruising the Trent-Severn Waterway.* Toronto: Kingswood House, 1959.

Arts and Design

Bartram, Ed. "Artist's Statement." Mira Godard Gallery, Toronto, 1997.

Bland, Salem. "Preferences." *Canadian Forum,* December 1929.

Bluewater Visions. Video. Owen Sound: Tom Thomson Memorial Art Gallery, 1992.

Comfort, Charles. "Georgian Bay Legacy." *Canadian Art,* Spring 1951.

Comstock, William P. *Bungalows, Camps and Mountain Houses.* 3rd ed. New York: William T. Comstock, 1924.

Crossley, Peter. "The Simple Life." *City and Country Home,* May 1989.

Fairley, Barker. "The Group of Seven." *Canadian Forum,* February 1925.

Forest and Stream, "A Sailing Dinghy," 15 April 1899.

Forrest, Diane. "A Work in Progress." *Cottage Life,* July-August 1996.

Fosbery, Ernest. "Landscape Painting in Canada." *Canadian Geographical Journal,* August 1930.

Harris, Lawren. "The Group of Seven in Canadian History." *Canadian Historical Association Report of Annual Meeting.* Toronto, 1948.

Hartman, John. Interview with Dick Gordon. "This Morning." Canadian Broadcasting Corporation. 10 September 1999.

Hubbard, Henry, and Theodora Kimball. *An Introduction to the Study of Landscape Design.* Rev. ed. New York: Macmillan, 1929.

Jackson, A.Y. "Interview with A.Y. Jackson." By Lawrence Sabbath. *Canadian Art,* July 1960.

Lismer, Arthur. "Canadian Art Should Interpret Environment." *Royal Architectural Institute of Canada Journal* 5, 29 (January 1928).

Macfarlane, David. "Curiosities." *Canadian Art,* Summer 1994.

Matyas, Joe. "A Painter's Refuge." *London Free Press,* 24 April 1997.

Mitchell, Michael. "Papa Duck's Tale." *City and Country Home,* June 1988.

Salinger, Jehanne Bietry. "Elizabeth Wyn Wood." *Canadian Forum,* May 1931.

Skoggard, Ross. "Summer on Georgian Bay." *City and Country Home,* June 1988.

Stone, Caroline. "The Works of Ed Bartram: Impressive Exhibit at MUN." *St. John's Daily News*, 21 August 1982.

Toronto Star, "Waterfront Living," 18 September 1999.

Underhill, Frank. "False Hair on the Chest." *Saturday Night,* 3 October 1936.

Fiction, Poetry, and Song

Asling-Riis, Stella E. *The Great Fresh Sea.* New York, 1931.

Atwood, Margaret. "True Trash." In *Wilderness Tips.* Toronto: McClelland and Stewart, 1997.

Béland, Madeleine. *Chansons des Voyageurs, Coureurs de Bois et Forestiers.* Quebec: Presses de L'Université Laval, 1982.

Brooks, Andrew. "Night on Georgian Bay." *National Book Week,* 1980.

Campbell, W.W. *Lake Lyrics and Other Poems.* St. John, NB: J. and A. McMillan, 1889.

Cook, Merrill H. *Shore Lines and Sand Songs.* Boston: Beacon Press, n.d.

Cowan, Hugh. *La Cloche: The Story of Hector MacLeod and His Misadventures in the Georgian Bay and the La Cloche Districts.* Toronto: Algonquin Historical Society, 1928.

Doerflinger, William M. *Shantymen and Shantyboys: Songs of the Sailor and Lumberman.* New York: Macmillan, 1951.

Du Vernet, Sylvia. *Beams from the Beacon: Poems of Georgian Bay.* Bracebridge, ON: Herald-Gazette Press, 1974.

Edgar, Mary S. *Wood-Fire and Candle-Light.* Toronto: Macmillan, 1945.

Fowke, Edith. *Lumbering Songs from the Northern Woods.* Austin: University of Texas Press/ American Folklore Society, 1970.

–. *Traditional Singers and Songs from Ontario.* Hatboro, PA: Folklore Associates; Don Mills, ON: Burns and MacEachern, 1965.

Govier, Katherine. *Angel Walk.* Toronto: Little, Brown, 1996.

–. *The Immaculate Conception Photography Gallery and Other Stories.* Toronto: Little, Brown, 1994.

Hodgins, James C. *The Wilderness Campers.* Toronto: Musson Book Co., 1921.

Irving, John. *A Prayer for Owen Meany.* New York: Ballantine Books, 1989.

Kroetsch, Robert. *Badlands.* Don Mills, ON: Paperjacks, 1975.

LePan, Douglas. *Far Voyages.* Toronto: McClelland and Stewart, 1990.

–. *Weathering It: Complete Poems 1948-87.* Toronto: McClelland and Stewart, 1987.

Lismer, Arthur. "To the Georgian Bay." *Canadian Forum,* May 1921.

Macfarlane, David. *Summer Gone.* Toronto: Alfred A. Knopf, 1999.

Montgomery, L.M. *The Blue Castle.* Toronto: McClelland and Stewart, 1926.

Ondaatje, Michael. *The English Patient.* Toronto: McClelland and Stewart, 1992.

Payne, Millicent. "July Storm on Georgian Bay." *Canadian Forum,* October 1921.

Pratt, E.J. *Brébeuf and His Brethren.* Toronto: Macmillan, 1940.

Rickaby, Franz. *Ballads and Songs of the Shanty-Boy.* Cambridge, MA: Harvard University Press, 1926.

Robinson, Percy. *Georgian Bay.* Foreword by A.Y. Jackson. Toronto: privately printed, 1966.

Scott, F.R., and A.J.M. Smith, eds. *The Blasted Pine.* Rev. ed. Toronto: Macmillan, 1967.

Slater, Patrick. *The Water-Drinker.* Toronto: Thomas Allen, 1937.

Solomon, Evan. *Crossing the Distance.* Toronto: McClelland and Stewart, 1999.

Thomas, Edward, and Mary Thomas. *Memories of Georgian Bay.* Collingwood: Privately printed, 1960.

Tudor, Stephen. *Hangdog Reef: Poems Sailing the Great Lakes.* Detroit: Wayne State University Press, 1989.

Wright, Richard B. *The Age of Longing.* Toronto: HarperCollins, 1995.

Young, Scott. *That Old Gang of Mine.* Toronto: Fitzhenry and Whiteside, 1982.

Interviews and Personal Correspondence

Baines, Chris, Past President, Georgian Bay Land Trust. Telephone interview. Toronto: 16 August 1999.

Band, John. Toronto: 3 June 1997.

Bartram, Ed. King City, ON: 3 January 1998.

Bates, Jill. Burlington, ON: letter received 3 March 1998.

Beck, Gregor. Toronto: 7 June 1997, and e-mail dated 12 October 2000.

Bird, Gerry, and Mike Chellew. Lakefield, ON: 23 June 1997.

Birnbaum, John, Executive Director, Georgian Bay Association. Toronto: 18 September 1999.

Cameron, John. MNR (Peterborough, ON): email dated 12 April 2000.

Campbell, Neil. Toronto: 19 and 25 May, 15 June 1997; 18 September 1999; 23 August 2000.

Cavanaugh, Chris. Telephone interview. London, ON: 31 January 1998.

Dawson, Dave. Aurora, ON: 6 June 1997.

Elliott, Ulla. Port Hope, ON: 2 June 1997.

Hall, Andrew. Toronto: 11 June 1997.

Hamelin, Andy and Jean. Midland, ON: 16 June 1997.

Hankinson, Margaret. London, ON: 9 February 1998.

Hanna, Dave. Toronto: 22 May 1997.

Hartman, John. La Fontaine, ON: 16 June 1997.

Hauch, W. Kuyler, Former Director, YMCA Kitchikewana. London, ON: 6 February 1998.

Higgins, Gary, Superintendant, Massasauga and Oastler Lake Provincial Parks. Telephone interview. Parry Sound, ON: 25 January 2000.

Hockin, Thomas. Toronto: 31 May 1997.

Hodgetts, David. Hamilton, ON: 26 June 1997.

Hodgetts, Ross. Spruce Grove, AB: taped comments received 6 August 1997.

Hughes, John. Toronto: 1 June 1997.

Hummel, Monte, Executive Director, World Wildlife Fund Canada. Toronto: 29 May 1997.

King, Frank. London, ON: 31 January 1998.

Lash, Tony and Marion. Toronto: 15 June 1997.

Laughlin, Brenda. Telephone interview. Township of the Archipelago Assessment Office, Parry Sound, ON: 25 November 1999.

Lawson, Tom. Port Hope, ON: 2 June 1997; Nares Inlet, ON: 12 August 1999.

Lord, John. Toronto: 28 May 1997.

MacGregor, Duncan, with Clem Sharp and Jan Trimble. Go Home Bay, ON: 10 August 1999.

Marston, Don. Camp Hurontario, ON: 20 July 1997.

McClelland, William. London, ON: 28 October 1997.

McDerment, Robert. Toronto: 15 July 1997.

McIntyre, Eric, MNR fisheries biologist. Parry Sound, ON: letter dated 9 February 1998.

McLeod, Ian. Toronto: 22 May 1997.

Michel, Gillian. London, ON: 5 March 1998.

Noble, Clark ("Knobby"). Toronto: 8 July 1997.
O'Brian, Peter. Toronto: 10 June 1997.
Otto, Mary Leigh. London, ON: 7 February 1998.
Powsey, Clive. Bomanville, ON: letter dated 27 January 1998.
Rossiter, Margaret. London, ON: 28 October 1997.
Somers, William. Toronto: 22 July 1997.
Stinson, Tim. Toronto: 5 June 1997.
Tada, Gunje and Tomi. Toronto: 26 June 1997.
Thomson, Rob. Charlottetown, PEI: email dated 12 July 1997.
Townley, Keith. Toronto: 22 May 1997.
Trusler, Jim, former MNR district geologist. Email dated 14 June 2000.
Will, Al, Executive Director, Ontario Sailing Association. Toronto: 19 June 1997.
Willoughby, Michael and Wendy. Toronto: 27 May 1997.
Wood, John. Toronto: 22 May 1997.

Galleries and Museums
Art Gallery of Ontario, Permanent Collection.
Ashton-Evicta Gallery, Toronto, "Clive Powsey," (2000).
London Regional Art and Historical Museum, Permanent Collection.
–. "Big North," works by John Hartman (2001).
–. "Tradition and Innovation in Canadian Sculpture," works by Elizabeth Wyn Wood and Emanuel Hahn (1997).
McMichael Canadian Art Collection, Permanent Collection.
Michael Gibson Gallery, "Recent Oil Paintings," works by Bruce Steinhoff (1998).
National Gallery of Canada, Permanent Collection.
Royal Ontario Museum, "Land Study, Studio View," works by Paul Kane (2000).
Sans Souci Museum (joint project of Huronia Museum and WPSM).
Thielsen Gallery, London, "New Prints," works by Ed Bartram (1997).
West Parry Sound Museum, Permanent Collection.

SECONDARY SOURCES

Landscape and Environment
Aitchison, Cara, Nicola E. MacLeod, and Stephen J. Shaw. *Leisure and Tourism Landscapes: Social and Cultural Geographies.* London: Routledge, 2000.
Alanen, Arnold R., and Robert Z. Melnick, eds. *Preserving Cultural Landscapes in America.* Baltimore: Johns Hopkins University Press, 2000.
Allen, Barbara, and Thomas J. Schlereth, eds. *Sense of Place: American Regional Cultures.* Lexington, KY: University Press of Kentucky, 1990.
Appleton, Jay. *The Experience of Landscape.* Rev. ed. London: John Wiley and Sons, 1996.
Baker, Alan R.H., and Gideon Biger, eds. *Ideology and Landscape in Historical Perspective.* Cambridge: Cambridge University Press, 1992.
Barnes, Trevor J., and James S. Duncan. *Writing Worlds: Discourse, Text and Metaphor in the Representation of Landscape.* London: Routledge, 1992.
Berleant, Arnold. *The Aesthetics of Environment.* Philadelphia: Temple University Press, 1992.
Boniface, Priscilla, and Peter J. Fowler. *Heritage and Tourism in the Global Village.* London: Routledge, 1993.

Buggey, Susan. "Cultural Landscapes in Canada." In Bernd von Droste, Harald Plachter, and Mechtild Rössler, eds., *Cultural Landscapes of Universal Value: Components of a Global Strategy.* New York: Gustav Fischer Verlag, 1995.

Buisseret, David, ed. *From Sea Charts to Satellite Images: Interpreting North American History through Maps.* Chicago: University of Chicago Press, 1990.

Clapp, B.W. *An Environmental History of Britain Since the Industrial Revolution.* London: Longman, 1994.

Conzen, Michael P., ed. *The Making of the American Landscape.* Boston: Unwin Hyman, 1990.

Cosgrove, Denis, ed. *Mappings.* London: Reaktion Books, 1999.

Cronon, William. *Nature's Metropolis: Chicago and the Great West.* New York: W.W. Norton, 1991.

–, ed. *Uncommon Ground: Toward Reinventing Nature.* New York: W.W. Norton, 1995.

Daniels, Stephen, and Denis Cosgrove, eds. *The Iconography of Landscape.* Cambridge: Cambridge University Press, 1988.

Duncan, James, and David Ley, eds. *Place/Culture/Representation.* London: Routledge, 1993.

Flader, Susan L., ed. *The Great Lakes Forest: An Environmental and Social History.* Minneapolis: University of Minnesota Press, 1983.

Gottlieb, Robert. *Forcing the Spring: The Transformation of the American Environmental Movement.* Washington, DC: Island Press, 1993.

Groat, Linda, ed. *Giving Places Meaning: Readings in Environmental Psychology.* London: Academic Press, 1995.

Groth, Paul, and Todd W. Bressi, eds. *Understanding Ordinary Landscapes.* New Haven, CT: Yale University Press, 1997.

Hall, C. Michael, and Stephen J. Page. "Tourism and Recreation in the Pleasure Periphery: Wilderness and National Parks," Chap. 7 in *The Geography of Tourism and Recreation: Environment, Place and Space.* London: Routledge, 2002.

Hays, Samuel. *Beauty, Health, and Permanence: Environmental Politics in the United States, 1955-1985.* Cambridge: Cambridge University Press, 1987.

Hiss, Tony. *The Experience of Place.* New York: Alfred A. Knopf, 1990.

Hough, Michael. *Out of Place: Restoring Identity to the Regional Landscape.* New Haven, CT: Yale University Press, 1990.

Hull, R. Bruce. "Image Congruity, Place Attachment and Community Design." *Journal of Architectural and Planning Research* 9, 3 (Autumn 1992).

Kemal, Salim, and Ivan Gaskell, eds. *Landscape, Natural Beauty and the Arts.* Cambridge: Cambridge University Press, 1993.

Jackson, John Brinckerhoff. *Discovering the Vernacular Landscape.* New Haven, CT: Yale University Press, 1984.

LaDow, Beth. *The Medicine Line: Life and Death on a North American Borderland.* New York: Routledge, 2001.

Light, Andrew, and Jonathan M. Smith, eds. *Philosophies of Place.* Lanham, MD: Rowman and Littlefield, 1998.

Locock, Martin, ed. *Meaningful Architecture: Social Interpretations of Buildings.* Aldershot, UK: Avebury Press, 1994.

McClelland, Linda Flint. *Building the National Parks: Historic Landscape Design and Construction.* Baltimore: Johns Hopkins University Press, 1998.

McGinnis, Michael Vincent, ed. *Bioregionalism.* New York: Routledge, 1999.

McNeill, J.R. *Something New Under the Sun: An Environmental History of the Twentieth-Century World.* New York: W.W. Norton, 2000.

Moodie, Jane. "Preparing the Waste Places for Future Prosperity? The New Zealand Pioneering Myth and Adaptation to Recent Change." Paper presented at "Landscapes of Memory: Oral History and the Environment," Annual Conference of the UK Oral History Society/University of Sussex, 16 May 1999.

Nash, Roderick. *Wilderness and the American Mind.* 3rd ed. New Haven, CT: Yale University Press, 1982.

Nassauer, Joan Iverson, ed. *Placing Nature: Culture and Landscape Ecology.* Washington, DC: Island Press, 1997.

Nelson, J.G., et al., eds. *Protected Areas and the Regional Planning Imperative in North America.* Calgary: University of Calgary Press; East Lansing, MI: Michigan State Press, 2002.

Norberg-Schulz, Christian. *Genius Loci: Towards a Phenomenology of Architecture.* London: Academy Editions, 1980.

Norton, William. *Explorations in the Understanding of Landscape: A Cultural Geography.* Westport, CT: Greenwood Press, 1989.

Nostrand, Richard L., and Lawrence E. Estaville. *Homelands: A Geography of Culture and Place across America.* Baltimore: Johns Hopkins University Press, 2001.

Oelschlaeger, Max. *The Idea of Wilderness, from Prehistory to the Age of Ecology.* New Haven, CT: Yale University Press, 1991.

Payne, Daniel G. *Voices in the Wilderness: American Nature Writing and Environmental Politics.* Hanover, NH: University Press of New England, 1996.

Penning-Rowsell, Edmund C., and David Lowenthal, eds. *Landscape Meanings and Values.* London: Allen and Union, 1986.

Pielou, E.C. *After the Ice Age: The Return of Life to Glaciated North America.* Chicago: University of Chicago Press, 1991.

Pierceall, Gregory M. *Residential Landscapes: Graphics, Planning, and Design.* Reston, VA: Reston Publishing, 1984.

Pyne, Stephen. *How the Canyon Became Grand.* New York: Viking, 1998.

Rowe, Stan. *Home Place: Essays on Ecology.* Edmonton: NeWest Press, 1990.

Ryden, Kent C. *Mapping the Invisible Landscape: Folklore, Writing, and the Sense of Place.* Iowa City: University of Iowa Press, 1993.

Sadler, Barry, and Allen Carlson, eds. *Environmental Aesthetics: Essays in Interpretation.* Victoria, BC: Western Geographical Press, 1982.

Schama, Simon. *Landscape and Memory.* New York: Alfred A. Knopf, 1995.

Sellars, Richard West. *Preserving Nature in the National Parks: A History.* New Haven, CT: Yale University Press, 1997.

Shields, Rob. *Places on the Margin: Alternative Geographies of Modernity.* London: Routledge, 1991.

Soulé, Michael E., and Gary Lease, eds. *Reinventing Nature? Responses to Postmodern Deconstruction.* Washington, DC: Island Press, 1995.

Stankey, George H. "Beyond the Campfire's Light: Historical Roots of the Wilderness Concept." *Natural Resources Journal* 29 (Winter 1989): 9-24.

Steinberg, Ted. *Down to Earth: Nature's Role in American History.* New York: Oxford University Press, 2002.

Thrower, Norman J.W. *Maps and Civilization: Cartography in Culture and Society.* Chicago: University of Chicago Press, 1996.

Tuan, Yi-Fu. *Topophilia: A Study of Environmental Perception, Attitudes, and Values.* Englewood Cliffs, NJ: Prentice-Hall, 1974.

Walker, Laurence C. *The North American Forests: Geography, Ecology, and Silviculture.* Boca Raton, FL: CRC Press, 1999.

White, Richard. *The Organic Machine: The Remaking of the Columbia River.* New York: Hill and Wang, 1995.

Whitfield, Peter. *New Found Lands: Maps in the History of Exploration.* London: The British Library, 1998.

Wilson, Alexander. *The Culture of Nature: North American Landscapes from Disney to the Exxon Valdez.* Toronto: Between the Lines, 1991.

Worster, Donald, ed. *The Ends of the Earth: Perspectives on Modern Environmental History.* Cambridge: Cambridge University Press, 1988.

–. *Nature's Economy: A History of Ecological Ideas.* 2nd ed. Cambridge: Cambridge University Press, 1994.

–. "Two Faces West: The Development Myth in Canada and the United States." In *Terra Pacifica: People and Place in the Northwest States and Western Canada.* Ed. Paul Hirt, 71-91. Pullman, WA: Washington State University Press, 1998.

–. *An Unsettled Country: Changing Landscapes of the American West.* Albuquerque, NM: University of New Mexico Press, 1994.

Canadian Environmental and Landscape Studies

Altmeyer, George. "Three Ideas of Nature in Canada, 1893-1914." *Journal of Canadian Studies* 11, 3 (August 1976): 21-36.

Andreae, Christopher. "Industry, Dereliction and Landscapes in Ontario." *Ontario History* 89, 2 (June 1997).

–. *Lines of Country: An Atlas of Railway and Waterway History in Canada.* Erin, ON: Boston Mills Press, 1997.

Angus, James T. "How the Dokis Protected Their Timber." *Ontario History* 81, 3 (September 1989).

Bella, Leslie. *Parks for Profit.* Montreal: Harvest House, 1987.

Berger, Carl. *Science, God, and Nature in Victorian Canada.* Toronto: University of Toronto Press, 1983.

Campbell, Claire. "'A New *Terra Incognita*': New Women in Canada's Outdoors, 1880-1920." Halifax: Department of History, Dalhousie University, 1995.

"Canadian Wilderness Journal." Segment on snakes of the Georgian Bay. Global. Aired 26 March 2000.

Clapp, R.A. "The Resource Cycle in Forestry and Fishing." *Canadian Geographer* 42, 2 (1998).

Coker, G.A., et al. *Morphological and Ecological Characteristics of Canadian Freshwater Fishes.* Canadian Manuscript Report of Fisheries and Aquatic Sciences 2554. Burlington, ON: Department of Fisheries and Oceans, 2001.

Davies, Stephen. "'Reckless Walking Must Be Discouraged': The Automobile Revolution and the Shaping of Modern Urban Canada to 1930." *Urban History Review,* 1989.

Dewar, Kenneth M. "Perceptions of the Canadian Wilderness: Literary and Visual Responses to the North Shore of Lake Superior, 1663-1926." MA thesis, York University, 1983.

Doern, G. Bruce, and Thomas Conway. *The Greening of Canada: Federal Institutions and Decisions.* Toronto: University of Toronto Press, 1994.

Farrell, Barbara, and Aileen Desbarats, eds. *Explorations in the History of Canadian Mapping.* Ottawa: Association of Canadian Map Libraries and Archives, 1988.

Fillmore, Stanley, and R.W. Sandilands. *The Chartmakers: The History of Nautical Survey-ing in Canada.* Toronto: NC Press, 1983.

Forkey, Neil. *Shaping the Upper Canadian Frontier: Environment, Society and Culture in the Trent Valley.* Calgary: University of Calgary, 2003.

Foster, Janet. *Working for Wildlife: The Beginning of Preservation in Canada.* 2nd ed. Toronto: University of Toronto Press, 1998.

Gaffield, Chad, and Pam Gaffield, eds. *Consuming Canada: Readings in Environmental History.* Toronto: Copp Clark, 1995.

Gentilcore, R. Louis, ed. *Historical Atlas of Canada.* Vol. 2, *1800-1891: The Land Trans-formed.* Toronto: University of Toronto Press, 1993.

–, and Kate Donkin. *Land Surveys of Southern Ontario.* Toronto: York University, 1973.

–, and C. Grant Head. *Ontario's History in Maps.* Toronto: University of Toronto Press, 1984.

Gillis, R. Peter. "Rivers of Sawdust: The Battle Over Industrial Pollution in Canada, 1865-1903." *Journal of Canadian Studies* 21, 1 (Spring 1986).

Gough, Joseph. *Fisheries Management in Canada, 1880-1910.* Canadian Manuscript Report of Fisheries and Aquatic Sciences 2105. Halifax: Department of Fisheries and Oceans, 1991.

Gray, Stephen L. *A Descriptive Forest Inventory of Canada's Forest Regions.* Information Report PI-X-122. Chalk River, ON: Canadian Forest Service, 1995.

The Great Lakes: An Environmental Atlas and Resource Book. 3rd ed. Toronto: Government of Canada; Chicago: US Environmental Protection Agency, 1995.

Halseth, Greg. *Cottage Country in Transition: A Social Geography of Change and Contention in the Rural-Recreational Countryside.* Montreal and Kingston: McGill-Queen's University Press, 1998.

Hart, E.J. *The Selling of Canada: The CPR and the Beginnings of Canadian Tourism.* Banff, AB: Altitude, 1983.

Hodgins, B.W., and Jamie Benidickson. *The Temagami Experience: Recreation, Resources, and Aboriginal Rights in the Northern Ontario Wilderness.* Toronto: University of Toronto Press, 1989.

–, and Bernadine Dodge, eds. *Using Wilderness: Essays on the Evolution of Youth Camping in Ontario.* Peterborough, ON: Frost Centre, 1992.

Howlett, Michael, and Jeremy Rayner. "Opening Up the Woods? The Origins and Future of Contemporary Canadian Forest Policy Conflicts." *National History* 1, 1 (Winter 1997).

Innis, Harold. *Essays in Canadian Economic History.* Ed. Mary Q. Innis. Toronto: University of Toronto Press, 1956.

Jasen, Patricia. *Wild Things: Nature, Culture, and Tourism in Ontario, 1790-1914.* Toronto: University of Toronto Press, 1995.

Johnston, Margaret. "The Canadian Wilderness Landscape as Culture and Commodity." *International Journal of Canadian Studies* 4 (Fall 1991): 127-44.

Killan, Gerald. *Protected Places: A History of Ontario's Provincial Park System.* Toronto: Dundurn Press, 1993.

Kuhlberg, Mark. "Ontario's Nascent Environmentalists: Seeing the Foresters for the Trees in Southern Ontario, 1919-1929." *Ontario History* 88, 2 (June 1996).

Labatt, Lori, and Bruce Litteljohn, eds. *Islands of Hope: Ontario's Parks and Wilderness.* Willowdale, ON: Firefly Books, 1992.

Ladell, John L. *They Left Their Mark: Surveyors and Their Role in the Settlement of Ontario.* Toronto: Dundurn Press, 1993.

Lambert, Richard S., and Paul Pross. *Renewing Nature's Wealth: A Centennial History of the Public Management of Lands, Forests and Wildlife in Ontario, 1763-1967.* Toronto: Department of Lands and Forests, 1967.

Loo, Tina. "Making a Modern Wilderness: Conserving Wildlife in Twentieth-Century Canada." *Canadian Historical Review* 82, 1 (2001): 91-120.

Lothian, W.F. *A History of Canada's National Parks.* Vol. 4. Ottawa: Parks Canada, 1981.

Lower, A.R.M. *The North American Assault on the Canadian Forest.* New Haven, CT: Yale University Press 1938. Rev. ed., New York: Greenwood Press, 1968.

MacEachern, Alan Andrew. "In Search of Eastern Beauty: Creating National Parks in Atlantic Canada, 1935-1970." PhD thesis, Queen's University, 1997.

–. *Natural Selections: National Parks in Atlantic Canada, 1935-1970.* Montreal: McGill-Queen's Press, 2001.

MacLaren, Ian. "Cultured Wilderness in Jasper National Park." *Journal of Canadian Studies* 34, 3 (Fall 1999): 3-53.

MacLellan, Duncan. "Shifting from the Traditional to the New Political Agenda: The Changing Nature of Federal-Provincial Environmental Relations." *American Review of Canadian Studies* 25, 2/3 (1995): 323-45.

Marsh, John, and Janice Fialkowski, eds. *Linking Cultural and Natural Heritage.* Peterborough, ON: Frost Centre for Canadian Heritage, 1995.

–, and Bruce W. Hodgins, eds. *Changing Parks: The History, Future and Cultural Context of Parks and Heritage Landscapes.* Toronto: Natural Heritage/Natural History, 1998.

Martin, Virgil. *Changing Landscapes of Southern Ontario.* Erin, ON: Boston Mills Press, 1988.

Marty, Sid. *A Grand and Fabulous Notion: The First Century of Canada's Parks.* Toronto: NC Press, 1984.

McIlwraith, Thomas. *Looking for Old Ontario: Two Centuries of Landscape Change.* Toronto: University of Toronto Press, 1997.

McNamee, Kevin. *The National Parks of Canada.* Toronto: Key Porter, 1998.

Moon, Barbara. *The Canadian Shield.* Toronto: Natural Science of Canada, 1970.

National Post, "When Waterfront Cottages Aren't," 4 April 2000.

Nelles, H.V. *The Politics of Development: Forests, Mines and Hydro-Electric Power in Ontario, 1849-1941.* Toronto: Macmillan, 1974.

Nelson, J.G. "Beyond Parks and Protected Areas: From Public and Private Stewardships to Landscape Planning and Management." *Environments* 21, 1 (1990): 23-34.

–. *An External Perspective on Parks Canada Strategies, 1986-2001.* Occasional Paper #2. Waterloo, ON: University of Waterloo/Parks Canada Liaison Committee, 1984.

–, and R.C. Scace. *The Canadian National Parks: Today and Tomorrow.* Proceedings of a conference organized by the National and Provincial Parks Association of Canada and the University of Calgary. Calgary, 1968.

Newell, Dianne, and Rosemary E. Ommer, eds. *Fishing Places, Fishing People: Traditions and Issues in Canadian Small-Scale Fisheries.* Toronto: University of Toronto Press, 1998.

Perera, Ajith H., et al. *Ecology of a Managed Terrestrial Landscape: Patterns and Processes of Forest Landscapes in Ontario.* Vancouver: University of British Columbia Press, 2000.

Rabe, Barry. "The Politics of Ecosystem Management in the Great Lakes Basin." *American Review of Canadian Studies* 27, 3 (Autumn 1997): 411-36.

Radforth, Ian. *Bushworkers and Bosses: Logging in Northern Ontario, 1900-1980.* Toronto: University of Toronto Press, 1987.

Read, Jennifer. "Addressing 'A Quiet Horror': The Evolution of Ontario Pollution Control Policy in the International Great Lakes, 1909-1972." PhD thesis, University of Western Ontario, 1999.

Rees, Ronald. *New and Naked Land: Making the Prairies Home.* Saskatoon: Western Producer Prairie Books, 1988.

"Refiguring Wilderness." *Journal of Canadian Studies* 33, 2 (Summer 1998).

Roach, Thomas R. "The Pulpwood Trade and the Settlers of New Ontario, 1919-1938." *Journal of Canadian Studies* 22, 3 (1987): 78-88.

Ross, Monique. *A History of Forest Legislation in Canada, 1867-1996.* Calgary: Canadian Institute of Resources Law, University of Calgary, 1997.

Scott, W.B., and E.J. Crossman. *Freshwater Fishes of Canada,* 2nd ed. Oakville, ON: Galt House, 1998.

Smith, Allan. "Farms, Forests and Cities: The Image of the Land and the Rise of the Metropolis in Ontario, 1860-1914." In David Keane and Colin Read, eds. *Old Ontario: Essays in Honour of J.M.S. Careless.* Toronto: Dundurn Press, 1990.

Theberge, John B., ed. *Legacy: The Natural History of Ontario.* Toronto: McClelland and Stewart, 1989.

Wall, Geoffrey, and John S. Marsh, eds. *Recreational Land Use: Perspectives on Its Evolution in Canada.* Ottawa: Carleton University Press, 1982.

Warecki, George M. *Protecting Ontario's Wilderness: A History of Changing Ideas and Preservation Politics, 1927-1973.* New York: Peter Lang, 2000.

Wood, J. David, ed. *Perspectives on Landscape and Settlement in Nineteenth Century Ontario.* Toronto: Macmillan, 1978.

–. *Making Ontario: Agricultural Colonization and Landscape Re-creation before the Railway.* Montreal and Kingston: McGill-Queen's University Press, 2000.

Wynn, Graeme, ed. *People, Places, Patterns, Processes: Geographical Perspectives on the Canadian Past.* Toronto: Copp Clark Pitman, 1990.

Zaslow, Morris. *Reading the Rocks: The Story of the Geological Survey of Canada, 1842-1972.* Toronto: Macmillan, 1975.

Zeller, Suzanne. "Mapping the Canadian Mind: Reports of the Geological Survey of Canada 1842-1863." *Canadian Literature* 131 (Winter 1991): 156-67.

History and Cultural Studies

Anderson, Benedict. *Imagined Communities: Reflections on the Origin and Spread of Nationalism.* Rev. ed. London: Verso, 1991.

Aron, Cindy S. *Working at Play: A History of Vacations in the United States.* New York: Oxford University Press, 1999.

Barnes, Thomas G. "Canada, True North: A 'Here There' or a Boreal Myth?" *American Review of Canadian Studies* 19, 4 (Winter 1989): 369-79.

Bartlett, Richard H. *Indian Reserves and Aboriginal Lands in Canada: A Homeland.* Saskatoon: University of Saskatchewan Native Law Centre, 1990.

Benidickson, Jamie. *Idleness, Water and a Canoe: Reflections on Paddling for Pleasure.* Toronto: University of Toronto Press, 1997.

Berger, Carl. *The Writing of Canadian History: Aspects of English-Canadian Historical Writing Since 1900.* 2nd ed. Toronto: University of Toronto Press, 1986.

Bodnar, John E. *Remaking America: Public Memory, Commemoration, and Patriotism in the Twentieth Century.* Princeton, NJ: Princeton University Press, 1992.

Bowden, Bruce, et al. *A History of Christian Island and the Beausoleil Band.* Vol. 3. Rev. ed. London, ON: Department of History, University of Western Ontario, 1990.

Cameron, Christina. "The Spirit of Place: The Physical Memory of Canada." *Journal of Canadian Studies* 35, 1 (Spring 2000): 77-94.

Careless, J.M.S. "Limited Identities in Canada." *Canadian Historical Review* 50 (1969): 1-10.

Carpenter, Carole Henderson. *Many Voices: A Study of Folklore Activities in Canada and Their Role in Canadian Culture.* Canadian Centre for Folk Culture Studies, Paper no. 26. Ottawa: National Museums of Canada, 1979.

Clayton, Daniel. *Islands of Truth: The Imperial Fashioning of Vancouver Island.* Vancouver: University of British Columbia Press, 1999.

Clifton, James A. *A Place of Refuge for All Time: Migration of the American Potawatomi into Upper Canada, 1830-1850.* Canadian Ethnology Service Paper no. 26. Ottawa: National Museums of Canada, 1975.

Conzen, Michael P., et al. *A Scholar's Guide to Geographical Writing on the American and Canadian Past.* University of Chicago Research Paper no. 235. Chicago: University of Chicago Press, 1993.

Coppock, J.T., ed. *Second Homes: Curse or Blessing?* Oxford: Pergamon Press, 1977.

Cross, Amy Willard. *The Summer House: A Tradition of Leisure.* Toronto: HarperCollins, 1992.

Crowe, Norman. *Nature and the Idea of a Man-Made World: An Investigation into the Evolutionary Roots of Form and Order in the Built Environment.* Cambridge, MA: MIT Press, 1995.

Dickason, Olive. *Canada's First Nations: A History of Canada's Founding Peoples from Earliest Times.* Toronto: McClelland and Stewart, 1996.

Eaton, Nicole, and Hilary Weston. *In a Canadian Garden.* Markham, ON: Viking Studio Books, 1989.

–. *At Home in Canada.* Toronto: Viking, 1995.

Ennals, Peter, and Deryck Holdsworth. *Homeplace: The Making of the Canadian Dwelling over Three Centuries.* Toronto: University of Toronto Press, 1998.

–. "Vernacular Architecture and the Cultural Landscape of the Maritime Provinces: A Reconnaissance." *Acadiensis* 10, 2 (1981): 86-106.

Evenden, Matthew. "The Northern Vision of Harold Innis." *Journal of Canadian Studies* 34, 3 (Fall 1999): 162-86.

Ferguson, Will. *Why I Hate Canadians.* Vancouver: Douglas and McIntyre, 1997.

Francis, Daniel. *National Dreams: Myth, Memory and Canadian History.* Vancouver: Arsenal Pulp Press, 1997.

Friesen, Gerald. "The Prairies as Region: The Contemporary Meaning of an Old Idea." Reprinted in Gerald Friesen, *River Road: Essays on Manitoba and Prairie History,* 165-82. Winnipeg: University of Manitoba Press, 1996.

Gardner, James B., and Peter S. LaPaglia, eds. *Public History: Essays from the Field.* Melbourne, FL: Krieger, 1999.

Gill, Brendan, and Dudley Witney. *Summer Places.* Toronto: McClelland and Stewart, 1978.

Glassberg, David. *Sense of History: The Place of the Past in American Life.* Amherst, MA: University of Massachusetts Press, 2001.

Gowans, Alan. *The Comfortable House: North American Suburban Architecture, 1890-1930.* Cambridge, MA: MIT Press, 1986.

Grant, Shelagh. "Myths of the North in the Canadian Ethos." *Northern Review* 3/4 (1989): 15-41.

Greer, Allan. *The People of New France.* Toronto: University of Toronto Press, 1997.

Hall, Roger, et al., eds. *Patterns of the Past: Interpreting Ontario's History.* Toronto: Dundurn Press, 1988.

Harris, Cole. "The Emotional Structure of Canadian Regionalism." In *The Challenges of Canada's Regional Diversity.* Vol. 5, The Walter L. Gordon Lecture Series 1980-1. Toronto, 1981, 9-30.

Hobsbawm, Eric, and Terence Ranger, eds. *The Invention of Tradition.* Cambridge: Cambridge University Press, 1983.

Hodgins, Bruce, and Margaret Hobbs, eds. *Nastawgan.* Toronto: Betelgeuse Books, 1985.

–, et al., eds. *Co-Existence? Studies in Ontario-First Nations Relations.* Peterborough, ON: Frost Centre for Canadian Heritage and Development Studies, 1992.

Jaenen, Cornelius J. "Gabriel Sagard: A Franciscan Among the Huron." In Ian K. Steele and Nancy L. Rhoden, eds., *The Human Tradition in Colonial America.* Wilmington, DE: Scholarly Resources, 1999, 37-48.

Jarvis, Eric. "The Georgian Bay Ship Canal: A Study of the Second Canal Age, 1850-1915." *Ontario History* 69, 2 (June 1977): 125-47.

Jennings, John, et al., eds. *The Canoe in Canadian Cultures.* Toronto: Natural Heritage/ Natural History, 1999.

Jury, Wilfrid, and Elsie McLeod Jury. *Sainte-Marie Among the Hurons.* Toronto: Oxford University Press, 1954.

Kalman, Harold. *A History of Canadian Architecture.* Vol. 2. Toronto: Oxford University Press, 1994.

Kapelos, George T. *Interpretations of Nature: Contemporary Canadian Architecture, Landscape and Urbanism.* Kleinburg, ON: McMichael Canadian Art Collection, 1994.

Kaufmann, Eric. "'Naturalizing the Nation': The Rise of Naturalistic Nationalism in the United States and Canada." *Comparative Studies in Society and History* 40, 4 (October 1998): 666-95.

Legget, Robert F. *Canals of Canada.* Vancouver: Douglas, David and Charles, 1976.

Leighton, Douglas. "The Historical Significance of the Robinson Treaties of 1850." Paper presented to the Annual Meeting of the Canadian Historical Association, Ottawa, June 1982.

–. "The Manitoulin Incident of 1863: An Indian-White Confrontation in the Province of Canada." *Ontario History* 69, 2 (June 1977): 113-24.

Lenskyj, Helen. "Common Sense and Physiology: North American Medical Views on Women and Sport, 1890-1930." *Canadian Journal of History of Sport* 21, 1 (May 1990): 49-64.

Lowenthal, David. *The Heritage Crusade and the Spoils of History.* Cambridge: Cambridge University Press, 1998.

MacLeod, D. Peter. "The Anishinabeg Point of View: The History of the Great Lakes Region to 1800 in Nineteenth-Century Mississauga, Odawa and Ojibway Historiography." *Canadian Historical Review* 73, 2 (June 1992): 194-210.

McKay, Ian. *The Quest of the Folk: Antimodernism and Cultural Selection in Twentieth Century Nova Scotia.* Montreal and Kingston: McGill-Queen's University Press, 1994.

McKillop, A.B. "Culture, Intellect, and Context: Recent Writing on the Cultural and Intellectual History of Ontario." *Journal of Canadian Studies* 24, 3 (1989): 7-31. Reprinted in Ajay Heble et al., eds. *New Contexts of Canadian Criticism.* Peterborough, ON: Broadview Press, 1997.

Morantz, Alan. *Where is Here? Canada's Maps and the Stories They Tell.* Toronto: Penguin Canada, 2002.

Morton, William. "Clio in Canada: The Interpretation of Canadian History." *University of Toronto Quarterly* 15, 3 (April 1946).

–. "The North in Canadian Historiography." *Transactions, Royal Society of Canada* 4, 8 (1970): 31-40.

–. "The Relevance of Canadian History." In *The Canadian Identity*. Madison, WI: University of Wisconsin Press, 1961.

Muise, D.A. "Who Owns History Anyway? Reinventing Atlantic Canada for Pleasure and Profit." *Acadiensis* 27, 2 (Spring 1998): 124-34.

Ontario Native Affairs Secretariat and Ministry of Citizenship. *Akwesasne to Wunnumin Lake: Profiles of Aboriginal Communities in Ontario*. Toronto: Queen's Printer, 1992.

Owram, Doug. *Promise of Eden: The Canadian Expansionist Movement and the Idea of the West, 1856-1900*. 1980. Reprint, Toronto: University of Toronto Press, 1992.

Parr, Joy, and Mark Rosenfeld, eds. *Gender and History in Canada*. Toronto: Copp Clark, 1996.

Ricketts, Shannon. "Cultural Selection and National Identity: Establishing Historic Sites in a National Framework, 1920-1939." *The Public Historian* 18, 3 (1996): 23-41.

Rogers, Edward S., and Donald B. Smith. *Aboriginal Ontario: Historical Perspectives on the First Nations*. Toronto: Dundurn Press, 1994.

Russell, Peter, ed. *Nationalism in Canada*. Toronto: McGraw-Hill, 1966.

Schmalz, Peter S. *The Ojibwa of Southern Ontario*. Toronto: University of Toronto Press, 1991.

Sioui, Georges. *Huron-Wendat: The Heritage of the Circle*. Vancouver: University of British Columbia Press, 1997.

Smith, Theresa A. *The Island of the Anishnaabeg: Thunderers and Water Monsters in the Traditional Ojibwe Life-World*. Moscow, ID: University of Idaho Press, 1995.

Taylor, C.J. *Negotiating the Past: The Making of Canada's National Historic Parks and Sites*. Montreal and Kingston: McGill-Queen's University Press, 1990.

Taylor, M. Brook. *Promoters, Patriots, and Partisans: Historiography in Nineteenth-Century English Canada*. Toronto: University of Toronto Press, 1989.

Tippett, Maria. *Making Culture: English-Canadian Institutions and the Arts before the Massey Commission*. Toronto: University of Toronto Press, 1990.

Tolpin, Jim. *The New Cottage Home*. Newtown, CT: Taunton Press, 1998.

Trigger, Bruce G. *The Children of Aataentsic: A History of the Huron People to 1660*. Reprint, Kingston and Montreal: McGill-Queen's University Press, 1987.

Upton, L.F.S. "The Origins of Canadian Indian Policy." *Journal of Canadian Studies* 8, 4 (November 1973): 51-61.

Vance, Jonathan F. *Death So Noble: Memory, Meaning, and the First World War*. Vancouver: UBC Press, 1997.

Viewpoints: One Hundred Years of Architecture in Ontario, 1889-1989. Kingston, ON: Ontario Association of Architects/Agnes Etherington Art Centre, 1989.

Vipond, Mary. "The Nationalist Network: English Canada's Intellectuals and Artists in the 1920s." *Canadian Review of Studies in Nationalism* 5 (1980): 32-52.

Voisey, Paul. "Rural Local History and the Prairie West." *Prairie Forum* 10, 2 (Autumn 1985): 327-38.

Wardhaugh, Robert. "Introduction: Tandem and Tangent." In Wardhaugh, ed., *Toward Defining the Prairies: Region, Culture, and History*. Winnipeg: University of Manitoba Press, 2001.

Regional and Local History

Angus, James T. *A Deo Victoria: The Story of the Georgian Bay Lumber Company, 1871-1942.* Thunder Bay: Severn Publications, 1990.

–. *A History of the Trent-Severn Waterway, 1833-1920.* Kingston and Montreal: McGill-Queen's University Press, 1988.

Baldwin, Norman S., et al. *Commercial Fish Production in the Great Lakes 1867-1977.* Technical Report no. 3. Ann Arbor, MI: Great Lakes Fishery Commission, 1979.

Barry, James. *Georgian Bay: The Sixth Great Lake.* 3rd ed. Erin, ON: Boston Mills Press, 1995. First published 1968 by Clarke, Irwin.

–. *An Illustrated History of the Georgian Bay.* Erin, ON: Boston Mills Press, 1992.

–. *Ships of the Great Lakes: 300 Years of Navigation.* Berkeley, CA: Howell-North Books, 1973.

Berst, A.H., and G.R. Spangler. *Lake Huron: The Ecology of the Fish Community and Man's Effects on It.* Technical Report no. 21. Ann Arbor, MI: Great Lakes Fishery Commission, 1973.

Berton, Pierre. *The Great Lakes.* Toronto: Stoddart, 1996.

Bogue, Margaret Beattie. *Fishing the Great Lakes: An Environmental History, 1783-1933.* Madison, WI: University of Wisconsin Press, 2000.

Boyer, Barbaranne. *Muskoka's Grand Hotels.* Erin, ON: Boston Mills Press, 1987.

Brazer, Marjorie Cahn. *The Sweet Water Sea: A Guide to Lake Huron's Georgian Bay.* Charlevoix, MI: Peach Mountain Press, 1984.

Brehm, Victoria, ed. *"A Fully Accredited Ocean": Essays on the Great Lakes.* Ann Arbor, MI: University of Michigan Press, 1998.

Byers, Mary. *Longuissa.* Erin, ON: Boston Mills Press, 1988.

Campbell, William A. *Northeastern Georgian Bay and Its People.* Britt, ON: 1982.

Chute, Janet E. "Ojibwa Leadership during the Fur Trade Era at Sault Ste. Marie." In *New Faces of the Fur Trade: Selected Papers of the Seventh North American Fur Trade Conference, Halifax, Nova Scotia 1995,* ed. Jo-Anne Fiske et al., 153-73. East Lansing, MI: Michigan State University Press, 1995.

Classen, H. George. "Georgian Bay Survey: Cradle of Canadian Hydrography." *Canadian Geographical Journal,* May 1963, 158-63.

Cognashene Book Corporation. *Cottage Histories of Cognashene.* Vol. 2. Cognashene Book Corporation, 1997.

Coombe, Geraldine. *Muskoka, Past and Present.* Toronto: McGraw-Hill Ryerson, 1976.

Croft, M.M. *Tall Tales and Legends of the Georgian Bay.* Owen Sound, ON: M.M. Croft, 1967.

Cruikshank, E. "An Episode of the War of 1812: The Story of the Schooner 'Nancy.'" *Ontario Historical Society Papers and Records.* Vol. 10, 1910, 75-126.

Cuthbertson, George A. *Freshwater: A History and a Narrative of the Great Lakes.* Toronto: Macmillan, 1931.

David, Jennifer, ed. *Wind, Water, Rock and Sky: The Story of Cognashene, Georgian Bay.* Toronto: The Cognashene Book Corporation, 1997.

Delaney, Paul J., and Andrew D. Nicholls. *After the Fire: Sainte-Marie Among the Hurons Since 1649.* Elmvale, ON: East Georgian Bay Historical Foundation, 1989.

Duke, A.H., and W.M. Gray. *The Boatbuilders of Muskoka.* Toronto: W.M. Gray, 1985.

East Georgian Bay Historical Journal 1-4 (1981-5).

Emery, Lee. *Review of Fish Species Introduced into the Great Lakes, 1819-1974.* Technical Report no. 45. Ann Arbor, MI: Great Lakes Fishery Commission, 1985.

Fleming, Keith. "Owen Sound and the CPR Great Lakes Fleet: The Rise of a Port, 1840-1912." *Ontario History* 76, 1 (March 1984).

Floren, Russell, et al. *Ghosts of the Bay: A Guide to the History of Georgian Bay.* Toronto: Lynx Images, 1994.

Foster, Gary, et al. *The Archaeological Investigations on Beausoleil Island, Georgian Bay Islands National Park, 1985.* Canada Parks Service Michrofiche Report Series MF 312, 1987.

Frost, Leslie. *Forgotten Pathways of the Trent.* Don Mills, ON: Burns and MacEachern, 1973.

Gough, Barry. *Fighting Sail on Lake Huron and Georgian Bay: The War of 1812 and its Aftermath.* Annapolis, MD: Naval Institute Press, 2002.

Gray, William M., and Timothy C. Du Vernet. *Wood and Glory: Muskoka's Classic Launches.* Toronto: Boston Mills Press, 1997.

Gutsche, Andrea, et al. *Alone in the Night: Lighthouses of Georgian Bay, Manitoulin Island and the North Channel.* Toronto: Lynx Images, 1996.

–, and Cindy Bisaillon. *Mysterious Islands: Forgotten Tales of the Great Lakes.* Toronto: Lynx Images, 1999.

Heidenreich, Conrad. *Huronia: A History and Geography of the Huron Indians, 1600-1650.* Toronto: Historical Sites Branch, Ontario Ministry of Natural Resources/McClelland and Stewart, 1971.

–. "Mapping the Great Lakes: The Period of Exploration, 1603-1700." *Cartographica* 17, 3 (1980): 32-64.

–. "Mapping the Great Lakes: The Period of Imperial Rivalries, 1700-1760." *Cartographica* 18, 3 (1981): 74-109.

Higgins, Robert. *The Wreck of the Asia: Ships, Shoals, Storms and a Great Lakes Survey.* Waterloo, ON: Escart Press, 1995.

Hultin, Neil C., and Warren U. Ober. "Captain Bayfield, Some Admiralty Clerks and the Naming of Islands." *Ontario History* 78, 1 (March 1986): 49-56.

Inland Seas. [Quarterly] Cleveland, OH: Great Lakes Historical Society, 1945-98.

Kuhlberg, Mark. "'We Are the Pioneers in This Business': Spanish River's Forestry Initiatives after the First World War." *Ontario History* 93, 2 (2001): 150-78.

Landon, Fred. *Lake Huron.* Indianapolis, IN: Bobbs-Merrill, 1944.

Lewis, G.M. "Mapping the Great Lakes between 1755 and 1795." *Cartographica* 17, 1 (1980).

Long, Charles. "Taxi's Waiting." *Cottage Life,* Spring 2000.

Mackay, Niall. *Over the Hills to Georgian Bay: A Pictorial History of the Ottawa, Arnprior and Parry Sound Railway.* Erin, ON: Boston Mills Press, 1981.

MacMahon, Paul. *Island Odyssey: A History of the San Souci Area of Georgian Bay.* Toronto: San Souci and Copperhead Association, 1990.

Madawaska Club. *Madawaska Club, Go-Home Bay: 1898-1923.* Midland: Madawaska Club, 1923.

–. *Madawaska Club, Go-Home Bay: 1898-1948.* Midland: Madawaska Club, 1948.

–. *Madawaska Club, Go-Home Bay: 1898-1973.* Toronto: Madawaska Club, 1973.

–. Notes from 1998 Cottage Tours.

–. *100 Years: Go Home Bay 1898-1998.* Madawaska Club at Go Home Bay, 1999.

Marchildon, Daniel. *Flying Low: A History of the Georgian Bay Scoot.* Midland, ON: Huronia Museum, 1994.

Mays, John Bentley. *Arrivals: Stories from the History of Ontario.* Toronto: Penguin Books, 2002.

McCannel, James. "Shipping out of Collingwood." Ontario Historical Society *Papers and Records.* Vol. 28, 1932, 16-24.

McCuaig, Ruth H. *Our Pointe au Baril.* Hamilton, ON: R.H McCuaig, 1984

McCullough, A.B. *The Commercial Fishery of the Canadian Great Lakes.* Studies in Archaeology, Architecture and History: Canadian Parks Service. Ottawa: Minister of Supply and Services Canada, 1989.

McKean, Fleetwood K. "Early Parry Sound and the Beatty Family." *Ontario History* 56, 3 (1964): 167-84.

Meers, David E. "The Use of a Communal Tourism/Recreation Resource: Case Study, The District of Muskoka." MA thesis, University of Western Ontario, 1990.

Mills, G.K. "The Nottawasaga River Route." Ontario Historical Society *Papers and Records.* Vol. 9, 1907, 40-8.

Morrison, James. "Upper Great Lakes Settlement: The Anishinabe-Jesuit Record." *Ontario History* 86, 1 (March 1994): 53-71.

Murray, Florence B., "Agricultural Settlement on the Canadian Shield: Ottawa River to Georgian Bay." In *Profiles of a Province,* ed. Edith G. Firth, 178-86. Toronto: Ontario Historical Society, 1967.

–, ed. *Muskoka and Haliburton 1615-1875: A Collection of Documents.* Toronto: Champlain Society/University of Toronto Press, 1963.

Newell, Diane. *Technology on the Frontier: Mining in Old Ontario.* Vancouver: University of British Columbia Press, 1986.

Orr, Sandra. "Captain Bayfield's Measure of Success." *Inland Seas* 55, 3 (1999): 224-32.

Parson, Helen E. "The Colonization of the Southern Canadian Shield in Ontario: The Hastings Road." *Ontario History* 79, 3 (September 1987): 263-73.

Peters, John H. "Commercial Fishing in Lake Huron, 1800 to 1915: The Exploitation and Decline of Whitefish and Lake Trout." MA thesis, University of Western Ontario, 1981.

Ratigan, William. *Great Lakes Shipwrecks and Survivals.* Grand Rapids, MI: Wm. B. Eerdmans, 1960.

Russell, Frances. *Mistehay Sakahegan: The Great Lake. The Beauty and the Treachery of Lake Winnipeg.* Winnipeg, MB: Heartland, 2000.

Rutherford, James H. "Early Navigation on the Georgian Bay." Ontario Historical Society *Papers and Records.* Vol. 18, 1920, 14-20.

Shanahan, David. "The Manitoulin Treaties, 1836 and 1862: The Indian Department and Indian Destiny." *Ontario History* 86, 1 (March 1994): 13-31.

Smith, Donald B. "The Dispossession of the Mississauga Indians: A Missing Chapter in the Early History of Upper Canada." *Ontario History* 73, 2 (1981): 67-87.

Spragge, George W. "Colonization Roads in Canada West, 1850-1867." *Ontario History* 49, 1 (1957): 1-18.

Tanner, Helen Hornbeck. *Atlas of Great Lakes Indian History.* Norman, OK: University of Oklahoma Press, 1987.

Taylor, Cameron. *Enchanted Summers: The Grand Hotels of Muskoka.* Toronto: Lynx Images, 1997.

Taylor, Monique. "'This Our Dwelling': The Landscape Experience of the Jesuit Missionaries to the Huron, 1626-1650." *Journal of Canadian Studies* 33, 2 (Summer 1998): 85-96.

Theberge, Clifford B., and Elaine Theberge. *The Trent-Severn Waterway: A Traveller's Companion.* Toronto: Samuel Stevens, 1978.

Troughton, Michael, and J. Gordon Nelson, eds. *The Countryside in Ontario: Evolution, Current Challenges and Future Directions.* Waterloo, ON: Heritage Resources Centre, University of Waterloo, 1998.

de Visser, John, and Judy Ross. *Georgian Bay.* Toronto: Stoddart, 1992.

Wall, Geoffrey. "Nineteenth-Century Land Use and Settlement on the Canadian Shield Frontier." In *The Frontier: Comparative Studies,* Vol. 1., ed. David Harry Miller and Jerome O. Steffen, 227-41. Norman, OK: University of Oklahoma Press, 1977.

Weller, G.R. "Hinterland Politics: The Case of Northwestern Ontario." *Canadian Journal of Political Science,* 10 December 1977, 727-54.

White, James. "Place Names in Georgian Bay." Ontario Historical Society *Papers and Records.* Vol. 11, 1913, 5-81.

Wightman, W.R. *Forever on the Fringe: Six Studies in the Development of the Manitoulin Island.* Toronto: University of Toronto Press, 1982.

Wilson, Scott D. "Henry Bayfield's Hydrographic Survey of Lake Huron." *Inland Seas* 48, 2 (1992): 117-28.

Arts and Literature

Armstrong, William. *Watercolour Drawings of New Ontario – From Georgian Bay to Rat Portage.* Thunder Bay, ON: Thunder Bay Art Gallery, 1996.

Arthur, John. *Spirit of Place: Contemporary Landscape Painting and the American Tradition.* Boston: Bulfinch Press/Little, Brown, 1989.

Atwood, Margaret. *Strange Things: The Malevolent North in Canadian Literature.* Oxford: Clarendon Press, 1995.

–. *Survival: A Thematic Guide to Canadian Literature.* Toronto: Anansi Press, 1972.

Bayer, Fern. *The Ontario Collection.* Markham, ON: Ontario Heritage Foundation/ Fitzhenry and Whiteside, 1984.

Bentley, D.M.R. "Charles G.D. Roberts and William Wilfred Campbell as Canadian Tour Guides." *Journal of Canadian Studies* 32, 2 (Summer 1997): 79-99.

–. *The Gay/Grey Moose: Essays on the Ecologies and Mythologies of Canadian Poetry, 1660-1990.* Ottawa: University of Ottawa Press, 1992.

Berger, John. *Ways of Seeing.* London: British Broadcasting Corporation/Penguin Books, 1972.

Bice, Megan. *Light and Shadow: The Work of Franklin Carmichael.* Kleinburg, ON: McMichael Canadian Art Collection, 1990.

Borcoman, James. *Goodridge Roberts: A Retrospective Exhibition.* Ottawa: National Gallery of Canada, 1969-70.

Bordo, Jonathan. "Jack Pine: Wilderness Sublime or the Erasure of the Aboriginal Presence from the Landscape." *Journal of Canadian Studies* 27, 4 (1992-3): 98-128.

Boulet, Roger. *The Canadian Earth: Landscapes of the Group of Seven.* Toronto: Cerebrus/ Prentice-Hall, 1982.

–. *Toni Onley: A Silent Thunder.* Toronto: Cerebrus/Prentice-Hall, 1981.

Boyanski, Christine, and John Hartman. *W.J. Wood: Paintings and Graphics.* Toronto: Art Gallery of Ontario, 1983.

Bridges, Marjorie Lismer. *Borders of Beauty: Arthur Lismer's Pen and Pencil.* Toronto: Red Rock, 1977.

Butlin, Susan. "Landscape as Memorial: A.Y. Jackson and the Landscape of the Western Front, 1917-1918." *Canadian Military Journal* 5, 2 (Autumn 1996): 62-70.

Campbell, Henry. *Early Days on the Great Lakes: The Art of William Armstrong.* Toronto: McClelland and Stewart, 1971.

Carroll, Jock. *The Life and Times of Greg Clark.* Toronto: Doubleday Canada, 1981.

Cole, Douglas. "Artists, Patrons and Public: An Enquiry into the Success of the Group of Seven." *Journal of Canadian Studies* 13, 2 (Summer 1978): 69-78.

–. "The History of Art in Canada." *Acadiensis* 10, 1 (Autumn 1980): 171-7.

Cook, Ramsay. "Landscape Painting and National Sentiment in Canada." In *The Maple Leaf Forever: Essays on Nationalism and Politics in Canada.* 2nd ed. Toronto: McClelland and Stewart, 1977.

Curtis, Andrea. "Testing the Waters." [Review of *Summer Gone.*] *Toronto Life,* August 1999.

Darroch, Lois. *Bright Land: A Warm Look at Arthur Lismer.* Toronto: Merritt, 1981.

Ed Bartram: 10 Years. Brantford, ON: Art Gallery of Brant, 1988.

Fairley, Barker. *12 Georgian Bay Sketches.* Toronto: Hugh Anson-Cartwright and Martin Ahvenus, 1972.

Fetherling, Douglas, ed. *Documents in Canadian Art.* Peterborough, ON: Broadview Press, 1987.

Flood, John. "Northern Ontario Art: A Study in Line Drawings." *Boréal* 9 (1977): 2-11.

Frye, Northrop. *Mythologizing Canada: Essays on the Canadian Literary Imagination.* Ed. Branko Gorup. Toronto: Legas, 1997.

–. "Sharing the Continent." Reprinted in Eli Mandel and David Taras, eds., *A Passion for Identity: An Introduction to Canadian Studies.* Toronto: Methuen, 1987.

Gatenby, Greg, ed. *The Wild Is Always There: Canada through the Eyes of Foreign Writers.* Toronto: Vintage Books, 1994.

Georgian Bay Moods: Paintings by Ivan Wheale. Sudbury, ON: Laurentian University Museum and Arts Centre, 1987.

Gibbon, Catherine, ed. *On the Edge: Artistic Visions of a Shrinking Landscape.* Erin, ON: Boston Mills Press, 1995.

Goodridge Roberts: Paintings from the 1950s and 60s. Toronto: Art Gallery of Ontario, 1980.

Grady, Wayne, ed. *Treasures of the Place: Three Centuries of Nature Writing in Canada.* Vancouver: Douglas and McIntyre, 1992.

Grande, John K. "Back to Bricks: Canadian Writing about Art Needs to Get Back to Basics." *Canadian Forum,* October 1993, 20-3.

Greenhill, Pauline. *True Poetry: Traditional and Popular Verse in Ontario.* Montreal and Kingston: McGill-Queen's University Press, 1989.

Harper, J. Russell. *Painting in Canada: A History.* 2nd ed. Toronto: University of Toronto Press, 1977.

Hartman, John. *Georgian Bay: Drawings by John Hartman.* Peterborough, ON: Broadview Press, 1989.

–. "William J. Wood, Etcher: 1877-1954." *East Georgian Bay Historical Journal,* 1981.

Hill, Charles C. *Art for a Nation: The Group of Seven.* Ottawa: National Gallery of Canada/ McClelland and Stewart, 1995.

Housser, F.B. *A Canadian Art Movement: The Story of the Group of Seven.* Toronto: Macmillan, 1926.

Hubbard, R.H. *Canadian Landscape Painting 1670-1930.* Madison, WI: University of Wisconsin Press, 1973.

Ivan Wheale: Rocks and Water. Toronto: Gallerie Dresdnere, 1980.

Jefferys, C.W. *Canada's Past in Pictures.* Toronto: Ryerson Press, 1943.

Keith, W.J. *Literary Images of Ontario.* Toronto: University of Toronto Press, 1992.

Klinck, Carl. *Canadian Anthology.* 1st ed. Toronto: Gage, 1966.

Landry, Pierre B. *The MacCallum-Jackman Cottage Mural Paintings*. Ottawa: National Gallery of Canada, 1990.

Larisey, Peter. *Light for a Cold Land: Lawren Harris's Work and Life – An Interpretation*. Toronto: Dundurn Press, 1993.

Litteljohn, Bruce, and Jon Pearce, eds. *Marked by the Wild: Literature Shaped by the Canadian Wilderness*. Toronto: McClelland and Stewart, 1973.

Liversidge, M.J.H. "Striking 'a Native Note': C.J. Jefferys and Canadian Identity in Landscape Painting." *British Journal of Canadian Studies* 9, 1 (1994): 64-71.

Lord, Barry. "Georgian Bay and the Development of the September Gale Theme in Arthur Lismer's Painting, 1912-1921." *National Gallery of Canada Bulletin* 5, 1-2 (1967): 28-38.

Lowrey, Carol, et al. *Visions of Light and Air: Canadian Impressionism, 1885-1920*. New York: Americas Society Art Gallery, 1995.

MacDonald, J.E.H. *The Tangled Garden*. Text by Paul Duval. Toronto, ON: Cerebrus/Prentice-Hall, 1978.

Macfarlane, David. Interview by Eleanor Wachtel. "The Arts Today." Canadian Broadcasting Corporation, 12 October 1999.

Mashel Teitelbaum: A Retrospective. Hamilton, ON: Art Gallery of Hamilton et al., 1992.

Mason, Bill. *Canoescapes*. Erin, ON: Boston Mills Press, 1995.

McDougal, Anne. *Anne Savage: The Story of a Canadian Painter*. Montreal: Harvest House, 1977.

McGregor, Gaile. *The Wacousta Syndrome: Explorations in the Canadian Landscape*. Toronto: University of Toronto Press, 1985.

McLeish, John A.B. *September Gale: A Study of Arthur Lismer and the Group of Seven*. Toronto: J.M. Dent and Sons, 1955.

Murray, Joan. *Confessions of a Curator: Adventures in Canadian Art*. Toronto: Dundurn Press, 1996.

–. *Northern Lights: Masterpieces of Tom Thomson and the Group of Seven*. Toronto: Key Porter, 1994.

Nasgaard, Roald. "Claude Breeze: Canadian Atlas and the Expressionist Landscape Tradition." *artscanada*, Spring 1974, 53-64.

–. *The Mystic North: Symbolist Landscape Painting in Northern Europe and North America, 1890-1940*. Toronto: Art Gallery of Ontario/ University of Toronto Press, 1984.

New, W.H. *A History of Canadian Literature*. London: Macmillan Education, 1989.

–. *Land Sliding: Imagining Space, Presence, and Power in Canadian Writing*. Toronto: University of Toronto Press, 1997.

Painting the Bay: Recent Work by John Hartman. Kleinburg, ON: McMichael Canadian Art Collection, 1993.

Panabaker, Frank. *Reflected Lights*. Toronto: Ryerson Press, 1957.

Pierce, L. Bruce. *Thoreau MacDonald: Illustrator, Designer, Observer of Nature*. Toronto: Norflex, 1971.

Prown, Jules David, et al. *Discovered Lands, Invented Pasts: Transforming Visions of the American West*. New Haven, CT: Yale University Press, 1992.

Public Archives Canada. *Elizabeth Simcoe*. Microfiche 9. Ottawa: 1978.

Rees, Ronald. *Land of Earth and Sky: Landscape Painting of Western Canada*. Saskatoon: Western Producer Prairie Books, 1984.

–. "Landscape in Art." *Dimensions of Human Geography: Essays on Some Familiar and Neglected Themes*. Ed. Karl Butzer. Chicago: University of Chicago, 1978.

Reid, Dennis. *A Concise History of Canadian Painting.* 2nd ed. Toronto: Oxford University Press, 1988.

–. *The MacCallum Bequest.* Ottawa: National Gallery of Canada, 1969.

–. *"Our Own Country Canada": Being an Account of the National Aspirations of the Principal Landscape Artists in Montreal and Toronto, 1860-1890.* Ottawa: National Gallery of Canada, 1979.

Silcox, David P. *Painting Place: The Life and Work of David B. Milne.* Toronto: University of Toronto Press, 1996.

Simpson-Housley, Paul, and Glen Norcliffe, eds. *A Few Acres of Snow: Literary and Artistic Images of Canada.* Toronto: Dundurn Press, 1992.

Staines, David, ed. *Beyond the Provinces: Literary Canada at Century's End.* Toronto: University of Toronto Press, 1995.

–. *The Canadian Imagination: Dimensions of a Literary Culture.* Cambridge, MA: Harvard University Press, 1977.

Stein, Alan. *Islands.* Parry Sound, ON: Church St. Press, 1994.

Stein, Roger B. *Seascape and the American Imagination.* New York: Clarkson N. Potter/ Whitney Museum of American Art, 1975.

Sweeney, J. Gray. *Great Lakes Marine Painting of the Nineteenth Century.* Muskegon, MI: Muskegon Museum of Art, 1983.

Teitelbaum: Georgian Bay Oils. Toronto: Loranger Gallery, 1979.

Thom, Ian M. *Angus Trudeau's Manitoulin.* Kleinburg, ON: McMichael Canadian Art Collection, 1986.

Tippett, Maria. *Art at the Service of War: Canada, Art and the Great War.* Toronto: University of Toronto Press, 1984.

–. *By A Lady: Celebrating Three Centuries of Art by Canadian Women.* Toronto: Viking, 1992.

–. *Stormy Weather: F.H. Varley, A Bibliography.* Toronto: McClelland and Stewart, 1998.

–, and Douglas Cole. *From Desolation to Splendour: Changing Perceptions of the British Columbia Landscape.* Toronto: Clarke, Irwin, 1977.

Tooby, Michael, ed. *The True North: Canadian Landscape Painting, 1896-1939.* London: Lund Humphries Publishers, 1991.

Warkentin, Germain, ed. *Stories from Ontario.* Toronto: Macmillan, 1974.

Westfall, William. "On the Concept of Region in Canadian History and Literature." *Journal of Canadian Studies* 15, 2 (Summer 1980): 3-15.

Wistow, David. *Landscapes of the Mind: Images of Ontario.* Toronto: Art Gallery of Ontario, 1986.

Wood, Susan J. "The Land in Canadian Prose, 1845-1945." PhD thesis, Carleton Univefrsity, 1988.

Zemans, Joyce. "Establishing the Canon: Nationhood, Identity and the National Gallery's First Reproduction Program of Canadian Art." *Journal of Canadian Art History* 16, 2 (1995): 6-35.

Index

Aboriginal peoples. *See* First Nations
acid rain, 127, 159, 190, 191
Adirondacks: park reserves, 168; resorts in, 87, 91; style of architecture, 91, 92
agency, 17
Agnew, Andrew, 55, 101
agriculture, 40, 41, 66, 99; missionaries and, 97; and seasonal employment, 67, 68; subsistence, 67, 68
airplanes, 121
Algoma, 68, 80, 84
Algonquian nations, 26, 67, 97, 100
Algonquin: Historical Society, 85; National Park, 146, 168, 179; uplands, 128; Wildlands League, 177
Algonquin people, 25, 178
American Forestry Congress, 168
Amerindians. *See* First Nations
Anishinabe, 12, 132, 134
anorthosite, 69
Appalachians, resorts in, 87
aquaculture, 167, 185
archipelago, 3; adaptation to, 148; aerial view, 118; appearance of, 3, 83-4, 84-5; beauty in, 137; "commons" vs private land, 94-5; defined, 3; difficulty of surveying, 37-9, 46-8; fishing clubs in, 87-8; fishing conditions in, 165; forestry on, 84-5; geology of, 128; as heritage landscape, 205; as imagined community, 148; invisibility of, 54; as landscape, 118; as littoral, 18, 53, 148, 188; mythical origin of, 106; number of islands, 41; number of islands in, 48; origin of, 105-6; parks in, 171-2; as political unit, 53; as shelter, 135; suitability for historical imagination, 106-7; township of, 52-3, 187; transformation through human presence, 18; trees of, 83-4; as wilderness, 181. *See also* islands
architecture: and granite, 129; and local ability, 91, 152; maritime orientation

of, 124-7; naturalism in, 91-2; schools of, 91; vernacular, 91, 152
Ardagh, A.G., 46
Armstrong, William, 56, 102, 119-20
art(s): Canadian experience in, 147; Canadian images in, 147, 155-6; criticism, 6; dangers depicted in, 132-3; documentary tradition in, 56-7; Georgian Bay in, 5, 113, 139, 145-7, 156, 160; Laurentian school, 155; military, 32; native perspective in, 146; and nature, 145; the Open in, 119; regional fragmentation, 156; regional groups, 156; and regional identity, 199; regional language in, 144; rock in, 127; subjective response to landscape, 57. *See also* literature
Asia, 41, 124, 149
Asling-Riis, Stella: *The Great Fresh Sea*, 85, 104, 144
Atwood, Margaret, xii, 106, 140, 147

back-to-nature movement, 44, 45, 85, 92, 104-5, 138, 147
Bartram, Edward, 127; *Precambrian Point #2*, plate 13
Baxter Township, 51, 52, 55, 126, 168
Bay of Quinte, 28, 72
Bayfield, Henry, 23, 24, 35, 36, 37-8, 48, 58, 59, 60, 123
Beatty, J.W., 145
Beausoleil Island: archaeological sites on, 109, 172; camps on, 173; DDT bombs on, 173, 203; in literature, 144; name, 59; National Park, 109, 121, 173-4; Ojibwa and, 26; parks on, 171; scoot on, 121; Sea Cadet camps on, 114; soil on, 67. *See also* Prince William Henry Island
Bellin, Nicolas: *Partie Occidentale de la Nouvelle France*, 29, 30
Benidickson, Jamie: *The Temagami Experience*, 8, 84